Fluid Dynamics Handbook

Fluid Dynamics Handbook

Edited by **Maria Forest**

New York

Published by NY Research Press,
23 West, 55th Street, Suite 816,
New York, NY 10019, USA
www.nyresearchpress.com

Fluid Dynamics Handbook
Edited by Maria Forest

© 2015 NY Research Press

International Standard Book Number: 978-1-63238-198-9 (Hardback)

Printed in the United States of America.

Contents

Preface

I am honored to present to you this unique book which encompasses the most up-to-date data in the field. I was extremely pleased to get this opportunity of editing the work of experts from across the globe. I have also written papers in this field and researched the various aspects revolving around the progress of the discipline. I have tried to unify my knowledge along with that of stalwarts from every corner of the world, to produce a text which not only benefits the readers but also facilitates the growth of the field.

Fluid dynamics is the sub-specialty of physics dealing with the study of flowing fluids. This book consists of various issues regarding fluid dynamics which will be of interest for scientists and researchers. Through the chapters included in this book, it intends to help its readers to advance their expertise in analyzing fluid dynamics as encountered in engineering fields.

Finally, I would like to thank all the contributing authors for their valuable time and contributions. This book would not have been possible without their efforts. I would also like to thank my friends and family for their constant support.

Editor

Fluid Dynamics of Gas – Solid Fluidized Beds

Germán González Silva[1], Natalia Prieto Jiménez[1] and Oscar Fabio Salazar
State University of Campinas
Brazil

1. Introduction

Fluidization refers to the contact between a bed of solids and a flow of fluid. As a result, the solid particles are transformed into a fluid-like behavior that can be used for different purposes. The fluidized bed reactor is one of the most important technologies for gas-solid heterogeneous operations chemical or petrochemical, considering catalytic or non catalytic processes (Kunii and Levenspiel 1991). The most important industrial applications include catalytic cracking, coal combustion and biomass combustion. One of the most relevant type of fluidized bed reactor is the ascendant flow reactor, which is also known as riser. The riser reactors consist of a tubular column in which both solid and gas flow upwards. The first fluidized bed gas generator was developed in Germany by Fritz Winkler in the 1920s. Later in the 1930s, the american petroleum industry started developing the fluidized bed technology for oil feedstock catalytic cracking, becoming the primary technology for such applications (Tavoulareas 1991).

Inside the riser reactor, solid particles have a wide range of residence time, which is a disadvantage that reduces the overall conversion and the selectivity of the chemical reactions. For that reason it has recently grown the interest in a new type of gas-solid circulating reactor known as downer. In this reactor the gas and the solid flow cocurrently downward, creating hydrodynamic features comparable to a plug flow reactor and allowing a better control over the conversion, the selectivity and the catalyst deactivation. The concept of downer reactor gas-solid appeared in the 1980s, with the first studies on the fluid dynamics of gas-solid suspensions (Kim and Seader 1983) and with the first downer reactors for patents developed by Texaco for the FCC process (Gross Benjamin and Ramage Michael P 1981; Niccum Phillip K and Bunn Jr Dorrance P 1983). In these studies it is observed that in the downer reactor has a uniform distribution of two-phase flow along the reactor, also observed that the contact time is very low, achieving a 20% decrease in the amounts of coke produced during the FCC process.

Applications, differences, advantages and disadvantages to these types of fluidized bed reactors can be found in various publications (Ancheyta 2010; Gonzalez, 2008; Yi Cheng et al. 2008; Crowe 2005; Wen-ching Yang 2003; Grace 1997; Gidaspow 1994; Geldart 1986)

2. Fluidization regimes and particle classification

Fluidization occurs when a gas or liquid is forced to flow vertically through a bed of particles at such a rate that the buoyed weight of the particles is completely supported by the drag force imposed by the fluid.

2.1 Flow regimes in fluidized beds

As the superficial gas velocity, U, is increased stepwise beyond the minimum fluidization velocity, it is observed different types of flow regimes. The principal ones are schematically shown in Figure 1. The flow regimes are listed by increasing value of U as follows:

- Bubble-free bed expansion
- Bubbling fluidization
- Slug flow
- Turbulent fluidization
- Fast fluidization and dense suspension upflow

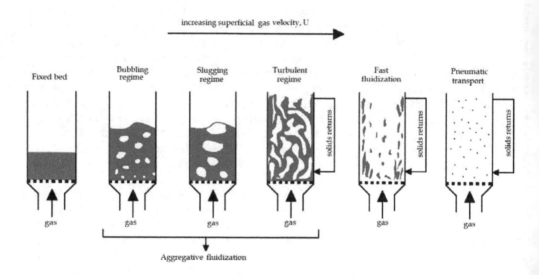

Fig. 1. Flow regimes of gas–solid fluidization.

The bubbling regime is one of the most studied flow regimes in gas-solid fluidization. Bubbles coalesce and break-up as fluid flow is increased. Finally, the bubbles become large enough to occupy a substantial fraction of the cross-section of the small diameter columns (Vejahati 2006). These large bubbles are called slug, as shown in the third column of Figure 1.

2.2 Particle classification

The behavior of solids fluidized by gases fall into four clearly recognizable groups, characterized by density difference $(\rho_s - \rho_f)$ and mean particle size. The features of the groups are: powders in group A exhibit dense phase expansion after minimum fluidization and prior to the commencement of bubbling; those in group B bubble at the minimum fluidization velocity; those in group C are difficult to fluidize at all and those in group D can form stable spouted beds (Geldart 1973). Desirable properties of particles and gas for fluidized bed are delineated in Table 1.

Property	Desirable Range
Particle Properties	
Mean diameter	50 μm to 1.6 mm
Size distribution	Neither too narrow or too broad, e.g., 90th to 10th decile ratio 5 to 25
Density	Wide range of values possible, but uniform from particle to particle
Shape	Rounded and with length to thickness ration no larger than ~3
Surface roughness	Smooth
Surface stickiness	Avoid sticky surfaces
Attrition resistance	Usually strong as possible
Hardness	Avoid resilience, but also excessive hardness
Gas Properties	
Density	No restriction, but higher value improves properties
Viscosity	No restriction
Relative humidity	Typically 10 to 90%

Table 1. Desirable properties of particles and gases for Gas-Solid fluidization (Jesse Zhu et al. 2005)

3. Experimental measurement techniques

For better understanding of these phenomena and to facilitate the solution of mathematical models is necessary to make an analysis of experimental data. This experimental analysis requires specialized measurement techniques are able to explain the flow field must also be automated to minimize human involvement in the process of collecting data.

The measurement techniques, to capture the important fluids dynamic behavior of the two-phase flow, can be classified as non-intrusive (**NMT**) and intrusive (**IMT**) techniques. The intrusive techniques are generally probes used to study local basic flow phenomena. Some of these are intended only as research instruments. The most common parameters that are measured with such probes are solids mass flows, radial and axial solids concentration, solids velocities, and distribution.

The particles can be deposited in the measuring device reducing its performance or causing malfunction. Besides this, the flow area reduction makes of the intrusive devices not the best solution. Non-intrusive techniques to characterize the flow within a fluidized bed are more desirable because it does not disturb the flow behavior. In the Table 2 and Table 3 classification techniques are included and recent successes have been achieved.

NMT		Ref for more details
Laser Doppler Anemometry (LDA)	LDA is a technology used to measure velocities of small particles in flows. The technique is based on the measurement of laser light scattered by particles that pass through a series of interference fringes (a pattern of light and dark surfaces). The scattered laser light oscillates with a specific frequency that is related to the velocity of the particles.	(C.H. Ibsen, T. Solberg, and B.H. Hjertager 2001; Claus H. Ibsen et al. 2002; Kuan, W. Yang, and Schwarz 2007; Lu, Glass, and Easson 2009; Vidar Mathiesen et al. 1999; Werther, Hage, and Rudnick 1996)
X-ray	Radiographic techniques based either based on electromagnetic radiation such as X and γ rays. The transmission of X-rays or γ-rays through a heterogeneous medium is accompanied by attenuation of the incident radiation, and the measurement of this attenuation provides a measure of the line integral of the local mass density distribution along the path traversed by the beam	(Franka and Heindel 2009; Newton, Fiorentino, and Smith 2001; Petritsch, Reinecke, and Mewes 2000; Tapp et al. 2003; C. Wu et al. 2008; Heindel, Gray, and Jensen 2008)
γ-ray		(Du, Warsito, and Fan 2005; Kumar, Moslemian, and Milorad P. Dudukovic 1995; Tan et al. 2007; Thatte et al. 2004; Veluswamy et al. 2011; H. G Wang et al. 2008)
Radioactive Particle Tracking (RPT)	Technique to measure velocity field and turbulent parameters of multiphase flow. This is based on the principle of tracking the motion of a single tracer particle as a marker of the solids phase. The tracer particle contains a radioactive element emitting γ-rays. This radiation is received by an ensemble of specific detector.	(Muthanna Al-Dahhan et al. 2005; S. Bhusarapu, M.H. Al-Dahhan, and Duduković 2006; Fraguío et al. 2009; Khanna et al. 2008; Larachi et al.; Vaishali et al. 2007)
Particle Image Velocimetry (PIV)	PIV measures whole velocity fields by taking two images shortly after each other and calculating the distance individual particles travelled within this time. The displacement of the particle images is measured in the plane of the image and used to determine the displacement of the particles	(van Buijtenen et al. 2011; Fu et al. 2011; He et al. 2009; Hernández-Jiménez et al.; Kashyap and Gidaspow 2011; Laverman et al. 2008; Sathe et al. 2010)

Table 2. Non-intrusive measurement techniques.

IMT		References
Pitot Tube	Mechanical method based on determination of momentum by means of differential pressure measurements	(Al-Hasan and Al-Qodah 2007; Bader, R., Findlay, J. and Knowlton, TM 1988; R.-C. Wang and Han 1999)
Fiber Optic Probe	This technique is commonly used as effective tools to measure the local porosity in fluidized beds.	(Fischer, Peglow, and Tsotsas 2011; Link et al. 2009; Meggitt 2010; Zhengyang Wang et al. 2009; Ye, Qi, and J. Zhu 2009; Zhou et al. 2010; Haiyan Zhu et al. 2008)
Capacitance Probe	This technique is used to measure the local dielectric constant of the gas-solid suspension, which is linked to the local volume fraction of solids	(A. Collin, K.-E. Wirth, and Stroeder 2009; Anne Collin, Karl-Ernst Wirth, and Ströder 2008; Demori et al. 2010; Guo and Werther 2008; Vogt et al. 2005; Wiesendorf 2000)

Table 3. Intrusive measurement techniques.

4. Computational fluid dynamics (CFD)

Computational Fluid Dynamics (CFD) is a technique which uses conservation principles and rigorous equations of fluid flow (Navier-Stokes) along with specialized turbulence models (k-ε, k-ω, SST among others). These models are more accurate and fundamentally more acceptable than empirical ones. The empirical models are approximations that assemble different phenomena to remove a number of unknown parameters. For this reason, these models are not reliable and therefore should not be generalized.

The CFD models can be divided into two groups: the *Eulerian-Eulerian* model in which the gas and solid phases are considered as two interpenetrating continuum flows; and the *Eulerian-Lagrangian* model that consider the gas as a fluid phase and the solids as discrete phase. The *Eulerian-Lagrangian* model calculates the trajectory of each individual particle using Newton's second law. The interaction between particles can be described by the potential energy or the dynamic of collisions. This method has the advantage of knowing exactly the particle trajectory and the system variables. However, this requires high computational effort, higher yet when gas and solid velocity fields are coupled.

4.1 Governing equations
Governing equations for *Eulerian-Eulerian* model are here presented in tensor notation.

4.1.1 Continuity equations
The gas and solid continuity equations are represented by:

$$\frac{\partial}{\partial t}\left(\alpha_g \rho_g\right)+\nabla \cdot \left(\alpha_g \rho_g \vec{v}_g\right)=0 \tag{1}$$

$$\frac{\partial}{\partial t}(\alpha_s \rho_s) + \nabla \cdot (\alpha_s \rho_s \vec{v}_s) = 0 \tag{2}$$

Where α, ρ and $\vec{v}$ are volume fraction, density and the vector velocity, respectively. No mass transfer is allowed between phases.

4.1.2 Momentum equations

The gas phase momentum equation may be expressed as:

$$\frac{\partial}{\partial t}(\alpha_g \rho_g \vec{v}_g) + \nabla \cdot (\alpha_g \rho_g \vec{v}_g \vec{v}_g) = -\alpha_g \nabla p + \nabla \cdot [\tau_g] + \alpha_g \rho_g \vec{g} + \beta(\vec{v}_s - \vec{v}_g) \tag{3}$$

p and $\vec{g}$ are fluid pressure and gravity acceleration. β is the drag coefficient between the phases g and s. The stress tensor is given by:

$$\tau_g = \alpha_g \mu_g \left[\nabla \vec{v}_g + (\nabla \vec{v}_g)^T \right] - \frac{2}{3} \alpha_g \mu_g \nabla \vec{v}_g \tag{4}$$

The solid phase momentum equation may be written as:

$$\frac{\partial}{\partial t}(\alpha_s \rho_s \vec{v}_s) + \nabla \cdot (\alpha_s \rho_s \vec{v}_s \vec{v}_s) = -\alpha_s G \nabla \alpha_s + \nabla \cdot [\tau_s] + \alpha_s \rho_s \vec{g} + \beta(\vec{v}_g - \vec{v}_s) \tag{5}$$

$$\tau_s = \alpha_s \mu_s \left[\nabla \vec{v}_s + (\nabla \vec{v}_s)^T \right] - \frac{2}{3} \alpha_s \mu_s \nabla \vec{v}_s \tag{6}$$

G is the modulus of elasticity given by:

$$G = \exp \left[C_G (\alpha_s - \alpha_{s,max}) \right] \tag{7}$$

Where $\alpha_{s,max}$ is the maximum solid volume fraction and β is the interface momentum transfer proposed by Gidaspow, (1994):

$$\begin{cases} \beta = 150 \dfrac{\alpha_s(1-\alpha_g)\mu_g}{\alpha_g d_p^2} + 1.75 \dfrac{\alpha_s \rho_g |\vec{v}_s - \vec{v}_g|}{d_p} & |\alpha_g \leq 0.8 \\[4mm] \beta = \dfrac{3}{4} C_D \dfrac{\alpha_s \alpha_g \rho_g |\vec{v}_s - \vec{v}_g|}{d_p} \alpha_g^{-2.65} & |\alpha_g > 0.8 \end{cases} \tag{8}$$

Where d_p and C_D are the particle diameter and the drag coefficient, based in the relative Reynolds number (Re_s)

$$C_D = \begin{cases} \dfrac{24(1+0.15 Re_s^{0.687})}{Re_s} & |Re_s \leq 1000 \\[4mm] 0.44 & |Re_s > 1000 \end{cases} \tag{9}$$

$$Re_s = \frac{\rho_g |\vec{v}_s - \vec{v}_g|}{\mu_g} \tag{10}$$

4.1.3 Energy equation
The gas and solid energy equations can be written as:

$$\frac{\partial}{\partial t}\left(\alpha_g \rho_g H_g\right) + \nabla \cdot \left(\alpha_g \rho_g \vec{v}_g H_g\right) = \nabla \cdot \left(\alpha_g \lambda_g \nabla T_g\right) + \gamma\left(T_s - T_g\right) + \alpha_g \rho_g \sum_r \Delta H_r \frac{\partial C_r}{\partial t} \qquad (11)$$

$$\frac{\partial}{\partial t}\left(\alpha_s \rho_s H_s\right) + \nabla \cdot \left(\alpha_s \rho_s \vec{v}_s H_s\right) = \nabla \cdot \left(\alpha_s \lambda_s \nabla T_s\right) + \gamma\left(T_g - T_s\right) \qquad (12)$$

Where
H = Specific enthalpy
T = Temperature
γ = Interface heat transfer coefficient: $\gamma = Nu\lambda \, / \, d_p$
λ = Thermal conductivity

4.2 Turbulence models
Turbulence is that state of fluid motion which is characterized by random and chaotic three-dimensional vorticity. When turbulence is present, it usually dominates all other flow phenomena and results in increased energy dissipation, mixing, heat transfer, and drag. The physical turbulence models provide the solution the closure problem in solving Navier – Stokes equations. While there are ten unknown variables (mean pressure, three velocity components, and six Reynolds stress components), there are only four equations (mass balance equation and three velocity component momentum balance equations). This disparity in number between unknowns and equations make a direct solution of any turbulent flow problem impossible in this formulation. The fundamental problem of turbulence modeling is to relate the six Reynolds stress components to the mean flow quantities and their gradients in some physically plausible manner.
The turbulence models are summarized in Table 4

Family group	Models	Description and advantages
Reynolds – Averaged Navier – Stokes (RANS)	Zero equation models	The most widely used models. Its main advantages are short computation time, stable calculations and reasonable results for many flows.
	One equation models	
	Two equation models $\kappa - \varepsilon$ $\kappa - \omega$	
Reynolds Stress Model (RSM)		Provides good predictions for all types of flows, including swirl, and separation. Longer calculation times than the RANS models.
Large Eddy Simulation (LES)	Smagorinsky-Lilly model	Provides excellent results for all flow systems. LES solves the Navier-Stokes equations for large scale motions of the flow models only the small scale motions.
	Dynamic subgrid-scale model	
	RNG – LES model	
	WALLE model	

Family group	Models	Description and advantages
Detached Eddy Simulation (DES)	The difficulties associated with the use of the standard LES models, has lead to the development of hybrid models (like that DES) that attempt to combine the best aspects of RANS and LES methodologies in a single solution strategy.
Direct Numerical Simulation (DNS)	The most exact approach to turbulence simulation without requiring any additional modeling beyond accepting the Navier–Stokes equations to describe the turbulent flow processes.

Table 4. Summary of turbulence models.

4.3 System discretization

The most important numerical methods used to approximate the partial differential equations by a system of algebraic equations in terms of the variables at some discrete locations in space and time (called "discretization method") are the Finite Volume (FV), the Finite Difference (FD) and the Finite Element (FE) methods. In this book, the finite volume method and the commercial software CFX® 12.0 were chosen; the solution domain is discretized in a computational mesh that can be structured or unstructured.

Finite volume (FV) method

The FV discretization method is obtained by integrating the transport equation around a finite volume. The general form of transport equations is given by:

$$\underbrace{\frac{\partial(\rho\phi)}{\partial t}}_{I} + \underbrace{\nabla \cdot (\rho\vec{v}\phi)}_{II} = \underbrace{\nabla \cdot (\Gamma_\phi \nabla\phi)}_{III} + \underbrace{S_\phi}_{IV} \tag{13}$$

i.   Transient term
ii. Convective term
iii. Diffusive term
iv. Source term

The transport equations are integrated in each computational cell using the divergence theorem over a given time interval Δt:

$$\int_{t}^{t+\Delta t} \left\{ \int_V \frac{\partial(\rho\phi)}{\partial t} dV + \oint \rho\phi\vec{v} \cdot d\vec{A} = \oint \Gamma_\phi \nabla\phi \cdot d\vec{A} + \int_v S_\phi dV \right\} dt \tag{14}$$

Linearization and interpolation techniques can be clarified considering the finite volume P shown in Figure 3.

In agreement with Figure 3 notation, diffusive term can be represented as

$$\oint \Gamma_\phi \nabla\phi \cdot d\vec{A} = \frac{\Gamma_\phi A_w}{h_w}(\phi_P - \phi_W) = D_w(\phi_P - \phi_W) \tag{15}$$

Fig. 2. Gas flow over a flat solid surface (left to right) experimental picture, refined mesh near the wall and contrast between experiment and discretization.

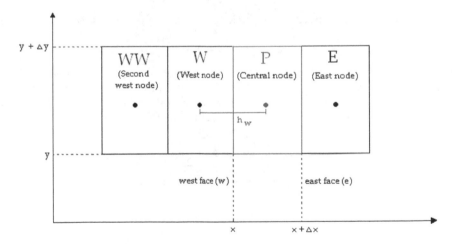

Fig. 3. Finite volume representation and notation.

4.4 Source term linearization

A generic source term may be written as

$$S_{\phi P} V_P = S_C^\phi + S_P^\phi \phi_P \tag{16}$$

Where $S_{\phi P}$ is the value of source term in the center of the cell P and V_P is the volume of computational cell centered on node P. The method to represent $S_{\phi P}$ was suggested by Patankar, 1980

$$S_{\phi P} = S_{\phi P}^* + \left(\frac{dS_{\phi P}}{d\phi}\right)^* \left(\phi_P - \phi_P^*\right) \tag{17}$$

This type of linearization is recommended since the source term decreases with increasing Φ. The source term coefficients are represented by:

$$S_C^\phi = \left[S_{\phi P}^* - \left(\frac{dS_{\phi P}}{d\phi}\right)^* \phi_P^*\right] V_P \tag{18}$$

$$S_P^\phi = \left(\frac{dS_{\phi P}}{d\phi}\right)^* V_P \tag{19}$$

4.4.1 Spatial discretization

The most widely used in CFD is first and second order Upwind methods. In the first order one, quantities at cell faces are determined by assuming that the cell-center values of any field variable represent a cell-average value and hold throughout the entire cell. The face value (Φ_w) are equal to the cell-center value of Φ in the upstream cell.

$$\oint \rho \phi \vec{v} \cdot d\vec{A} = \rho v_w A_w \phi_w = C_w \phi_w \tag{20}$$

Where, C_w is the west face convective coefficient. A_w can be represented by:

$$A_w = MAX\left(C_w, 0\right) + D_w \tag{21}$$

In the second order one, quantities at cell faces are computed using a multidimensional linear reconstruction approach (Jespersen and Barth 1989). In this approach, higher-order accuracy is achieved at cell faces through a Taylor series expansion of the cell-centered solution about the cell centroid. Thus, the face value Φ_w is computed using the following expression:

$$\phi_w = \frac{3}{2}\phi_W - \frac{1}{2}\phi_{WW} = \phi_W + \frac{1}{2}\left(\phi_W - \phi_{WW}\right) \tag{22}$$

The east face coefficient and matrix coefficient are shown below

$$\phi_e = \frac{3}{2}\phi_P - \frac{1}{2}\phi_W \tag{23}$$

$$A_w = MAX\left(C_w, 0\right) + \frac{1}{2}MAX\left(C_e, 0\right) + D_w \tag{24}$$

4.4.2 Temporal discretization

Temporal discretization involves the integration of every term in the differential equations over a time step Δt. A generic expression for the time evolution of a variable Φ is given by

$$\frac{\partial \phi}{\partial t} = F(\phi) \tag{25}$$

Where the function F incorporates any spatial discretization. The first-order accurate temporal discretization is given by

$$\frac{\phi^{n+1} - \phi^n}{\Delta t} = F(\phi) \tag{26}$$

And the second-order discretization is given by

$$\frac{3\phi^{n+1} - 4\phi^n + \phi^{n-1}}{2\Delta t} = F(\phi) \tag{27}$$

5. Case studies

In order to give a better introduction with regards to the simulation of fluidized beds, in this chapter there are presented three case studies that were carried out by using a CFD software package.

The case studies were carried out using simulations in dynamic state. These simulations were set up taking into account the average value of the Courant number, which is recommended to be near 1. Besides this, it was used a constant step time, in this way was possible to have numerical stability during the execution of each of the simulations.

5.1 Cases 1 and 2

Lab scale riser reactor (Samuelsberg and B. H. Hjertager 1996; V Mathiesen 2000). Riser height, 1 m; riser diameter, 0.032 m. Experimental data and LES - Smagorinsky simulations were compared for three velocities with initial particle bed, 5cm.

5.1.1 Mesh parameters and boundary conditions
- Control volumes number: 100.000
- $\Delta x = 2$ mm
- Matrix determinant > 0.5 and minimum angle > 50°

The boundary conditions for both cases are shown in Table 5 and Table 6.

In addition, tests were made with a 500.000 control volume mesh with same block distribution (the description of volume distribution in the meshes, are presented in Table 7). Obtaining similar results with the 100.000 control volume mesh. Both meshes are shown in Figure 4.

In	Gas velocity = 0.36; 1.42 m/s
	Particle mass flow equal to the output
Out	*Opening* = atmospheric pressure
Wall	Particles = *free slip* and *No slip*
	Gas = no slip
Initial height	Bed height = 0,05 m
Particles	60 μm; 1600 kg/m3

Table 5. Boundary conditions for the Case 1.

In	Gas velocity = 1 m/s
	Particle mass flow equal to the output
Out	*Opening* = atmospheric pressure
Wall	Particles = *No slip*
	Gas = No slip
Initial height	Bed height = 0.05 m
Particles	120 μm, 2400 kg.m^{-3}

Table 6. Boundary conditions for the Case 2.

Mesh	dx/dp	Volumes Number	Δ/dx
I	15	99900	0.05
II	10	467313	0.08

Table 7. Volume discretization of the meshes.

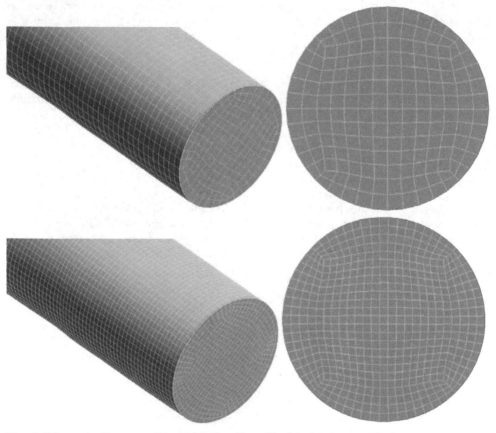

Fig. 4. Schematic diagram of the Table 7 meshes. Up: Mesh I. Down: Mesh II

Numeric calculations performed (Vreman, Geurts, and Kuerten 1997; Chow and Moin 2003) showed that the required values to obtain an accurate numerical solution, it is necessary to use a ratio $\Delta/dx \leq 0.25$ for the second order spatial scheme, and a ratio $\Delta/dx < 0.5$ for the sixth order scheme.

The values of Δ/dx presented in Table 7 are within the range recommended in the literature (Chow and Moin 2003; Agrawal et al. 2001; van Wachem 2000; Ahmed and Elghobashi 2000; Vreman, Geurts, and Kuerten 1997).

Figure 5 presents the solid volume fraction time evolution for the mesh II with superficial velocity 1 m/s. At the beginning, the solids present in the riser are forced to flow in the upward direction, similar to a plug flow.

When the bed of solids starts to expand, it is observed high solid particle concentration at the center of the tube and near the walls (Figure 5). This reordering of solid particles is a counteraction in order to offer a lower resistance to the gas flow. This type of flow regime is known as pre-fluidized bed.It is important to mention that one of most relevant characteristics of the fluidization is the high contact area between the solid particles and the fluid. In this way, a cubic meter of particles of 100 micron contains a superficial area of around 30000 m2. The advantage of this high surface area is reflected in a high mass and heat transfer rates between the solid and the fluid.

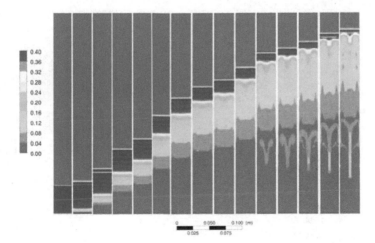

Fig. 5. Evolution of the volume fraction field in a fluidized bed at 0, 11, 35, 70, 90, 132, 165, 185, 198, 220, 242, 264, 275, 290, and 317 ms.

Figure 6 shows the similarity between results presented by Miller and Gidaspow (1992). Here it is represented the regions of high and low solid concentration. Near the walls velocity is negative and near the center velocity is positive.

The annular-core behavior is something that detrimental in the units of Fluid Catalytic Cracking (FCC), since big fraction of the oil is converted in a region where the catalyst works less efficient. In addition to this, the particles that flow at center core are expose to bigger concentrations of oil compounds, which is something that produces faster deactivation of the catalyst. One the strategies to solve this issue is to inject pressurized gas in perpendicular direction to the flow in the reaction zone. Another solution is to include rings connected to walls, with the purpose of redirecting the solids from the wall towards the center.

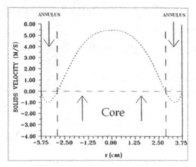 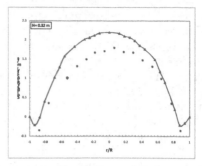

Fig. 6. Comparison of solid phase velocity profile presented by Miller and Gidaspow (1992) with the CFD simulations (-▲-) and experimental data performed by Samuelsberg and B. H. Hjertager (1996) (●).

To get an impression regarding the flow behavior inside the column, the time averaged solid volume fraction is plotted at different column heights, 0.16 m, 0.32 m and 0.48 m (Figure 7). Here it can be observed the strong tendency of the solid particles to be near the wall.

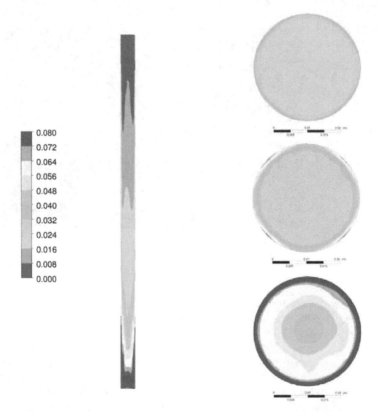

Fig. 7. Axial profile of the solid phase volume fraction fields in the center (left) and radial profiles at 0.48 m, 0.32 m, 0.16 m (right up to down). Superficial velocity 0.36 m s⁻¹

5.2 Case 3

Pilot plant scale riser reactor (Bader, R., Findlay, J. and Knowlton, TM 1988). Riser height: 13 m, riser diameter 0.3 m. Entrance with angle 60°, gas superficial velocity 3.7 m and solids flux 98 kg/(s.m^2) as shown in Figure 8.

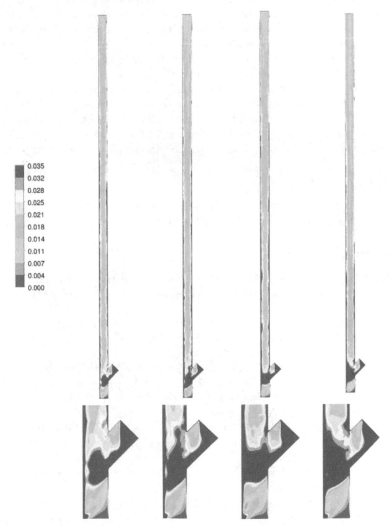

Fig. 8. Solids volumetric fraction in the center of the riser. Simulation time 15 sec. Left to right: LES Smagorinsky, LES WALE, LES Dynamic model, Detached Eddy Simulation (DES).

In the Figure 8 can be observed that the solid particles enter to the reactor uniformly distributed, after a short distance these particles start falling due to the gravity and they start flowing over the wall of the inclined pipe. After this, the solids fall into a turbulent zone where they get mixed. Some of the particles will continue falling over the vertical wall opposite to the entrance. The core-annular zone is formed at some height in the middle of the column.

6. Conclusions

Computational fluids dynamics is a very powerful tool understanding the behavior of multi phase in engineering applications.
Large eddy simulation (LES) turbulence method provides a very detailed description of two phase flow, which makes it suitable for simulation models that are validated with experimental data. By applying the LES method, it is possible to characterize different regions of a fluidized bed (core-annulus). LES can be considered as a valuable method for development and validation of closure models that include additional phenomena like heat exchange, mass transfer and chemical reactions.
It is important to constantly monitor the simulation, using parameters such as the Courant number, creating a function that calculates the maximum and average number of the control volume courant. The average value is recommended that is near or less than unity.
Finally, it is important to comment that success in the validation of experimental data depends on the appropriate choice of the experimental technique used to measure variables.

7. Acknowledgements

The author G. Gonzalez is grateful to PETROBRAS and the National Council for Scientific and Technological Development (CNPq) for the financial support to this research.

8. References

Agrawal, Kapil, Peter N Loezos, Madhava Syamlal, and Sankaran Sundaresan. 2001. "The role of meso-scale structures in rapid gas solid flows." *Journal of Fluid Mechanics* 445 (01) (October): 151-185.

Ahmed, A. M., and S. Elghobashi. 2000. "On the mechanisms of modifying the structure of turbulent homogeneous shear flows by dispersed particles." *Physics of Fluids* 12 (11): 2906. doi:10.1063/1.1308509.

Al-Dahhan, Muthanna, Milorad P. Dudukovic, Satish Bhusarapu, Timothy J. O'hern, Steven Trujillo, and Michael R. Prairie. 2005. Flow Mapping in a Gas-Solid Riser via Computer Automated Radioactive Particle Tracking (CARPT). http://www.osti.gov/energycitations/servlets/purl/881590-Kfq80v/.

Al-Hasan, M., and Z. Al-Qodah. 2007. Characteristics of gas-solid flow in vertical tube. In *9th International Symposium on Fluid Control Measurement and Visualization 2007, FLUCOME 2007*, 1:264-271.

Ancheyta, Jorge. 2010. *Modeling and simulation of catalytic reactors for petroleum refining*. Oxford: Wiley-Blackwell.

Bader, R., Findlay, J. and Knowlton, TM. 1988. Gas/ Solid Flow Patterns in a 30.5-cm-Diameter Circulating Fluidized Bed. In *Circulating fluidized bed technology II: proceedings of the Second International Conference on Circulating Fluidized Beds, Compiègne, France, 14-18 March 1988*, by Prabir Basu and Jean François Large. Pergamon Press.

Bhusarapu, S., M.H. Al-Dahhan, and M.P. Duduković. 2006. "Solids flow mapping in a gas-solid riser: Mean holdup and velocity fields." *Powder Technology* 163 (1-2): 98-123.

van Buijtenen, Maureen S., Willem-Jan van Dijk, Niels G. Deen, J.A.M. Kuipers, T. Leadbeater, and D.J. Parker. 2011. "Numerical and experimental study on multiple-spout fluidized beds." *Chemical Engineering Science* 66 (11) (June 1): 2368-2376. doi:16/j.ces.2011.02.055.

Collin, A., K.-E. Wirth, and M. Stroeder. 2009. "Characterization of an annular fluidized bed." *Powder Technology* 190 (1-2): 31-35.

Collin, Anne, Karl-Ernst Wirth, and Michael Ströder. 2008. "Experimental characterization of the flow pattern in an annular fluidized bed." *The Canadian Journal of Chemical Engineering* 86 (3) (June 1): 536-542. doi:10.1002/cjce.20056.

Crowe, Clayton T. 2005. *Multiphase Flow Handbook*. 1st ed. CRC Press, September 19.

Cheng, Yi, Changning Wu, Jingxu Zhu, Fei Wei, and Yong Jin. 2008. "Downer reactor: From fundamental study to industrial application." *Powder Technology* 183 (3) (April 21): 364-384. doi:16/j.powtec.2008.01.022.

Chow, Fotini Katopodes, and Parviz Moin. 2003. "A further study of numerical errors in large-eddy simulations." *Journal of Computational Physics* 184 (2) (January 20): 366-380. doi:doi: 10.1016/S0021-9991(02)00020-7.

Demori, M., V. Ferrari, D. Strazza, and P. Poesio. 2010. "A capacitive sensor system for the analysis of two-phase flows of oil and conductive water." *Sensors and Actuators, A: Physical* 163 (1): 172-179.

Du, Bing, W. Warsito, and Liang-Shih Fan. 2005. "ECT Studies of Gas–Solid Fluidized Beds of Different Diameters." *Industrial & Engineering Chemistry Research* 44 (14) (July 1): 5020-5030. doi:10.1021/ie049025n.

Fischer, C., M. Peglow, and E. Tsotsas. 2011. "Restoration of particle size distributions from fiber-optical in-line measurements in fluidized bed processes." *Chemical Engineering Science* 66 (12) (June 15): 2842-2852. doi:16/j.ces.2011.03.054.

Fraguío, M.S., M.C. Cassanello, S. Degaleesan, and M. Dudukovic. 2009. "Flow regime diagnosis in bubble columns via pressure fluctuations and computer-assisted radioactive particle tracking measurements." *Industrial and Engineering Chemistry Research* 48 (3): 1072-1080.

Franka, Nathan P., and Theodore J. Heindel. 2009. "Local time-averaged gas holdup in a fluidized bed with side air injection using X-ray computed tomography." *Powder Technology* 193 (1) (July 10): 69-78. doi:16/j.powtec.2009.02.008.

Fu, Y., T. Wang, J.-C. Chen, C.-G. Gu, and F. Xu. 2011. "Experimental investigation of jet influence on gas-solid the two-phase crossflow in a confined domain." *Shiyan Liuti Lixue/Journal of Experiments in Fluid Mechanics* 25 (1): 48-53+64.

Geldart, D. 1986. *Gas fluidization technology*. Chichester, New York: Wiley.

Geldart, D. 1973. "Types of gas fluidization." *Powder Technology* 7 (5) (May): 285-292. doi:16/0032-5910(73)80037-3.

Gidaspow, Dimitri. 1994. *Multiphase flow and fluidization: continuum and kinetic theory descriptions*. Boston: Academic Press.

Gonzalez,, S.G. 2008. Modeling and simulation of cocurrent downflow reactor (Downer). Master's thesis, Campinas, Brazil: State University of Campinas.

Grace, John. 1997. *Circulating fluidized beds*. 1st ed. London;: Blackie Academic & Professional.

Gross Benjamin, and Ramage Michael P. 1981. FCC Reactor With A Downflow Reactor Riser. April 14.

Guo, Q., and J. Werther. 2008. "Influence of a gas maldistribution of distributor design on the hydrodynamics of a CFB riser." *Chemical Engineering and Processing: Process Intensification* 47 (2): 237-244.

He, Y., N. G. Deen, M. van Sint Annaland, and J. A. M. Kuipers. 2009. "Gas–Solid Turbulent Flow in a Circulating Fluidized Bed Riser: Experimental and Numerical Study of Monodisperse Particle Systems." *Industrial & Engineering Chemistry Research* 48 (17): 8091-8097. doi:10.1021/ie8015285.

Heindel, Theodore J., Joseph N. Gray, and Terrence C. Jensen. 2008. "An X-ray system for visualizing fluid flows." *Flow Measurement and Instrumentation* 19 (2) (April): 67-78. doi:16/j.flowmeasinst.2007.09.003.

Hernández-Jiménez, F., S. Sánchez-Delgado, A. Gómez-García, and A. Acosta-Iborra. "Comparison between two-fluid model simulations and particle image analysis & velocimetry (PIV) results for a two-dimensional gas-solid fluidized bed." *Chemical Engineering Science* In Press, Corrected Proof. doi:16/j.ces.2011.04.026. http://www.sciencedirect.com/science/article/pii/S0009250911002685.

Ibsen, C.H., T. Solberg, and B.H. Hjertager. 2001. "Evaluation of a three-dimensional numerical model of a scaled circulating fluidized bed." *Industrial and Engineering Chemistry Research* 40 (23): 5081-5086.

Ibsen, Claus H., Tron Solberg, Bjørn H. Hjertager, and Filip Johnsson. 2002. "Laser Doppler anemometry measurements in a circulating fluidized bed of metal particles." *Experimental Thermal and Fluid Science* 26 (6-7): 851-859. doi:16/S0894-1777(02)00196-6.

Jespersen, Dennis C, and Timothy J Barth. 1989. "The design and application of upwind schemes on unstructured meshes." *AIAA paper* 89 (89-0366): 1–12.

Kashyap, Mayank, and Dimitri Gidaspow. 2011. "Measurements of Dispersion Coefficients for FCC Particles in a Free Board." *Industrial & Engineering Chemistry Research* 50 (12) (June 15): 7549-7565. doi:10.1021/ie1012079.

Khanna, Pankaj, Todd Pugsley, Helen Tanfara, and Hubert Dumont. 2008. "Radioactive particle tracking in a lab-scale conical fluidized bed dryer containing pharmaceutical granule." *The Canadian Journal of Chemical Engineering* 86 (3) (June 1): 563-570. doi:10.1002/cjce.20073.

Kim, J. M, and J. D Seader. 1983. "Pressure drop for cocurrent downflow of gas-solids suspensions." *AIChE Journal* 29 (3) (May 1): 353-360. doi:10.1002/aic.690290302.

Kuan, B., W. Yang, and M.P. Schwarz. 2007. "Dilute gas-solid two-phase flows in a curved 90° duct bend: CFD simulation with experimental validation." *Chemical Engineering Science* 62 (7): 2068-2088.

Kumar, Sailesh B., Davood Moslemian, and Milorad P. Dudukovic. 1995. "A [gamma]-ray tomographic scanner for imaging voidage distribution in two-phase flow systems." *Flow Measurement and Instrumentation* 6 (1): 61-73. doi:16/0955-5986(95)93459-8.

Kunii, D, and O Levenspiel. 1991. *Fluidization engineering.* 2nd ed. Boston Mass.: Butterworth-Heinemann.

Larachi, Faical, M H Al-Dahhan, M P Duduković, and Shantanu Roy. "Optimal design of radioactive particle tracking experiments for flow mapping in opaque multiphase reactors." *Applied radiation and isotopes including data instrumentation and methods for use in agriculture industry and medicine* 56 (3): 485-503.

Laverman, Jan Albert, Ivo Roghair, Martin van Sint Annaland, and Hans Kuipers. 2008. "Investigation into the hydrodynamics of gas–solid fluidized beds using particle image velocimetry coupled with digital image analysis." *The Canadian Journal of Chemical Engineering* 86 (3) (June 1): 523-535. doi:10.1002/cjce.20054.

Link, J.M., W. Godlieb, P. Tripp, N.G. Deen, S. Heinrich, J.A.M. Kuipers, M. Schönherr, and M. Peglow. 2009. "Comparison of fibre optical measurements and discrete element simulations for the study of granulation in a spout fluidized bed." *Powder Technology* 189 (2): 202-217. doi:16/j.powtec.2008.04.017.

Lu, Y., D.H. Glass, and W.J. Easson. 2009. "An investigation of particle behavior in gas-solid horizontal pipe flow by an extended LDA technique." *Fuel* 88 (12): 2520-2531.

Mathiesen, V. 2000. "An experimental and computational study of multiphase flow behavior in a circulating fluidized bed." *International Journal of Multiphase Flow* 26 (3) (March): 387-419. doi:10.1016/S0301-9322(99)00027-0.

Mathiesen, Vidar, Tron Solberg, Hamid Arastoopour, and Bjørn H Hjertager. 1999. "Experimental and computational study of multiphase gas/particle flow in a CFB riser." *AIChE Journal* 45 (12) (December 1): 2503-2518. doi:10.1002/aic.690451206.

Meggitt, B.T. 2010. Fiber Optics in Sensor Instrumentation. In *Instrumentation Reference Book (Fourth Edition)*, 191-216. Boston: Butterworth-Heinemann.

Miller, Aubrey, and Dimitri Gidaspow. 1992. "Dense, vertical gas-solid flow in a pipe." *AIChE Journal* 38 (11) (November 1): 1801-1815. doi:10.1002/aic.690381111.

Newton, D., M. Fiorentino, and G.B. Smith. 2001. "The application of X-ray imaging to the developments of fluidized bed processes." *Powder Technology* 120 (1-2): 70-75.

Niccum Phillip K, and Bunn Jr Dorrance P. 1983. Catalytic Cracking System. March 23.

Patankar, Suhas. 1980. *Numerical heat transfer and fluid flow*. Washington; New York: Hemisphere Pub. Corp; McGraw-Hill.

Petritsch, Georg, Nicolas Reinecke, and Dieter Mewes. 2000. Visualization Techniques in Process Engineering. In *Ullmann's Encyclopedia of Industrial Chemistry*. Wiley-VCH Verlag GmbH & Co. KGaA.

Samuelsberg, A., and B. H. Hjertager. 1996. "An experimental and numerical study of flow patterns in a circulating fluidized bed reactor." *International Journal of Multiphase Flow* 22 (3) (June): 575-591. doi:16/0301-9322(95)00080-1.

Sathe, M.J., I.H. Thaker, T.E. Strand, and J.B. Joshi. 2010. "Advanced PIV/LIF and shadowgraphy system to visualize flow structure in two-phase bubbly flows." *Chemical Engineering Science* 65 (8): 2431-2442.

Tan, H.-T., G.-G. Dong, Y.-D. Wei, and M.-X. Shi. 2007. "Application of γ-ray attenuation technology in measurement of solid concentration of gas-solid two-phase flow in a FCC riser." *Guocheng Gongcheng Xuebao/The Chinese Journal of Process Engineering* 7 (5): 895-899.

Tapp, H.S., A.J. Peyton, E.K. Kemsley, and R.H. Wilson. 2003. "Chemical engineering applications of electrical process tomography." *Sensors and Actuators, B: Chemical* 92 (1-2): 17-24.

Tavoulareas, E S. 1991. "Fluidized-Bed Combustion Technology." *Annual Review of Energy and the Environment* 16 (1) (November): 25-57. doi:10.1146/annurev.eg. 16.110191.000325.

Thatte, A. R., R. S. Ghadge, A. W. Patwardhan, J. B. Joshi, and G. Singh. 2004. "Local Gas Holdup Measurement in Sparged and Aerated Tanks by γ-Ray Attenuation Technique." *Industrial & Engineering Chemistry Research* 43 (17): 5389-5399. doi:10.1021/ie049816p.

Vaishali, S., S. Roy, S. Bhusarapu, M.H. Al-Dahhan, and M.P. Dudukovic. 2007. "Numerical simulation of gas-solid dynamics in a circulating fluidized-bed riser with geldart group B particles." *Industrial and Engineering Chemistry Research* 46 (25): 8620-8628.

Vejahati, F. 2006. CFD simulation of gas-solid bubbling fluidized bed. Master's thesis, University of Regina.

Veluswamy, Ganesh K., Rajesh K. Upadhyay, Ranjeet P. Utikar, Geoffrey M. Evans, Moses O. Tade, Michael E. Glenny, Shantanu Roy, and Vishnu K. Pareek. 2011. "Hydrodynamics of a Fluid Catalytic Cracking Stripper Using γ-ray Densitometry." *Industrial & Engineering Chemistry Research* 50 (10) (May 18): 5933-5941. doi:10.1021/ie1021877.

Vogt, C., R. Schreiber, G. Brunner, and J. Werther. 2005. "Fluid dynamics of the supercritical fluidized bed." *Powder Technology* 158 (1-3): 102-114.

Vreman, Bert, Bernard Geurts, and Hans Kuerten. 1997. "Large-eddy simulation of the turbulent mixing layer." *Journal of Fluid Mechanics* 339 (May): 357-390. doi:10.1017/S0022112097005429.

van Wachem. 2000. Derivation, Implementation, and Validation of Computer Simulation Models for Gas-Solid Fluidized Beds. Ph.D. Thesis, Delft University of Technology.

Wang, H. G, W. Q Yang, P. Senior, R. S Raghavan, and S. R Duncan. 2008. "Investigation of batch fluidized-bed drying by mathematical modeling, CFD simulation and ECT measurement." *AIChE Journal* 54 (2) (February 1): 427-444. doi:10.1002/aic.11406.

Wang, R.-C., and Y.-C. Han. 1999. "Momentum dissipation of jet dispersion in a gas-solid fluidized bed." *Journal of the Chinese Institute of Chemical Engineers* 30 (3): 263-271.

Wang, Zhengyang, Shaozeng Sun, Hao Chen, Qigang Deng, Guangbo Zhao, and Shaohua Wu. 2009. "Experimental investigation on flow asymmetry in solid entrance region of a square circulating fluidized bed." *Particuology* 7 (6): 483-490. doi:16/j.partic.2009.07.004.

Werther, J., B. Hage, and C. Rudnick. 1996. "A comparison of laser Doppler and single-fibre reflection probes for the measurement of the velocity of solids in a gas-solid circulating fluidized bed." *Chemical Engineering and Processing: Process Intensification* 35 (5) (October): 381-391. doi:16/0255-2701(96)80018-3.

Wiesendorf, Volker. 2000. *The Capacitance Probe: A Tool for Flow Investigations in Gas-solids Fluidization Systems.* Shaker Verlag GmbH, Germany, September 8.

Wu, C., Y. Cheng, M. Liu, and Y. Jin. 2008. "Measurement of axisymmetric two-phase flows by an improved X-ray-computed tomography technique." *Industrial and Engineering Chemistry Research* 47 (6): 2063-2074.

Yang, Wen-ching. 2003. *Handbook of fluidization and fluid-particle systems.* New York: Marcel Dekker.

Ye, S., X. Qi, and J. Zhu. 2009. "Direct Measurements of Instantaneous Solid Flux in a CFB Riser using a Novel Multifunctional Optical Fiber Probe." *Chemical Engineering & Technology* 32 (4) (April 1): 580-589. doi:10.1002/ceat.200800361.

Zhou, Hao, Guiyuan Mo, Jiapei Zhao, Jianzhong Li, and Kefa Cen. 2010. "Experimental Investigations on the Performance of a Coal Pipe Splitter for a 1000 MW Utility Boiler: Influence of the Vertical Pipe Length." *Energy & Fuels* 24 (9): 4893-4903. doi:10.1021/ef1007209.

Zhu, Haiyan, Jesse Zhu, Guozheng Li, and Fengyun Li. 2008. "Detailed measurements of flow structure inside a dense gas-solids fluidized bed." *Powder Technology* 180 (3): 339-349. doi:16/j.powtec.2007.02.043.

Zhu, Jesse, Bo Leckner, Yi Cheng, and John Grace. 2005. Fluidized Beds. In *Multiphase Flow Handbook*, ed. Clayton Crowe, 20052445:5-1-5-93. CRC Press, September.

2

An Experimental and Computational Study of the Fluid Dynamics of Dense Cooling Air-Mists

Jesús I. Minchaca M., A. Humberto Castillejos E.* and F. Andrés Acosta G.
Centre for Research and Advanced Studies – CINVESTAV, Unidad Saltillo
Mexico

1. Introduction

Spray cooling of a hot body takes place when a dispersion of fine droplets impinges upon its surface to remove a large amount of heat by evaporation and convection (Deb & Yao, 1989). In metallurgical processes such as continuous casting of steel (Camporredondo et al., 2004) the surface temperature, T_w, of the hot steel strand exceeds considerably the saturation temperature, T_s, of the cooling liquid (water), i.e., T_w-T_s ranges between ~600 to 1100°C. These harsh temperature conditions have traditionally called for the use of high water impact fluxes (w, L/m²s) to remove the heat arriving to the surface as a result of the solidification of the liquid or semi-liquid core of the strand. The boundary between dilute and dense sprays has been specified at w= 2 L/m²s (Deb & Yao, 1989, Sozbir et al., 2003). In modern continuous casting machines the w found are well above this value. Most of the impingement area of the spray or mist jets will have w> 10 L/m²s, with regions where w can be as large as ~110 L/m²s. Heat treatment of alloys requiring the rapid removal of large amounts of heat also makes use of dense sprays or mists (Totten & Bates, 1993).

Sprays and air-mists are dispersions of drops produced by single-fluid (e.g., water) and twin-fluid (e.g. water-air) nozzles, respectively. In sprays, the energy to fragment the water into drops is provided by the pressure drop generated across the narrow exit orifice, while in air-mists nozzles a high speed air-stream breaks the water-stream generating fine, fast-moving droplets (Lefebvre, 1989; Nasr et al., 2002). In air-mist nozzles with internal mixing and perpendicular inlets for the fluids, as those shown in Fig. 1, the water splatters against a deflector surface and the resulting splashes are further split by the shear forces exerted by the axial air-stream, which also accelerates the drops as they move along the mixing chamber toward the exit port. Thus, the liquid emerges in the form of drops with different sizes and velocities and with a non-uniform spatial distribution (Hernández et al., 2008).

In addition to w, the size, d_d, and velocity, u, of the drops in dense air-mists play a crucial role in the cooling of highly superheated surfaces (Bendig et al., 1995; Jenkins et al., 1991; Hernández et al., 2011). This behavior stresses the important relationship between the heat transfer process and the droplet impact or deformation and break-up behavior. Since, for a specified fluid those two parameters, d_d and u, determine the local impingement Weber number (We_{zs}= $\rho_d u_{zs}^2 d_d/\sigma$), which in general has been agreed to characterize the impact behavior (Wachters & Westerling, 1966; Araki & Moriyama, 1981; Issa & Yao, 2005). As the

---

* Corresponding Author

impingement Weber number increases the drops tend to deform more widely, break more profusely, stay closer to the surface and agitate more intensively the liquid film formed by previous drops. Thus, it is clear that knowledge of the local parameters characterizing free mist-jets is needed to arrive to a quantitative description of the fluid dynamic interaction of drops with a surface and of the boiling-convection heat transfer that would result.

Experimentally, the water impact flux has been the parameter most frequently determined, using a patternator (Camporredondo et al., 2004; Puschmann & Specht, 2004). The drop size distribution in mists has often been measured by: (a) laser diffraction (Jenkins et al., 1991; Bul, 2001), (b) phase Doppler particle analysis, PDPA (Bendig et al., 1995; Puschmann & Specht, 2004) and (c) particle/droplet image analysis, PDIA (Minchaca et al., 2011). The last two methods allow the simultaneous determination of the droplet velocity and hence of the correlation between both parameters. To the best knowledge of the authors only PDIA has been used for the characterization of dense sprays and mists. Particle image velocimetry, PIV, has been employed for measuring the velocity of drops in dense mists, but the technique did not allow the simultaneous determination of size (Hernández et al., 2008). Recent works have presented a detailed experimental characterization of the local variation of w, d_d and u obtained with typical air-mist nozzles, operating over a wide range of conditions of practical interest (Minchaca et al., 2011; Hernández et al., 2011).

The phenomena involved in the atomization of a liquid stream are very complex and therefore the generation of drops and their motion are generally treated separately. Knowledge of the influence of the fluid physical properties, nozzle design and operating conditions on atomization is crucial to generate drops with the size distribution that would perform better the task for which they are intended. The best well-known method for modeling drop size distributions is the empirical method (Babinski & Sojka, 2002). This consists in fitting a curve to data collected over a wide range of nozzles and operating conditions. In the case of nozzles with internal mixing and 90° intersecting streams of air and water, the number and volume frequency distributions of drop size have been adequately modeled by log-normal and Nukiyama-Tanasawa, NT, distribution functions (Minchaca et al., 2011), respectively. The statistical parameters of the distributions have been correlated with the water and air inlet pressures allowing the prediction of different characteristic mean diameters, over a wide range of operating conditions. Alternative modeling approaches are the maximum entropy and the discrete probability function methods (Babinski & Sojka, 2002).

Two-phase flow models generally treat the continuous phase (e.g., air) in an Eulerian frame of reference while the disperse phase (e.g., water droplets) is considered by either one of two approaches: (a) Eulerian representation, which treats it as a continuum whose characteristics (e.g., velocity, concentration, etc) are declared and updated at grid cells shared with the continuous phase, and (b) Lagrangian representation, where the drops characteristics (e.g., position, velocity, concentration, etc) are tracked along their path-lines (Crowe et al., 1998). The Eulerian-Eulerian approach is best suited for flows of monodisperse or narrow size range drops. But models have been developed to handle efficiently polydisperse sprays by describing the distribution of sizes through the moments of the droplet distribution function (Beck & Watkins, 2002). The Eulerian-Lagrangian approach can handle more efficiently a large range of particle sizes and give more details of the behavior of individual particles and of their interaction with walls. Both approaches use submodels to represent phenomena such as droplet break-up, droplet-droplet collisions, droplets-wall interaction, etc.

A two-dimensional (2-D) transient Eulerian-Lagrangian model was developed to describe the motion of air and drops in a domain that included the nozzle chamber, the free jet and the impingement region (Hatta et al., 1991a, 1991b). The researchers considered mono-disperse drops with sizes of 1 and 10 μm and found that the motion of both phases depended strongly on the particle size. More recently, a computational model was developed to calculate the in-flight and impingement motion of air and droplets with a size distribution (Issa & Yao, 2005). The rebounding of multiple drops from the surface was simulated by extending empirical information regarding the variation of the normal coefficient of restitution of single droplets with the impingement We_{zs}. The authors claimed that large drops with high momentum tended to impinge closer to the stagnation point, whereas smaller drops tended to collision farther away because they were entrained by the air. In another study, the equation of motion for drops projected horizontally in quiescent air was solved considering sizes ranging from 100 to 1000 μm and velocities of 20 m/s and 50 m/s (Ciofalo et al., 2007). It was found that drops smaller than 100 μm would experience large deflections due to gravity, and would never reach a plane beyond 0.25 m.

The sprays and mists that have been studied experimentally and computationally are far apart from those used in important metallurgical processes. In recent studies the authors presented a 3-D computational fluid dynamic (CFD) Eulerian-Lagrangian model for free dense air-mist jets (Hernández et al., 2008). However, since new and rigorous experimental information has been generated the model has been refined in regard to the size distribution imposed at the nozzle orifice. The experimental information generated in this work has also enabled to carry out a detailed validation of the model. The model predicts very well the correlation between drop velocity and particle size, the velocity and trajectory of the drops and the water impact density as a function of the nozzle operating conditions, over the whole range of practical interest.

2. Experimental methods and conditions

A schematic of the experimental set-up used for measuring the mist parameters is displayed in Figure 1. It consists of: (a) a patternator for measuring water impact density distribution, (b) a particle/droplet image analysis, PDIA, system for acquiring and analyzing the images of fine moving droplets to determine their size and velocity and (c) a water and air supply system for the nozzle.

To determine w the nozzle was oriented horizontally and this parameter was evaluated collecting the drops entering tubes with an area a, to measure the total volume of water v accumulated during a period of time t in the bottles connected to the tubes. The collecting tubes were arranged forming a grid and their diameter and spacing are given in Figure 1. Hence, the local water impact flux at a position x-y-z was calculated according to the following expression,

$$w(x,y,z) = \frac{v(x,y,z)}{(a\cos\gamma)\,t} \tag{1}$$

where (cos γ) is the direction cosine of the angle formed between the nozzle axis and the line connecting the centers of the nozzle orifice and of a given tube, i.e., (a cos γ) gives the projected area of a tube perpendicular to the direction of motion of the drops. The accuracy of the measured w distributions was verified by integrating w over the impingement area to compare it with the total water flow rate, W. In general, the computed W had an error smaller than ±10 %.

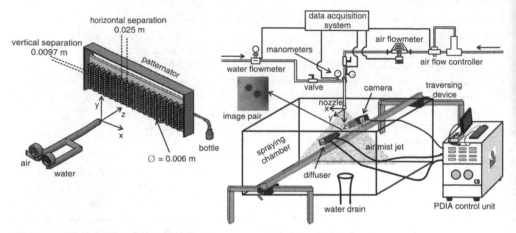

Fig. 1. Schematic of experimental setup

The PDIA system (VisiSizer N60V, Oxford Laser Ltd. Didcot, United Kingdom) schematically illustrated in Fig. 1 is a spatial multiple counting apparatus that captures instantaneously (i.e., in 4 ns) shadow images of the droplets moving through a thin (~400 µm) sampling volume and analyzes them in real time (Minchaca et al., 2011). A dual head Nd:YAG laser sends light pulses (15 mJ at 532 nm) through a fluorescent diffuser to illuminate the region of interest from behind while a high resolution camera placed in front captures the shadow images of the objects passing in between. The disposition of these elements is illustrated in the figure. Operating in dual pulse mode the laser and camera are triggered to capture image pairs separated by a time interval of 1.7 µs, the figure displays a single pair extracted from superposed frames. The analysis of single and superposed frames allows, respectively, the simultaneous determination of the size and velocity of the drops appearing. The criteria employed for the consideration of single drops and drop pairs have been described elsewhere and were validated by off-line analysis of single and superposed frames (Minchaca, 2011; Minchaca et al., 2011). Lenses with two magnifications (2×, 4×) were employed to resolve the whole spectrum of drop sizes. With each magnification 1000 frames were captured to obtain samples with over 5500 drops that ensured statistical confidence limits of 95 % (Bowen & Davies, 1951). The magnifications allowed resolving drops with sizes ranging from 5 µm to 366 µm and velocities of up to 185 m/s. The field of view with both magnifications was 2.561×2.561 mm², which allowed combination of the samples obtained from both to carry out statistical analysis of the data. The calibration (i.e., µm/pixel) provided for the camera, lens and magnifications used was validated measuring standard circles in a reticule and standard line spacings in a grating and the agreement was better than 0.5 %. The traversing rail shown in the figure moved the diffuser and camera to 7 prescribed x-positions (0.0013, 0.030, 0.059, 0.088, 0.116, 0.145 and 0.174 m), while the y and z_s positions where maintained constant at 0 m and 0.175 m, respectively. Differently from the measurements with the patternator the measurements with the PDIA system were carried out with the nozzle oriented vertically downward, but it was experimentally verified that the distributions of d_d and u obtained with both orientations were not significantly different.

The water for the pneumatic nozzle was supplied from a reservoir using an immersion pump instrumented with a digital turbine flow-meter, a valve and a digital pressure gauge. A compressor provided the air and this line was instrumented with an automatically controlled valve, to minimize flow rate variations, a mass flow-meter and a digital manometer. The results reported in this article are for a Casterjet 1/2-6.5-90 nozzle (Spraying Systems Co., Chicago, IL), whose operating diagram is displayed in Figure 2. The conditions investigated are indicated by the triangles drawn in the plot, and it is seen that they correspond to constant W with different air inlet pressures, p_a, and vice versa.

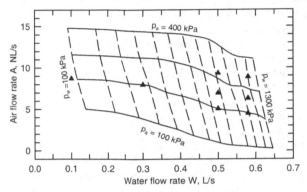

Fig. 2. Measured operating diagram of a Casterjet 1/2-6.5-90 nozzle

3. Mathematical model and computational procedure

3.1 System considered and assumptions

The 3-D system domain considered in the model is shown in Fig. 3(a), it includes the two-phase free-jet issuing from a pneumatic nozzle and the surrounding environment; the mixing chamber is excluded from the analysis. Since the visualization of the jets and the

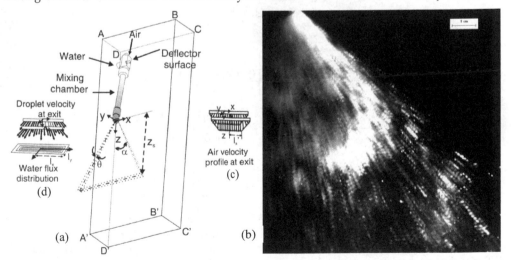

Fig. 3. (a) Schematic of system considered and computational domain, (b) quadruple exposure PIV image of drops in the neighborhood of the nozzle orifice, (c) schematic of assumed air-velocity profiles at nozzle exit and (d) schematic of assumed drop velocity profiles and water flux distribution at nozzle exit

measurements of their impact footprints indicated double symmetry over the x–z and y–z planes the computational domain involved just one quarter of the physical domain, as seen in the figure. Additionally, since the air- and water-flow rates were stable it was assumed that on a time average basis, the flow characteristics of the two-phase jet could be simulated in steady-state conditions. For their treatment the continuous air-phase was considered in an Eulerian frame of reference and the discrete droplets were regarded in a Lagrangian frame. The assumptions for the model were: (a) the liquid emerges from the nozzle as drops. This is supported by PIV observations done close of the nozzle orifice, as that displayed in Figure 3(b). This figure shows a quadruple-exposure photograph with trails of 4-images of droplets. Additionally and in agreement with PDIA observations the drops are assumed spherical; (b) the size distribution of the drops exiting the orifice is equal to the distribution measured at a distance $z = z_s$ (i.e., at the typical working distance of a given nozzle). This is reasonable since drop coalescence and break-up are rare events. The low volume fraction, α_d, of the drops prevents coalescence and the PDIA images, taken at different positions in the free mists, rarely show droplet break-up; (c) drops of all the specified sizes leave the orifice at the terminal velocity reached in the mixing chamber while dragged by the air. Calculations indicate that this would be the case for individual drops and since α_d is low the assumption would seem reasonable for the dilute multi-drop system moving within the chamber; in the mixing chamber $\alpha_d < 0.08$; (d) the droplets in the jet do not interact with each other and only interact with the air through interfacial drag, and (e) the air and the droplets are at room temperature and condensation and vaporization are negligible.

3.2 Governing equations

Under the considerations just described, the governing equations for the motion of the air are: the continuity equation (2), the Navier-Stokes equations (3) and the turbulence transport equations (4) and (5), which are expressed as follows,

$$\frac{\partial U_i}{\partial x_i} = 0 \tag{2}$$

$$\rho U_i \frac{\partial U_j}{\partial x_i} = -\frac{\partial p}{\partial x_j} + \frac{\partial}{\partial x_j}\left[(\mu + \mu_t)\left(\frac{\partial U_i}{\partial x_j} + \frac{\partial U_j}{\partial x_i}\right)\right] + S_i \tag{3}$$

$$\rho U_i \frac{\partial k}{\partial x_i} = \frac{\partial}{\partial x_i}\left[\left(\mu + \frac{\mu_t}{\sigma_k}\right)\frac{\partial k}{\partial x_i}\right] + \mu_t\left(\frac{\partial U_i}{\partial x_j} + \frac{\partial U_j}{\partial x_i}\right)\frac{\partial U_i}{\partial x_j} - \rho\varepsilon \tag{4}$$

$$\rho U_i \frac{\partial \varepsilon}{\partial x_i} = \frac{\partial}{\partial x_i}\left[\left(\mu + \frac{\mu_t}{\sigma_\varepsilon}\right)\frac{\partial \varepsilon}{\partial x_i}\right] + f_1 C_1 \mu_t \frac{\varepsilon}{k}\left(\frac{\partial U_i}{\partial x_j} + \frac{\partial U_j}{\partial x_i}\right)\frac{\partial U_i}{\partial x_j} - \rho f_2 C_2 \frac{\varepsilon^2}{k} \tag{5}$$

S_i in Eq. (3) is the source term expressing the i-direction momentum transferred between the air and the drops in a given cell of the fixed Eulerian grid over a Lagrangian time step. It is equal to the change in the momentum (only due to interfacial drag) of the drops following all the trajectories traversing a cell over that time step (Crowe et al., 1977) and it is given as,

$$S_i = \frac{\pi \rho_d}{6 v_{cell}} \sum_{k=1}^{n_{cell}} (N_d\, d_d^3\, (u_{i,out} - u_{i,in}))_k \tag{6}$$

The meaning of the symbols appearing in the equations is given in Section 7. The turbulence of the air was treated by the k-ε model for low Reynolds flows of Lam-Bremhorst modified by Yap, 1987. The constants and functions appearing in Eqs. (4) and (5) are listed in Table 1. Also, under the considerations done in Sec. 3.1 the equation of motion for individual drops in the mist and under the effects of aerodynamic drag and gravity is expressed as,

$$\frac{du_i}{dt} = \frac{3}{4} C_D \frac{\rho}{\rho_d d_d} |U_i - u_i| (U_i - u_i) + (1 - \frac{\rho}{\rho_d}) g_i \tag{7}$$

The drag coefficient C_D was assumed to vary with the particle Reynolds number, Re_d, according to the expressions given in Table 1. The trajectory of the drops was computed from the variation with time of the components of the position-vectors, according to,

$$\frac{dx_i}{dt} = u_i \tag{8}$$

Constants and functions involved in the turbulence model
$C_1= 1.44$; $C_2= 1.92$; $C_d= 0.09$; $\sigma_k= 1.0$; $\sigma_\varepsilon= 1.3$
$f_\mu = (1-\exp(-0.0165 Re_{z'}))^2 \left(1+\dfrac{20.5}{Re_t}\right)$; $f_1 = 1 + \left(\dfrac{0.05}{f_\mu}\right)^3$; $f_2 = 1 - \exp(-Re_t^2)$ where,
$\mu_t = \rho f_\mu C_d \dfrac{k^2}{\varepsilon}$; $Re_{z'} = \dfrac{\sqrt{k}z'}{v}$; $Re_t = \dfrac{k^2}{v\varepsilon}$
Drag coefficient expressions
Stokes law region, $Re_d< 2$ $C_D = 24/Re_d$
Intermediate region, $2 \leq Re_d \leq 500$ $C_D = 10/\sqrt{Re_d}$
Newton's law region, $500 \leq Re_d \leq 2\times10^5$ $C_D = 0.44$
where $Re_d = \dfrac{d_d|U_i - u_i|}{v}$

Boundary conditions at nozzle exit		
Drop-phase		Air-phase
$u_{x,k} = u_{z,t} \sin(\alpha\, x/l_x)$ $u_{y,k} = u_{z,t} \sin(\theta\, y/l_y)$ $u_{z,k} = [u_{z,t}^2 - u_{x,k}^2 - u_{y,k}^2]^{1/2}$	$0 \leq x \leq l_x'$	$U_x= 0$, $U_y= 0$, $U_z= U_{z,max}$
	$l_x' \leq x \leq l_x$	$U_x= U_{z,max}(\tan \alpha_x)$, $U_y= 0$ $U_z= U_{z,max}(l_x - x)/(l_x - l_x')$ $\alpha_x = \alpha\,(x - l_x')/(l_x - l_x')$ $U_{z,max}= (A+W)/\,[2l_y(l_x + l_x')]$ *
	$0 \leq x \leq l_x$	$k_o = 0.01 U_{z,max}^2$; $\varepsilon_o = 2k_o^{1.5}(l_x +l_y)/(4l_x l_y)$

Physical properties
$\rho= 1.02$ kg/m³; $\mu= 1.8\times10^{-5}$ Pa s; $\rho_d= 998$ kg/m³
Nozzle dimensions and parameters
$l_x= 0.01$ m; $l_y= 0.00325$ m; $l_x'= 0.00585$ m
$z_s= 0.175$ m; $\alpha= 45°$; $\theta= 10°$

*A is computed at local conditions: 25°C, 86 kPa.

Table 1. Auxiliary equations, properties and dimensions

3.3 Boundary and initial conditions

At the boundaries of the calculation domain shown in Fig. 3(a), Eqs. (2) through (5) describing the turbulent motion for the air-phase were solved imposing the following conditions:

- Ambient conditions at boundaries ABCD, A'B'C'D', ABB'A' and BCC'B', specified as,

$$P = P_{amb} \; ; k = e = 0 \tag{9}$$

To approach these conditions, the boundaries were located far away from the jet.

- Symmetry conditions at the boundaries ADD'A', DCC'D', given as:

$$U_j = \frac{\partial U_i}{\partial x_j} = \frac{\partial k}{\partial x_j} = \frac{\partial \varepsilon}{\partial x_j} = 0 \tag{10}$$

where j represents the index for the coordinate normal to the respective symmetry plane.

- Non-penetration and non-slip conditions were specified at the external wall of the nozzle,

$$U_i = k = \varepsilon = 0 \tag{11}$$

- Air velocity profiles as those shown in Fig. 3(c) were specified at the nozzle orifice, which corresponds to an internal boundary. Along the x-direction these profiles were uniform over the length l_x' of the flat hollow portion of the flanged orifice, in the rest of the orifice the profiles decreased to zero varying in angle from 0 deg to α deg at the edge; the distributions were the same throughout the whole thickness (y-)direction of the orifice. These velocity profiles were suggested by the geometry of the flanged orifice and are supported by the results presented in Section 4. The expressions describing the profiles are listed in Table 1, together with the expressions for the turbulence kinetic energy and the dissipation rate of turbulence kinetic energy at this boundary, k_o and ε_o, respectively.

- Positions and velocities were specified to the droplets as initial conditions for the solution of their motion (7) and trajectory (8) equations. For doing this, a first step was to decide a series of criteria to distribute throughout the orifice drops of different size and velocity in a random fashion that reflected that the water flux profile decreases from its center to its edges. To do this, the orifice was simulated as a grid of ports, k, releasing drops satisfying the diameter distribution measured at z= z_s (according with the assumption indicated in Section 3.1) for the particular set of nozzle operating conditions under consideration. For deciding the number of ports assigned to each drop size category it is important to establish what type of distribution to use, number or volume frequency? Figure 4 shows both size distributions measured for a representative set of operating conditions W and p_a. The number frequency distribution shows that droplets smaller than 25 μm account for a large number percentage of the drops (82.85 % of them), but that they represent only 6.82 % of the volume of the drops in the sample. Since the number of ports that can be used cannot be excessively large the assignment of the ports according to the number frequency distribution would leave many sizes unrepresented. The volume frequency distribution does not present this disadvantage and was chosen to designate the number of ports for each size category. The size assigned to each port was done through a random number generator to simulate the stochastic emergence of drops with different characteristics from distinct sites of the orifice. The drops with volume $v_{d,k} = (\pi d_d^3)_k/6$ exiting the ports k with a number frequency, $N_{d,k}$, had to satisfy the water-flow rate, W, according to the following expression,

$$W = \sum_{k=1}^{n_T} N_{d,k} v_{d,k} \tag{12}$$

Furthermore, based on the form of the mist footprint obtained with a patternator the water flow was assumed to be distributed in the orifice according to an obelisk-shaped distribution (Camporredondo et al., 2004), as that shown schematically in Figure 3(d).

With the criteria given, the initial conditions for Eqs. (7) and (8) were assigned to each port of the nozzle orifice, such that at t = 0 the position and velocity of the drop are specified as follows,

$$u_i = (u_i)_k \text{ and } x_i = (x_i)_k \tag{13}$$

As mentioned in Sec. 3.1, the initial velocity, $(u_i)_k$, of the drops was prescribed by assuming that the drops exit with the terminal velocity that they reach in the mixing chamber (Minchaca et al., 2010). As suggested by the observed drop trajectories (Hernández et al., 2008), according to the position assigned to the drop the angle of the velocity varied from 0 deg to α deg in the x-direction and from 0 deg to θ deg in the y-direction. A schematic representation of the velocity vectors of the drops is displayed in Fig. 3(d), and the expressions for the $u_{x,k}$, $u_{y,k}$ and $u_{z,k}$ velocity components are given in Table 1. Also, the physical properties of the fluid, the dimensions of the orifice and the angles of expansion of the jet are given in the table.

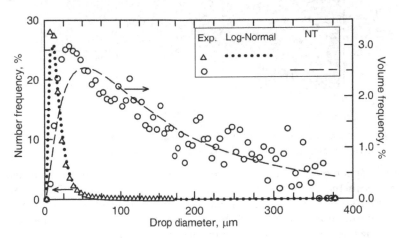

Fig. 4. Numeric and volume frequency drop size distributions measured for W= 0.58 L/s and p_a= 279 kPa. The log-normal and Nukiyama-Tanasawa, NT, distributions fitted to the respective data are included

3.4 Solution procedure

The Eulerian and Lagrangian equations of the model were solved using the control volume method and the particle tracking facility implemented in Phoenics. The mesh used had 128×25×99 control volumes to achieve mesh independent results and the number of ports was 100×12 in the x-y directions. The convergence criterion specified a total residual for all the dependent variables ≤ 10⁻³.

4. Experimental and computational results and discussion

4.1 Drop velocity and size correlation

As indicated in Sec. 2 the drop size and velocity measurements were done at 7 sampling volumes spaced along the x-direction at y= 0 and z= z_s. Figure 5 shows the correlation between drop velocity and diameter for the set of measurements done over all the sampling volumes. Droplets of all sizes exhibit a wide range of velocities when they arrive to the measuring positions. It is noticed that as the droplet diameter decreases their velocities exhibit a broader range, many small drops arrive at the measuring axis with small velocities and this causes a weak positive correlation between droplet velocity and diameter, i.e., the results denote a slight trend in the velocities to be larger as the size of the drops increases. Correlation coefficients, for several conditions and positions, were evaluated quantitatively in another work and confirm the weak positive correlation appreciated for the particular case illustrated in the figure (Minchaca et al., 2011).

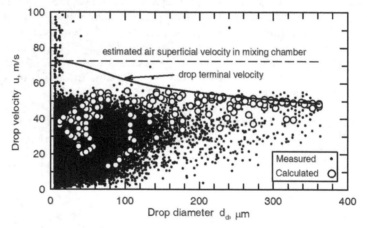

Fig. 5. Measured and calculated velocities for drops of different sizes reaching the x-axis at y= 0 and z= z_s. The results are for W= 0.58 L/s and p_a= 279 kPa

Figure 5 also shows results of computed velocities for drops arriving at the measuring axis. In agreement with the experiments, the results of the model reveal that the small drops exhibit a broader spectrum of velocities than the larger ones, causing the development of a weak positive correlation of velocity with size. Considering that the model assumes that all drops of each size leave the nozzle at the particular terminal velocity that they reach in the mixing chamber, as a result of the drag exerted by the air, the results of the figure indicate that the smaller droplets are more susceptible to lose their momentum while moving in the mist jet interacting with the air. This is evidenced by comparing the dispersion results with the calculated terminal velocity curve included in the figure. From the nozzle exit the drops follow ballistic nearly rectilinear trajectories and drops of the same size, which according to the model exit at the same velocity, will decelerate more when leaving from external than from internal positions of the nozzle orifice, so that they travel in the periphery of the jet interacting with the quiescent environment. This statement is supported by the experimental and computational results displayed in Fig. 6, which shows the variation of the normal and tangential velocity components of drops of different size traversing the sampling volumes located at the different x-positions. The regression curves fitted to the experimental and computational results show an excellent agreement. The dispersion exhibited by the computed results displayed in Figs. 5 and 6 is smaller than the experimental due to the very different number of drop trajectories traversing the sampling volumes, thousands in the experimental case versus a few tens in the computational case. Despite of this the model is able to represent very well the trend in the behavior of the actual system.

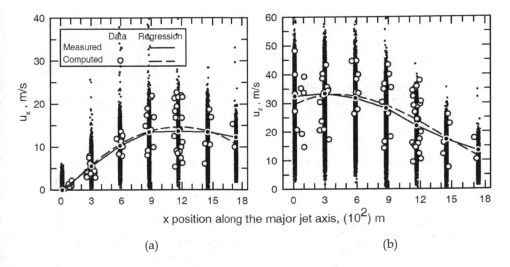

(a)   (b)

Fig. 6. Measured and computed velocity components as a function of x-position for drops of all sizes: (a) tangential velocity component and (b) normal velocity component. The results are for $W= 0.58$ L/s and $p_a= 279$ kPa

4.2 Effect of nozzle operating conditions on the velocity of the drops
4.2.1 Effect of water flow rate at constant air inlet pressure
In the application of air-mist nozzles for the cooling of surfaces at high temperature it is common to vary the water flow rate maintaining constant the air inlet pressure. This procedure would be equivalent to move along the curves of constant p_a appearing in the operating diagram of Figure 2. The reason behind this is that the spray cooling intensity is commonly associated only to the flux of water impinging upon the hot surface, when actually there is another mist parameter that plays an important role and this is the velocity of the drops (Hernández et al., 2011). Experimentally, it has been found that the velocity of the drops increases with W up to a certain value, but once this value is exceeded the opposite effect takes place and the drop velocity decreases markedly (Minchaca et al., 2011). With the increase in W at constant p_a the drops generated by the nozzle become larger (Minchaca et al., 2011) and the air flow rate, A, gets smaller as indicated by Figure 2. Both factors will alter the terminal velocities that the drops will reach at the nozzle exit and also their behavior in the free jet.

The multivariate effects that the droplet velocity experiences when changing W at constant p_a are complex. Therefore, it was important to examine the predictions of the CFD model in this regard. Figure 7 shows experimental and computational profiles of the normal and tangential volume weighed mean velocity components defined as,

$$u_{z,v} = \sum_{i=1}^{N} u_{z,i} d_{d,i}^3 / \sum_{i=1}^{N} d_{d,i}^3 \; ; \; u_{x,v} = \sum_{i=1}^{N} u_{x,i} d_{d,i}^3 / \sum_{i=1}^{N} d_{d,i}^3 \tag{14}$$

for four different water flow rates and a constant p_a= 205 kPa. It is seen that both, experimental and computational results, indicate that the increase in W from 0.1 to 0.3 L/s causes an increase in the normal and tangential velocity components and that further increase leads to a decrease in the velocities. The drop velocities obtained with W equal to 0.30 L/s and 0.58 L/s are considerably different, being substantially smaller for the higher W. This behavior could be one of the factors of why the heat transfer does not augment considerably when W and hence w do it (Montes et al., 2008). The computed velocities are somewhat larger than the experimental because the volume frequency distribution of sizes, chosen to establish the model, generates a greater number of large drops than small drops.

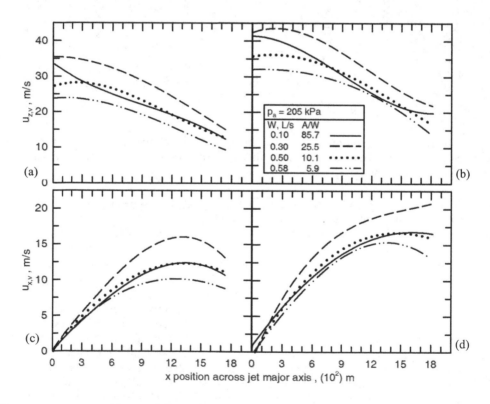

Fig. 7. Measured and computed volume weighed mean velocity components as a function of x-position for different W and a constant p_a. Normal velocity components: (a) measured, (b) computed. Tangential velocity components: (c) measured, (d) computed

4.2.2 Effect of air inlet pressure at constant water flow rate

The computed and measured x- and z-components of the volume weighed mean velocity are shown in Fig. 8, for conditions involving a constant W and different p_a. The agreement between computed and experimental results is quite reasonable and the curves show that as p_a increases both velocity components become larger. This behavior suggests that if the drops were to impinge over a surface, a more intimate contact would take place as p_a increases; this as a consequence of the higher impingement Weber numbers that would result. The larger tangential velocity component of the drops hints a faster renewal of the liquid on the surface as p_a increases. In fact, heat transfer experiments have shown a substantial increase in the heat flux with the increase in p_a at constant W (Montes et al., 2008; Hernández et al., 2011), suggesting that the change in the fluid dynamic behavior of the drops with the increase in p_a favors heat transfer. The phenomena occurring during the impingement of dense air-mist jets with solid surfaces is being investigated. The effect of p_a on the intensity of heat extraction could have important implications to achieve water savings during cooling operations. Similar to the results in Fig. 7, the computed velocities in Fig. 8 are somewhat larger than the experimental because the volume frequency distribution of sizes tends to generate a greater number of large drops than small drops.

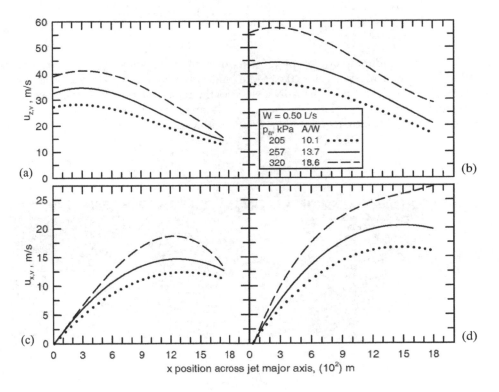

Fig. 8. Measured and computed volume weighed mean velocity components as a function of x-position for different p_a and constant W. Normal velocity components: (a) measured, (b) computed. Tangential velocity components: (c) measured, (d) computed

4.3 Effect of air inlet pressure on droplet volume fraction and water impact flux

As mentioned in Sec. 1 the mists have been classified in dense or dilute according to the value of the water impact density. However, little has been investigated about the actual mist density defined as the volume of liquid of the drops per unit volume of space; which is equivalent to the local liquid volume fraction, α_d. This parameter would give an indication of how critical could be to the model the assumption that the drops in the free jet do not interact as a consequence that they are far apart from each other. Although, the direct experimental measurement of this parameter is difficult the validity of the computational estimation of α_d can be tested by its relation with the water impact density. The local water volume fraction can be defined by the following expression,

$$\alpha_d = \frac{\sum\limits_{k=1}^{n_{cell}}(N_d v_d \Delta t)_k}{V_{cell}} \qquad (15)$$

and the water impact density can be evaluated as,

$$w(x,y) = \frac{\sum\limits_{k=1}^{n_A}(N_d v_d)_k}{A(x,y)} \qquad (16)$$

Figure 9 shows computed contour plots of α_d over the x-z symmetry plane of mist jets generated with a constant water flow rate and different air nozzle pressures. It is appreciated that as the air inlet pressure decreases the region close to the nozzle exhibits a higher liquid fraction. This behavior arises from the larger size and smaller velocities of the drops generated as p_a decreases. For conservation of mass this last factor would imply that

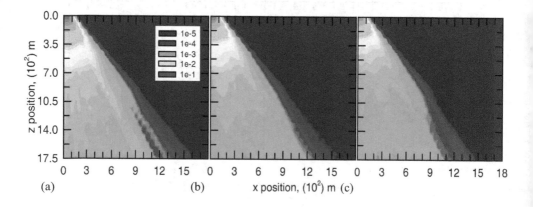

Fig. 9. Computed contour maps of α_d over the main symmetry plane of the mist jet for a W= 0.50 L/s and p_a of: (a) 205 kPa, (b) 257 kPa and (c) 320 kPa

the drops travel more closely spaced as p_a decreases, leading this to a higher particle packing. This behavior continues up to z= z_s where regions of higher liquid volume fraction are more widely spread in the case of smaller p_a.

The predicted liquid fraction contours indicate that the increase in p_a at constant W causes a redistribution of the liquid in the free jet that could affect the water impact density. In previous w measurements no effect was detected (Hernández et al., 2008). Thus, it was decided to refine the patternator to try to reveal if there was an influence of p_a on the water impact density for a constant W. Figure 10 shows measured and computed water impact density maps for p_a of 257 kPa and 320 kPa with W= 0.50 L/s. It is seen that both results agree very well and indicate an increase of the water impact density with the increase in p_a, in the central region. Based on the model, this result points out that although at higher p_a the drops tend to travel more widely spaced from each other, having a lower volume fraction, their higher velocities causes them to arrive more frequently to the collecting cells of the patternator or equivalently to the virtual impingement plane (in the case of the model), leading this to higher w in the central region of the footprint. The differences observed in the figure between experimental and computed results are mainly in the sizes of the footprint. This discrepancy arises because the model considers the nominal value of the expansion semi-angle α and a semi-angle θ= 10 deg. However, the w maps and the visualization of the jets indicate that the actual angles were slightly larger than the nominal values.

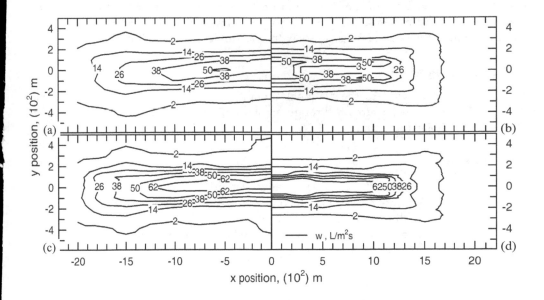

Fig. 10. Contour maps of w at z= z_s for a W= 0.50 L/s with p_a= 257 kPa: (a) experimental, (b) computational and with p_a= 320 kPa: (c) experimental, (d) computational

5. Summary and conclusions

A comprehensive experimental and computational study of the fluid dynamics of air-mist jets generated under conditions of interest in the cooling of surfaces found in metallurgical processes and in other high temperature processes was carried out. The work analyzes the variation of droplet velocity and water impact density as a function of droplet size and nozzle operating conditions.

The rigorous determination of statistically meaningful samples of droplet size allowed to establish that a relatively small numeric proportion (~0.17) of 'large' drops (between ~25 μm and 370 μm) are responsible for transporting a large fraction of the water in the mist. The smaller drops transport only a small proportion of the liquid volume. This fact, which is typical of the different nozzles and operating conditions, suggested that the description of the drop sizes mainly responsible for carrying the water arriving at a given distance from the nozzle orifice would be crucial for formulating a 3-D turbulent Eulerian-Lagrangian model for simulating the dynamics of fan-shaped air-mists. Thus, the present model considers the volume frequency distribution of sizes, instead of the number frequency distribution, to prescribe the inlet ports assigned to the different sizes. This specification, together with those for the distributions of the velocity of the air and drops and of the water volume flux at the orifice, was critical in the results of the model.

The model predicts very well the correlation between drop velocity and size, indicating that the finest drops tend to decelerate rapidly when traveling in the periphery of the mist. Also, it gives an accurate description of the influence that the variation in water flow rate at constant air inlet pressure, and of the variation in air inlet pressure at constant water flow rate, has on the velocity of the drops and on the water impact density distribution. The results on the fluid dynamics of free mist jets are being very useful to develop models for simulating the interaction of dense mists with solid surfaces.

6. Acknowledgments

The authors are grateful to the National Council of Science and Technology of Mexico (CONACYT) for financial support through grant No. 57836. JIMM wish to thank CONACYT for his Ph.D scholarship grant.

7. Nomenclature

a	Area of collector tube in patternator, m^2
A	Air flow rate at normal conditions (i.e., 0°C, 101.3 kPa), NL s^{-1}, or ambient conditions (25°C, 86 kPa), L s^{-1}
$A(x,y)$	Local area in impact plane centered around coordinates x, y.
C_1, C_2, C_d	Constants in the turbulence model
C_D	Drag coefficient
d_d	Drop diameter, m
f_μ, f_1, f_2	Functions defined in Table 1
g	Acceleration due to gravity, m s^{-2}
k, k_o	Turbulence kinetic energy; at nozzle orifice, m^2 s^{-2}

l_x, l_y, l_x'	Half length; half width nozzle orifice; half length of hollow portion of nozzle orifice, m
n	Number of drop trajectories
n_p	Port number or number of ports
N_d	Number frequency of drops, s^{-1}
p, P	Pressure in Ecs. (3); in Eq. (9), kPa
p_a, p_w	Air-; water nozzle inlet pressures, kPa
$Re_d, Re_t, Re_{z'}$	Reynolds number defined in Table 1
S	Source term for momentum transfer interaction between the drops and the air, $m\ s^{-2}$
t	Time, s
T_s	Saturation temperature of water, °C
T_w	Surface temperature, °C
u	Velocity of drops, $m\ s^{-1}$
$u_{x,v}, u_{z,v}$	Tangential; normal volume weighed mean velocity, $m\ s^{-1}$
u_{zs}	Normal drop velocity at $z=z_s$, $m\ s^{-1}$
U	Velocity of the continuous phase (air), $m\ s^{-1}$
$U_{z,max}$	Velocity of air phase defined in Table 1
v	Volume of water collected in bottles of patternator, L; volume, m^3
v_d	Volume of drop, m^3
w	Water impact flux or water impact density, $L\ m^{-2}s^{-1}$
W	Water flow rate, $L\ s^{-1}$
We_{zs}	Impinging droplet Weber number
x	Coordinate, m
x, y, z	Rectangular coordinates, m
z_s	Setback distance of nozzle tip from plane of interest, m

Greek symbols

α, θ	Jet expansion half angles in x and y directions, deg
α_d	Volume fraction of drops, dimensionless
γ	Angle defined in Eq. (1)
Δt	Time interval, s
$\varepsilon; \varepsilon_o$	Dissipation rate of turbulence kinetic energy; at nozzle orifice, $m^2\ s^{-3}$
μ, μ_t	Continuous-phase molecular; turbulent dynamic viscosity, Pa s
ν	Kinematic viscosity, $m^2\ s^{-1}$
ρ, ρ_d	Continuous; discontinuous-phase density, $kg\ m^{-3}$
σ	Surface tension of drop phase, $N\ m^{-1}$
$\sigma_k, \sigma_\varepsilon$	Laminar and turbulent Schmidt numbers for k and ε

Subscripts

amb	Ambient conditions, P = 86 kPa, T = 25 °C
cell	Discretization cell
i, j	Indexes for coordinate directions
in, out	Input, output to control volume
k	Ports or trajectories
max	Maximum

t	terminal
T	Total
x, y, z	Coordinates directions

8. References

Araki, K. & Moriyama, A. (1981). Theory on Deformation Behavior of a Liquid Droplet Impinging onto Hot Metal Surface. *Transactions of the Iron and Steel Institute of Japan*, Vol. 21, No. 8, (n.d. 1981), pp. 583–590, ISSN 0021-1583.

Babinsky, E. & Sojka, P. (2002). Modeling Drop Size Distributions. *Progress in Energy and Combustion Science*, Vol. 28, No. 4, (February 2002), pp. 303-329.

Beck, J. & Watkins, A. (2002). On the Development of Spray Submodels Based on Droplet Size Moments. *Journal of Computational Physics*, Vol. 182, No. 2, (November 2002), pp. 586-621.

Bendig, L., Raudensky, M. & Horsky, J. (1995). Spray parameters and heat transfer coefficients of spray nozzles for continuous casting, *Proceedings of Seventy-eighth Steelmaking Conference*, Warrendale, PA, USA, April 2-5, 1995.

Bowen, I. & Davies, G. (1951). Technical Report ICT 28. *Shell Research Ltd.*, London, 1951.

Bul, Q. (2001). Development of a Fan Spray Nozzle for Continuous Caster Secondary Cooling. *AISE Steel Technology*, Vol. 78, No. 5, (May 2001), pp. 35-38.

Camporredondo, J., Castillejos, A., Acosta, F., Gutiérrez, E. & Herrera, M. (2004). Analysis of Thin Slab Casting by the Compact-Strip Process: Part 1. Heat Extraction and Solidification. *Metallurgical and Materials Transactions B*, Vol. 35B, No. 3, (June 2004), pp. 541-560.

Ciofalo, M., Caronia, A., Di Liberto, M. & Puleo, S. (2007). The Nukiyama Curve in Water Spray Cooling: Its Derivation from Temperature-time Histories and Its Dependence on the Quantities that Characterize Drop Impact. *International Journal of Heat and Mass Transfer*, Vol. 50, No. 25-26, (December 2007), pp. 4948-4966.

Crowe, C., Sharma, M. & Stock, D. (1977). The Particle-Source-In-Cell (PSI-CELL) Model for Gas-Droplet Flows. *Journal of Fluids Engineering*, Vol. 99, No. 2, (June 1977), pp. 325-332.

Crowe, C., Schwarzkopf, J., Sommerfeld, M. & Tsuji, Y. (1998). *Multiphase Flows with Droplets and Particles*, CRC Press LLC, ISBN 0-8493-9469-4. Boca Raton, FL, USA.

Deb, S. & Yao, S. (1989). Analysis on Film Boiling Heat Transfer of Impacting Sprays. *International Journal of Heat and Mass Transfer*, Vol. 32, No. 11, (March 1989), pp. 2099–2112.

Hatta, N., Fujimoto, H., Ishii, R. & Kokado, J. (1991a). Analytical Study of Gas-Particle Two-Phase Free Jets Exhausted from a Subsonic Nozzle. *ISIJ International*, Vol. 31, No. 1, (n.d. 1991), pp. 53-61, ISSN 0915-1559.

Hatta, N., Fujimoto, H. & Ishii, R. (1991b). Numerical Analysis of a Gas-Particle Subsonic Jet Impinging on a Flat Plate. *ISIJ International*, Vol. 31, No. 4, (n.d. 1991), pp. 342-349, ISSN 0915-1559.

Hernández, I., Acosta, F., Castillejos, A. & Minchaca, J. (2008). The Fluid Dynamics of Secondary Cooling Air-Mist Jets. *Metallurgical and Materials Transactions B*, Vol. 39B, No. 5, (October 2008), pp. 746-763.

Hernández, C., Minchaca, J., Castillejos, A., Acosta, F., Zhou, X. & Thomas, B. (2011). Measurement of Heat Flux in Dense Air-Mist Cooling: Part II. The Influence of Mist Characteristics on Heat Transfer. Submitted for publication.

Issa, R. & Yao, S. (2005). Numerical Model for Spray-Wall Impaction and Heat Transfer at Atmospheric Conditions. *Journal of Thermophysics and Heat Transfer*, Vol. 19, No.4, (October–December 2005), pp.441-447.

Jenkins, M., Story, S. & David, R. (1991). Defining Air-mist Nozzle Operating Conditions for Optimum Spray Cooling Performance, *Proceedings of the Nineteenth Australasian Chemical Engineering Conference*, Newcastle, New South Wales, Australia, September 12–20, 1991.

Lefebvre, A. (December 1989). *Atomization and Sprays*, Taylor & Francis Group, ISBN 9780891166030. Boca Raton, FL., USA.

Minchaca, J., Castillejos, A., Acosta, F. & Murphy, S. (2010). Fluid Dynamics of Thin Steel Slab Continuous Casting Secondary Cooling Zone Air Mists, *Proceedings of Twenty-second ILASS-Americas Conference on Liquid Atomization and Spray Systems*, Cincinnati, OH, USA, May 2010.

Minchaca, J. (2011). *Estudio de las Distribuciones de Velocidad y Tamaño de Gota en Rocíos y Nieblas de Agua usados en la Colada Continua del Acero*. Ph.D. Thesis in preparation, CINVESTAV- Saltillo, Coahuila, México.

Minchaca, J., Castillejos, A. & Acosta, F. (2011). Size and Velocity Characteristics of Droplets generated by Thin Steel Slab Continuous Casting Secondary Cooling Air-mist Nozzles. *Metallurgical and Materials Transactions B*, Vol. 42B, No. 3, (June 2011), pp. 500-515.

Montes, J., Castillejos, A., Acosta, F., Gutiérrez, E. & Herrera, M. (2008). Effect of the Operating Conditions of Air-mists Nozzles on the Thermal Evolution of Continuously Cast Thin Slabs. *Canadian Metallurgical Quarterly*, Vol. 47, No. 2, (n.d. 2008), pp. 187-204.

Nasr, G., Yule, A. & Bendig, L. (Eds.). (September 2002). *Industrial Sprays and Atomization: Design, Analysis and Applications*, Springer, ISBN 1-85233-611-0, Great Britain, UK.

Puschmann, F. & Specht, E. (2004). Transient Measurement of Heat Transfer in Metal Quenching with Atomized Sprays. *Experimental Thermal and Fluid Science*, Vol. 28, No. 6, (June 2004), pp. 607–615.

Sozbir, N., Chang, Y. & Yao, S. (2003). Heat Transfer of Impacting Water Mist on High Temperature Metal Surfaces. *Transactions of the ASME, Serie C: Journal of Heat Transfer*, Vol. 125, No. 1, (n.d. 2003), pp. 71-74, ISSN 0022-1481.

Totten, G. & Bates, C. (Eds.). (January 1993). *Handbook of Quenchants and Quenching Technology*, ASM International, The Materials Information Society, ISBN 087170448X, Clinton, OH., USA.

Wachters, L. & Westerling, N. (1966). The Heat Transfer from a Hot Wall to Impinging Water Drops in the Spheroidal State. *Chemical Engineering Science*, Vol. 21, No. 11, (November 1966), pp.1047-1056.

Yap, C. (n.d. 1987). *Turbulent Heat and Momentum Transfer in Recirculating and Impinging Flows*. Ph.D. Thesis, University of Manchester, Manchester, UK.

3

Direct Numerical Simulations of Compressible Vortex Flow Problems

S.A. Karabasov[1] and V.M. Goloviznin[2]
[1]University of Cambridge Department of Engineering
[2]Moscow Institute of Nuclear Safety, Russian Academy of Science
[1]UK
[2]Russia

1. Introduction

Vortical flows are one of the most fascinating topics in fluid mechanics. A particular difficulty of modelling such flows at high Reynolds (Re) numbers is the diversity of space and time scales that emerge as the flow develops.

For compressible flows, in particular, there are additional degrees of freedom associated with the shocks and acoustic waves. The latter can have very different characteristic amplitudes and scales in comparison with the vorticity field. In case of high Re-number flows, the disparity of the scales becomes overwhelming and instead of Direct Numerical Simulations (DNS) less drastically expensive Large Eddy Simulations (LES) are used in which large flow scales are explicitly resolved on the grid and the small scales are modelled. For engineering applications, examples of unsteady vortical flows include the interaction of wakes and shocks with the boundary layer in a transonic turbine and vorticity dissipation shed due to the temporal variations in blade circulation that can have a profound loss influence and affect the overall performance of a turbomachine (e.g., Fritsch and Giles, 1992; Michelassi et al, 2003). Another example is dynamics and acoustics of high-speed jet flows that is affected by the jet inflow conditions such as the state of the boundary layer at the nozzle exit (e.g., Bogey and Bailly, 2010). The computational aspects involved in the modelling of such complex flows, typically, include the issues of high-resolution numerical schemes, boundary conditions, non-uniform grids and the choice of subgrid scale parameterization in case of LES modelling.

Stepping back from this complexity to more idealised problems, two-dimensional (2D) vortex problems are a key object for testing different modelling strategies. Such reduced-order systems play an important role in the understanding of full-scale flow problems as well as in benchmarking of computational methods.

One example of such important idealised systems is isolated vortices, their interaction with acoustic waves and also nonlinear dynamics when interacting with each other. In particular, such vortical systems are a classical problem in the theory of sound generation and scattering by hydrodynamic non-uniformities (e.g., Kreichnan, 1953; Howe, 1975)

The structure of the chapter is the following. In part I, an outline of unsteady computational schemes for vortical flow problems is presented. In part II, the test problem of a stationary inviscid vortex in a periodic box domain is considered and a few numerical solutions obtained with unsteady Eulerian schemes are discussed. Part III is devoted to the sound scattering by a slowly decaying velocity field of a 2D vortex. In part IV, the canonical problem of 2D leapfrogging vortex pairs is considered and numerical solutions based on the Eulerian and Lagrangian approach are discussed.

2. Numerical methods for solving unsteady flow problems sensitive to vortex dynamics

Numerical dissipation and dispersion are typical drawbacks of the Eulerian computational schemes (e.g., Hirsch, 2007). These drawbacks are partially overcome in the Lagrangian and mixed Eulerian-Lagrangian methods, which describe flow advection by following fluid particles, rather than by considering fixed coordinates on the Eulerian grid (e.g., Dritschel et al, 1999). A remarkable property of the Lagrangian methods is that they are exact for linear advection problems with a uniform velocity field, therefore, in principle, their accuracy is limited only by the accuracy of solving the corresponding Ordinary Differential Equations (ODEs), rather than by the accuracy of solving the full Partial Differential Equations (PDEs), which is the case for the Eulerian schemes. This class of methods can be very efficient for simulations, which involve multiple contact discontinuities, e.g., in the context of multi-phase flows and strong shock waves (e.g., Margolin and Shashkov, 2004). However, for the problems where vorticity plays an important role, the standard Lagrangian-type methods have to be adjusted, after not many Lagrangian steps, by some ad-hoc 'repair' or 'contour surgery' procedure. The 'repair' procedure can be actually viewed as a special kind of numerical dissipation that is needed to stabilise the numerical solution.

For the Eulerian schemes, one of the frequently used approaches for improving the numerical dissipation and dispersion properties is based on using central schemes of high-order spatial approximation. The optimized schemes employ a non-conservative form of the governing equations, and, typically, use large computational stencils to replicate the spectral properties of the linear wave propagation in the (physical) space-time domain (Lele, 1992; Tam and Webb, 1993; Bogey and Bailly, 2004). By construction, such methods are particularly efficient in handling linear wave phenomena. The optimized finite-difference methods were developed to overcome typical problems of spectral and pseudo-spectral methods by handling non-periodic boundary conditions and large flow gradients which they handle with the use of hyper diffusion.

On the other hand, there is another popular approach, based on the conservation properties of the governing equations, that forms the basis for the so-called shock-capturing schemes. This is the family of methods based on the quasi-linear hyperbolic conservation laws (Roe, 1986; Toro, 2001; LeVeque, 2002). For improving the numerical properties in this approach, either a second-order or higher 'variable-extrapolation', or 'flux-extrapolation' techniques are used, such as in Method for Upwind Scalar Conservation Laws (MUSCL, Kolgan, 1972; B.van Leer, 1979), for enhancing linear wave properties of the solution away from the large-solution gradients discontinuities. The time stepping is usually treated separately from the spatial approximation and one popular method for time integration is multi-stage Runge-Kutta schemes (e.g, Hirsch, 2007).

To eliminate spurious oscillations of the solution obtained with the second- or higher-order schemes, in the vicinity of the discontinuities, local non-linear limiter functions are suggested, as, for example, in Totally Variation Diminishing (TVD) schemes (Boris et al.,

1975). By enforcing the TVD property on the solution, the limiter functions introduce implicit numerical dissipation. If the numerical dissipation gets too strong, artificial anti-diffusion terms are added to make the method less dissipative (Harten et al., 1987). The non-oscillatory methods are very attractive for computing flows with shocks. For weakly non-linear flow problems, however, the shock-capturing TVD schemes tend to introduce too much dissipation and for vortical flows, especially in acoustics sensitive applications, the limiters are recommended to switch off (e.g., Colonius and Lele, 2004), i.e., selectively use the non-oscillatory methods only for strong discontinuities.

One notable exception is the so-called Compact Accurately Adjusting high-Resolution Technique (CABARET) (Karabasov and Goloviznin, 2009). CABARET is the extension of Upwind Leapfrog (UL) methods (Iserlis, 1986; Roe, 1998; Kim, 2004; Tran and Scheurer, 2002) to non-oscillatory conservative schemes on staggered grids with preserving low dissipative and low dispersive properties. CABARET is an explicit conservative finite-difference scheme with second-order approximation in space and time and it is found very efficient in a number of Computational Fluid Dynamics (CFD) problems, (Karabasov and Goloviznin, 2007; Karabasov et al, 2009). In comparison to many CFD methods, CABARET has a very compact stencil which for linear advection takes only one computational cell in space and time. The compactness of the computational stencil results in the ease of handling boundary conditions and the reduction of CPU cost. For non-linear flows, CABARET uses a low-dissipative conservative correction method directly based on the maximum principle.

For collocated-grid schemes, the mainstream method of reducing numerical dissipation is to upgrade them to a higher order (typically, by extending its computational stencil). There is a broad range of recommendations on the subject, starting from Essentially or Weighted Essentially Non-Oscillatory schemes (ENO and WENO) (Liu et al, 1994) to Discontinuous Galerkin methods (Cockburn and Shu, 2001). All these methods show significant improvements in terms of preserving the linear flow properties, if compared with the conventional second-order schemes.

For illustration of numerical properties of different Eulerian schemes, Fig 1 shows the comparison of phase speed error and the non-dimensional group speed as a function of grid resolution for several semi-discrete central finite differences. E2, E4, E6 denote standard central differences of the second, fourth and sixth-order, respectively, DRP denotes the fourth order Dispersion Relation Preserving scheme by Tam and Webb; and LUI stands for the sixth order pentadiagonal compact scheme of Pade-type. CABARETx stands for the CABARET dispersion characteristic at various Courant number CFL=x. All solutions are shown as a function of the grid refinement parameter, $N_\lambda = \pi/(k\,h)$ and the non-dimensional wavenumber, $k\,h$, respectively. Note that the solutions for the second-order discretization are typical of the 'low-order' shock-capturing methods, e.g., the Roe MUSCL scheme, with the limiter switched off. Higher-order central schemes of the 4th and the 6th order are analogues to the high-order shock-capturing methods, such as WENO, in the smooth solution region. The results for two pseudo-spectral optimised dispersion schemes are also shown.

Note that the dispersion errors of semi-discrete schemes correspond to exact integration in time, which neglects the possible increase of dispersion error due to inaccuracies in time marching. For most Courant numbers and for a wide range of grid resolution (7-20 points per wavelength) the dispersion error of the CABARET scheme remains below that of the conventional and optimised fourth-order central finite differences and close to that of the six-order central schemes. Away from the optimal Courant number range (e.g., for

CFL=0.1), the CABARET dispersion error is similar to that of the conventional fourth-order scheme. Fig1b shows that the numerical group speed of central finite-difference schemes on coarse grids is negative that leads to spurious wave reflection and sets the limit to the minimum grid resolution if numerical backscatter is to be avoided (Colonius and Lele, 2004). In comparison with the central schemes, the CABARET group speed remains in the physically correct direction for all wavenumbers, i.e., the non-physical backscatter is always absent.

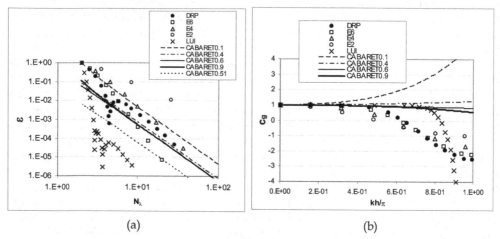

(a)   (b)

Fig. 1. Linear wave properties of several spatial finite-difference schemes: (a) phase errors and (b) normalised group speeds.

3. Steady vortex solution in a finite domain

Let's first consider a steady problem of isolated compressible Gaussian vortex in a square periodic domain. The vortex is specified in the centre of the box domain, as a perturbation to a uniform background flow with zero mean velocity $(\rho_\infty, \mathbf{u}_\infty, p_\infty) = (1,0,1)$:

$$\rho' = \rho_\infty \left[\left(1 - \frac{(\gamma-1)}{4\alpha\gamma} \varepsilon^2 \exp\{2\alpha(1-\tau^2)\} \right)^{\frac{1}{\gamma-1}} - 1 \right], u' = \varepsilon\tau \exp\{\alpha(1-\tau^2)\} \sin\theta,$$

$$v' = -\varepsilon\tau \exp\{\alpha(1-\tau^2)\} \cos\theta, p' = p_\infty \left[\left(1 - \frac{(\gamma-1)}{4\alpha\gamma} \varepsilon^2 \exp\{2\alpha(1-\tau^2)\} \right)^{\frac{\gamma}{\gamma-1}} - 1 \right], \quad (1)$$

$$\tau = r/L; r = \sqrt{(x-x_0)^2 + (y-y_0)^2}; \theta = \tan^{-1}((y-y_0)/(x-x_0)), \varepsilon = 0.3, L = 0.05, \alpha = 0.204.$$

To simplify the treatment of external boundary conditions, the box size is set 20 times as large as the vortex radius, L so that the vortex induced velocity vanishes at the boundaries. The vortex field corresponds to a steady rotation that is a stable solution of the governing compressible Euler equations (e.g., Colonius at al, 1994). The characteristic space scale of the problem is the vortex core radius L. It is also useful to introduce the time scale based on the vortex circulation time $T = 2\pi L / \varepsilon \approx 1.047$.

The analytical solution of the problem is trivial: at all time moments the solution remains equal to the initial conditions. From the viewpoint of unsteady computational schemes, however, preserving the vortex solution on a fixed Eulerian grid that is not specifically tailored to the initial vortex shape tends to be a challenge.

To illustrate the point we consider numerical solutions of this problem obtained with two high-resolution Eulerian methods mentioned in the introduction. These are the Roe-MUSCL scheme with and without TVD limiter (MinMod) and the CABARET method. The former method is based on the third-order MUSCL variable extrapolation in characteristic variables and the third order Runge-Kutta scheme for in time. The latter is based on a staggered space-time stencil and is formally second order. Note that the MinMod limiter used with the Roe MUSCL scheme is more robust for vortical flow computations in comparison with more 'compressive' limiters, e.g., SuperBee, that are better tailored for 1-D shock-tube problems. This is because the former is less subjected to the 'stair-casing' artefacts in smooth solution regions (e.g., see Hirsch, 2007). The Euler equations with the initial conditions (1) are solved on several uniform Cartesian grids: (30x30), (60x60), (120x120) and (240x240) cells. These correspond to the grid density of 1.5, 3, 6 and 12 grid spacings per the vortex core radius, respectively. Figs 2 show the grid convergence of the vorticity solution obtained with the CABARET method at control time t=100. The shape and the peak of the vortex is well preserved on all grids including the coarsest one. For qualitative examination, the kinetic energy integral has been computed $K(t) = \sum_{x,y} \rho u_i u_i$, as a fraction of its initial value $K(0)$. The

relative error $\varepsilon(t) = 1 - K(t) / K(0)$ of this nonlinear problem at t =100 shows approximately a linear decay with the grid size: it is 0.011 for grid (30x30), 0.0061 for grid (60x60), and 0.003 for grid (120x120). For the Roe-MUSCL scheme, the solution of the vortex problem is much more challenging. The activation of the TVD limiter leads to a notable solution smearing, which builds up with time, and which affects even the solution on the fine grid (120x120) (fig.3a). It is, therefore, tempting to deactivate the TVD limiter since in the case considered there are no shocks involved. Without the limiter, the Roe-MUSCL scheme initially preserves the vortex shape well (as in fig.3b). However, after a few vortex circulation times, spurious oscillations that correspond to the nonphysical propagation direction of the short scales (cf. fig.1b) grow until they completely contaminate the vortex solution (fig3c).

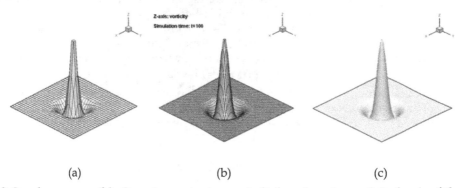

(a)   (b)   (c)

Fig. 2. Steady compressible Gaussian vortex in a periodic box domain: vorticity levels of the CABARET solution at time t=100 on (a) grid (30x30), (b) grid (60x60) , and (c) grid (120x120).

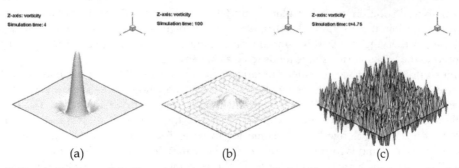

(a)   (b)   (c)

Fig. 3. Steady compressible Gaussian vortex in a periodic box domain: vorticity levels of the Roe-MUSCL solution on grid (240x240) cells with (a) MinMod limiter at time t=100, (b) MinMod limiter at time t=4, and (c) with the limiter deactivated at time t=5.

4. Sound scattering by a steady vortex

We next consider an isolated viscous-core vortex from Colonius at al (1994) that corresponds to a slowly decaying velocity field of constant circulation $\Gamma_\infty > 0$

$$v_r = 0$$
$$v_\theta = \frac{\Gamma_\infty}{2\pi r}(1 - \exp(-\alpha(r/L)^2))$$
$$\alpha = 1.256431, \ M_{max} = (v_\theta)_{max} / c_\infty$$
(2)

where c_∞ is acoustic far-field sound and L is the vortex core radius. The density and pressure field satisfy the usual isentropic relation and the steady tangential momentum equation:

$$p \cdot \rho^{-\gamma} = const;$$
$$\frac{\delta p}{\delta r} = \rho \frac{v_\theta^2}{r}$$
(3)

The vortex is specified in the centre of an open square Cartesian domain, three sides of which are open boundaries and the forth one corresponds to an incident acoustic wave that is monochromatic with frequency f. and normal to the boundary. The incident wave boundary condition is imposed at distance $R=10L$ from the vortex centre which is offset from the centre of the square computational box domain of linear size $40L$.

The velocity perturbations of the incident acoustic wave are several orders of magnitude as small as the maximum velocity of the vortex, $u'=1.e-5 \left(v_\theta\right)_{max}$. The problem has two length and time scales associated with the vortex circulation and the acoustic wave. The case of long acoustic wavelength $\lambda=2.5L$ is considered first.

The solution of the acoustic wave scattered by the vortex is sought in the form of the scattered wave component

$$p' = p - p^{(a)} - \delta p, \quad \delta p = (p^{(v)} - p_\infty)$$
(4)

where p is the full pressure field obtained as the solution of the sound wave interaction with the vortex, $p^{(a)}$ is the solution that corresponds to the acoustic wave propagating in the free space without any hydrodynamic perturbation, $p^{(v)}$ is the steady solution vortex without any incoming acoustic wave, and p_∞ is the pressure at the far field. Note, that from the numerical implementation viewpoint it is preferable to compute the scattered solution in form (5) instead of using $\delta p = (p_{t=0} - p_\infty)$ in order to account for a small systematic approximation error of the round vortex on a rectilinear Cartesian grid.

Colonius et al (1994) obtain the benchmark solution to this problem by using the 6-th order Pade-type compact finite-difference scheme in space and 6-th order Runge-Kutta integration in time with the grid density of 7-8 grid points per vortex radius. The reference solution corresponds to the Navier-Stokes equations at Reynolds number 10^5 integrated over four acoustic wave time periods in the open computational domain with well-tailored numerical boundary conditions to minimise numerical reflections from the boundaries.

It is interesting to compare the reference solution with the results obtained with the CABARET scheme and the third-order Roe-MUSCL-Runge-Kutta method from the previous section. To reduce the numerical dissipation error of the latter, the MinMod limiter has been deactivated. For CABARET, the complete formulation including the nonlinear flux correction is used. For the sake of comparison, the vortex with core Mach number $M_{max} = 0.25$ is considered. Characteristic-type nonreflecting boundary conditions and grid stretching close to the open boundaries are used to minimise artificial reflections.

Fig.4 shows the computational problem configuration and the distribution of the root-means-square (r.m.s.) of the scattered pressure fluctuations for the CABARET solution, where the vortex centre corresponds to the origin of the system of coordinates.

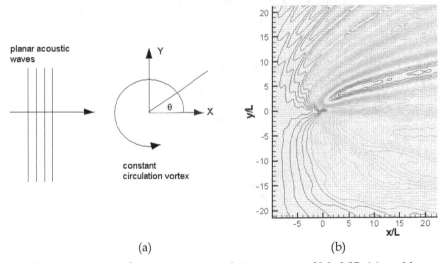

(a)   (b)

Fig. 4. Sound wave scattering by a non-zero circulation vortex of M=0.25: (a) problem configuration, (b) computed r.m.s of the scattered pressure field of the CABARET solution on coarse grid of 2.5 cells per vortex core radius.

The main emphasis of this subsection is the effect of non-uniform hydrodynamic flow on sound scattering, hence, the numerical solutions for the scattered pressure field intensity

$r.m.s.(p') = f(R,\theta)$ (acoustic pressure directivity) at a large distance from the vortex centre $R=10L$ are considered. Figs.5 show the acoustic pressure directivity $r.m.s.(p') = f(R,\theta)$ with respect to the polar angle defined anti-clock-wise from the positive x-direction. The comparison of the CABARET solution with the reference solution of Colonius et al (1994) is shown in Fig.5a. To monitor the grid convergence, the Euler equations are solved on a Cartesian grid whose resolution is gradually increasing: (100,100), (200x200) и (400x400) cells (2.5, 5 and 10 grid points per vortex radius, respectively). For CABARET, the r.m.s. distributions on the two finer grids virtually coincide.

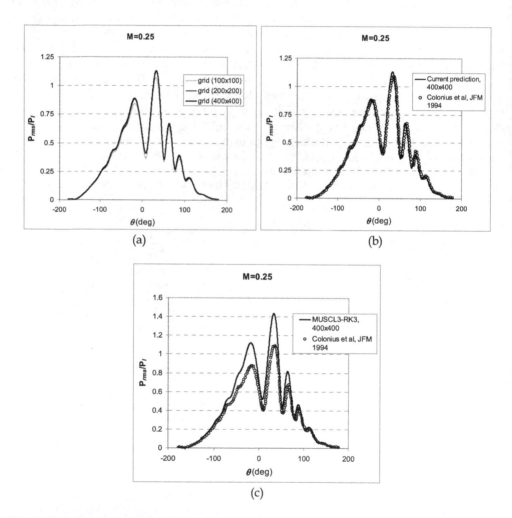

Fig. 5. Sound wave scattering by a vortex: (a) grid convergence of the scattered pressure r.m.s. field of the CABARET solution, (b) comparison of the fine-grid CABARET results with the reference solution of Colonius et al (1994), (c) comparison of the fine-grid 3rd order Roe/MUSCL results with the reference solution of Colonius et al (1994).

For the Roe-MUSCL scheme, the comparison with the reference solution on the finest grid (400x400) is shown in Fig.5c. In comparison to the CABARET results (fig.5a), for the Roe-MUSCL scheme there is some 30% overprediction of the peak sound directivity that is associated with numerical dispersion. Clearly, the acoustic peak corresponds to the downstream vortex direction where the sound waves spend more time inside the strongest vortex-induced hydrodynamic field and which direction is more sensitive to the linear dispersion error of the numerical scheme.

The next case considered is the high frequency acoustic wave imposed as the inflow boundary condition. It is well known (e.g., Kinsler and Frey, 2000) that in the high-frequency limit $\lambda \ll L$ the Euler equations can be reduced to the ray-theory equations. The latter, for example, describe the effect of focusing and defocusing of acoustic rays as they pass through a non-uniform medium. In particular, the focusing of acoustic rays creates caustics which loci can be found from the solution of eikonal equation (Georges, 1972). On the other hand, caustic locations correspond to the most intense root-mean-square (r.m.s) fluctuations of the pressure field that can be obtained directly from solving the Euler equations.

To illustrate this numerically, let's consider the incident acoustic wave at a high-frequency wavenumber $\lambda = 0.076\ L$ and solve the Euler equations with the CABARET method. For this calculation, the computational grid with the resolution of 7-8 cells per acoustic wavelength that corresponds to (1000x1200) grid cell points is used. Fig.6a shows the scattered pressure r.m.s. field obtained from the Euler solution, where the loci of the caustics bifurcating into two branches, as obtained in Colonius at al (1994), are shown. The centre of the vortex corresponds to the origin of the coordinate system. The caustics branches outline the acoustic interference zone that develops behind the vortex. Fig.6b shows the pressure r.m.s. directivity, $r.m.s.(p') = f(R,\theta)$ of the computed solution at distance R=L from the vortex centre. Two grid resolutions are considered, 7 and 14 grid cells per acoustic wavelength. The polar angle variation corresponds to the top half of the computational domain which intersects one of the caustics bifurcation point at $\sim 70^0$ relative to the incident wave direction.

For the solution grid sensitivity study, the scattered pressure r.m.s. solutions are computed with two grid densities, as shown in Figs 6c,d. It can be seen the main features such as the caustic point location and the peak amplitudes are well captured on both grids. The scattered acoustic pressure solution component is then further used to compute the trajectories of sound rays. The trajectories are defined as the normal to the scattered pressure r.m.s. fronts. In particular, from this vector field, the maximum angle of the acoustic ray deflected by the vortex can be compared with the ray-tracing solution. According to the ray theory, the maximum deflection angle scales linearly with the vortex Mach number.

Fig. 7 shows the maximum deflection angles obtained from the Euler calculation (Euler) and the reference values obtained from the ray-tracing solutions. All solutions are in a good agreement and follow the linear trend expected. In particular, the Euler solution almost coincides with the eikonal solution of Tucker and Karabasov (2009) that corresponds to the same computational domain size. The slight disagreement with the other ray-tracing solutions is likely to be caused by the differences in the domain size, i.e., the proximity of inflow boundary conditions, as discussed by Colonius et al, 1994.

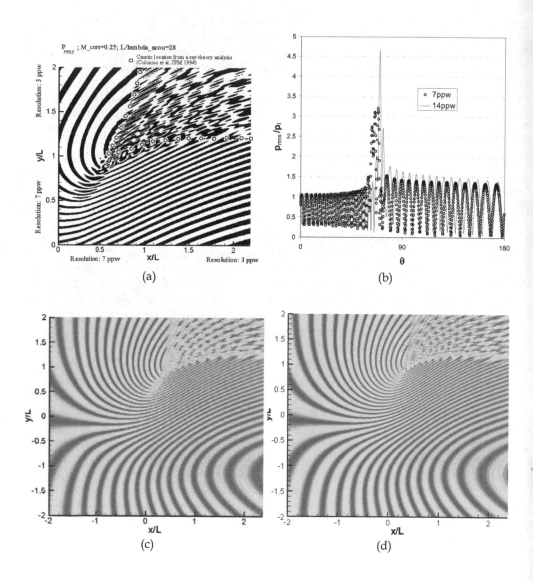

Fig. 6. Euler solution of the sound scattering by a vortex at high frequency: (a) pressure r.m.s. field where the loci of caustic branches are shown with the open symbols, (b) pressure r.m.s. directivity in the top half of the domain for the grid resolutions 7 and 14 cell per acoustic wavelength (7 ppw and 14 ppw); the scattered pressure r.m.s. for vortex core Mach number $M_{max}=0.295$ with the grid density of (c) 7 cells per acoustic wavelength and (d) 14 cells per acoustic wavelength.

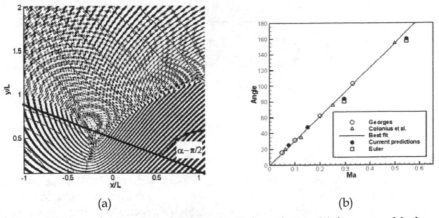

(a)   (b)

Fig. 7. Extracting sound ray trajectories from the Euler solution (a) for vortex Mach number 0.55 and (b) comparing the extracted maximum ray deflection angle as a function of various vortex Mach numbers with several ray-tracing solutions.

5. Dynamics of counter-rotating vortices

As the final example, we consider the test problem of interacting counter-rotating vortices that involves both their nonlinear dynamics and, as a by-product, sound generation. For small viscosity, the direct simulation of vortex dynamics and acoustics by solving the compressible Navier-Stokes equations on a Eulerian grid is a challenging problem because of the thin vorticity filaments that are generated as the process evolves in time. These are difficult to capture because of numerical dissipation-dispersion problems mentioned in the introduction. In the literature, examples of flow simulations have Reynolds number, as defined based on the velocity circulation, in the range of 1000-4000 (e.g., Inoue, 2002). Eldridge (2007) manages to accurately compute the problem of dynamics and acoustics of counter-rotating vortex pairs at a high Reynolds number, Re=10000 with the use of a Lagrangian vortex particle method. In the latter, the governing fluid flow equations are solved in a non-conservative form and the advected vortex solution is regularly reinitialised on a Eulerian grid to reduce the complexity of thin vorticity filaments and stabilise the solution. In the present subsection, the problem of counter-rotating vortices is solved on a fixed Eulerian grid for the range of Reynolds numbers, Re=5000-10000 with the conservative Navier-Stokes CABARET method.

Fig. 8 shows the problem setup. Four viscous-core counter-rotating vortices are initiated in an open domain. Each of the vortices has a constant velocity circulation at infinity $\pm\Gamma, \Gamma > 0$ and a Gaussian distribution of the vorticity with the core radius r_0

$$\xi = \frac{1.25\Gamma}{\pi r_0^2} e^{-1.25\left(\frac{r}{r_0}\right)^2} \tag{5}$$

where $\Gamma = 0.24\,\pi\,\delta\,a_\infty$ that corresponds to the vortex core Mach number $M_0 = 0.3$ and $\delta_x = \delta_y = \delta = \frac{10}{3} r_0$ is half-distance between the adjacent vortex centres. In non-dimensional variables, the flow parameters at infinity are taken to be $p_\infty = 1, \rho_\infty = 1$ and $\delta = 0.2$. The initial location of the centre of mass of the system corresponds to x=0.

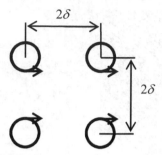

Fig. 8. Problem configuration for two counter-rotating vortex pairs.

The vortex system undergoes a jittering motion with one vortex pair sleeping through the other and taking turns. The centre of mass travels at a positive subsonic speed, which corresponds to the right horizontal direction in Fig.8. After each sleep-through, the distance between the vortices decreases until they finally coalesce and continue the movement as a single core. By using a point-vortex approximation with neglecting viscous effects, the dynamics of the vortex system before coalescence can be described by the classical analytical solution of Hicks (1922). This solution gives the following expressions for the centre mass velocity $\overline{U}$ and the slip through period T_p of the vortex system:

$$.\overline{U} = \frac{\delta_y \cdot \Gamma}{4\pi\delta_x^2}\left[\frac{k^2 E(k)}{E(k)-(1-k^2)K(k)}\right], T_p = \frac{32\pi\delta_x^2}{\Gamma k^2(1+k)}\left(E(k)-(1-k^2)K(k)\right) \qquad (6)$$

where $k=1/[1+(\delta_y/\delta_x)^2]=0.5$ and $K(k), E(k)$ are the complete elliptic integrals of the first and second kind. For the specified parameters of the model, the centre mass velocity and slip-through period are

$$T_p = \frac{54.46\delta^2}{\Gamma} = 12.21, \ \overline{U} = \frac{0.1437 \cdot \Gamma}{\delta} = 0.1282 . \qquad (7)$$

The velocity of the centre mass corresponds to the Mach number $M_\infty = 0.1083$.
Because of the problem symmetry with regard to axis y=0, one-half of the computational domain is considered with the symmetry boundary condition. The problem is solved in the inertial frame of reference which velocity with regard to the absolute frame equals the velocity of the centre of mass, $\overline{U}$. The latter is available from the analytical point-vortex solution. Because of this choice the centre of mass is approximately stationary in the reference coordinate system. The latter is helpful for minimising the size of the computational domain.
The computational domain of size $440\delta \times 220\delta$ (axially times vertically) is covered by a Cartesian grid that has a uniform grid spacing in the central block ($60\delta \times 30\delta$). Exponential grid stretching is applied near the outer boundaries to reduce numerical reflections (Fig.9). Three grid resolutions are considered: 6, 9 and 12 grid cells per vortex core radius.
Initial conditions in conservation variables are computed in the following way. By combining the initial vorticity distribution (5) with the solenoidal velocity field condition, the velocity field is computed from solving the Laplace equation for velocity potential. The

resulting velocity field is substituted to the momentum equations in conservation variables that are then integrated numerically with the use of the isentropic flow relation between pressure and density.

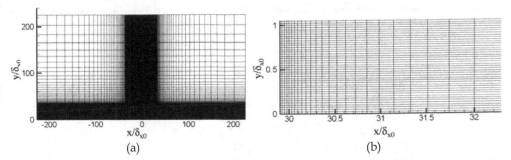

(a)   (b)

Fig. 9. Computational grid: (a) full domain view and (b) zoom in the beginning of the grid stretching zone.

Fig.10 shows the time evolution of vorticty field of the system as the vortex pairs sleep through each other. The grid density is 6 grid cells per vortex radius and the Reynolds number of the CABARET simulation is 9400. This particular Reynolds number is chosen as the best match for the reference Lagrangian particle solution of Eldridge (2007) that corresponded Re=10000. The 5% difference in the Reynolds numbers may be attributed to the differences in numerical approximation of viscous terms in the momentum and energy equations in the conservative CABARET and the non-conservative vortex particle method.

For Re=9400, the vortex pairs merge after 4 sleep-through events. The sleep-though events correspond to Fig10(c),(d),(e),(f). For Re=5000, the coalescence happens earlier in comparison with the Re=9400 case: for the lower Re-number the vortices coalesce already after 3 sleep-throughs. This change can be compared with results for the fully inviscid follow case calculation that was conducted with the same CABARET Euler method. In the latter case the vortices manage to undergo 6 sleep-through events before coalescence. The capability of the CABARET model to capture the qualitative differences between the Re=9400 and the fully inviscid solution is noted as a good indication of how low-dissipative the method is.

To zoom into the flow details, Fig.11 compares instantaneous vorticity contours obtained from the CABARET solution at two grid resolutions, 6 and 12 cells per core radius, with the reference Lagrangian particle solution from Eldridge (2007) at one time moment. This time moment corresponds to the last vortex sleep-through before the coalescence. At this time the vortices are very close and strongly interact with each other through fine vorticity filaments. For all three solutions, the same contour levels are plotted that show a good agreement down to a small detail.

It is also interesting to compare the 2D solution computed with the results of the vortical structure visualization obtained experimentally by Bricteux et al 2011 for a 3D high-Reynolds number jet. In this experimental work a moving window technique is used in the framework of Particle Image Velocimetry (PIV) method to visualise vortex paring in the jet shear layer. Fig.12 shows the results of the visualisation in the jet symmetry plane that appear qualitatively very similar to the results of the 2D simulation (cf. fig.10a,b,d,g).

For a quantitative comparison, the centre mass velocity of the vortex system and the values of first few vortex slip-through periods are compared with the reference analytical solution for point vortices in inviscid flow.

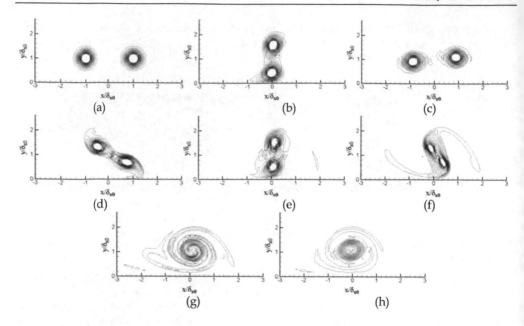

Fig. 10. Vorticity distribution of leapfrogging vortices at several consecutive time moments: $a_\infty t / \delta = 0$ (a), 18 (b), 36 (c), 54 (d), 72 (e), 90 (f), 108 (g), 120 (h).

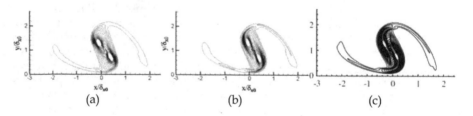

Fig. 11. Vorticity distribution of the leapfrogging vortex pairs at $a_\infty t / \delta = 90$ for (a) CABARET solution with the grid density of 6 cells per vortex core radius, (b) CABARET solution on the grid with 12 cells per vortex core radius, and (c) the reference vortex particle method solution from Eldridge (2007).

Fig. 12. PIV of vorticity distribution of leapfrogging vortex rings in the symmetry plane of a high Reynolds number jet obtained with a moving window technique by Bricteux et al 2011.

Grid spacing, h	$r_0 / 6$	$r_0 / 9$	$r_0 / 12$
$\overline{u}$	0.1271	0.1277	0.1279
$\left(T_p\right)_1$	10.85	10.9	10.9
$\left(T_p\right)_2$	7.9	8.1	8.2

Table 1. Integral characteristics of the vortex system as obtained from the numerical solution

Here $(T_p)_1$ is the time period between the first and the second vortex sleep-through and $(T_p)_2$ is the time period before the second and the third sleep-through. The agreement for the meanflow velocity between the point-vortex theory value (0.1282) and the numerical values on 3 different grids is within 1%. For the sleep-through period, the values obtained on different grids are converged within 0.5%. The numerically predicted time period, however, is 10-15% shorter in comparison with the point-vortex theory (12.21). This discrepancy is within the order of accuracy the point-vortex model, $(r_0/\delta)^2 = 0.16$ and thus characterises how non-compact the viscous vortex core is in comparison with the distance between the adjacent vortex centres.

In addition to the near-filed, the far-field pressure field has been computed on a circle control surface at distance of 20δ from the vortex centre of vortices. The control surface is located in the same reference system as the centre of mass that moves at a small subsonic speed, $M_\infty = 0.1083$ with respect to the absolute frame. Fig.13 shows the pressure signals obtained at the control points corresponding to 30^0 and 90^0 angle to the flow direction for Re=5000 and 9400 on the grids of different resolution.

The pressure fluctuations are defined with the reference to the pressure field value at infinity, $p_\infty = 1$. The peaks of the pressure signatures correspond to the vortex sleep-through events and the number of the peaks corresponds to the total number of vortex sleep-throughs, respectively. The phase of intense vortex interaction during the vortex pairing is followed by a "calming" period that corresponds to the vortex roll-up after the coalescence. In comparison with the pre-coalescence time history that is dominated by large-time scales the post-coalescence signal is dominated by small-time-scale events.

For the higher Re-number case, the amplitude of the last acoustic "burst" that corresponds to $a_\infty t / \delta \sim 110$ has some 20% higher amplitude in comparison with other peaks. This loud acoustic event corresponds to the last vortex sleep-through, which takes place at $a_\infty t / \delta \sim 90$ and which is well-captured on the grids of different resolution. After the vortex coalescence, the increase of Reynolds number from 5000 to 9400 also leads to a notable prolongation of small-scale acoustic fluctuations in the post-coalescence phase. These effects may be associated with the small spatial structures that are generated shortly before the vortices coalesce (e.g., fig.10f,g) and which are more sensitive to viscous dissipation.

For the pre-coalescence period of vortex evolution, the numerical solutions that correspond to the grids of different resolution are converged within 1-2% for both Reynolds numbers. For the post-coalescence time history, the grid convergence for the high Reynolds-number case, Re=9400, slows down in comparison with the Re=5000 case. For both Re-number cases, however, the CABARET solution on the grid resolution 12 cells per vortex radius appears adequate to capture the fine pressure field fluctuations well.

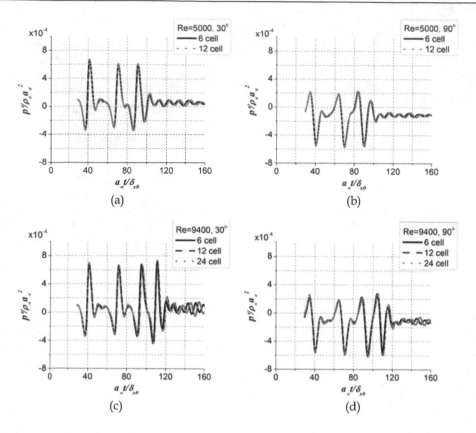

Fig. 13. Acoustic pressure signals at different observer angles to the flow: 30° (a),(c) and 90° (b),(d) for Re=5000 (a),(b) and Re=9400 (c),(d).

6. Conclusion

The computational of compressible vortical flows is challenging because of the multi-scale phenomena involved. Computational approaches and numerical methods for the solution of compressible vortical flow problems have been discussed. In particular, the key elements of a successful computational method have been outlined that include low numerical dissipation and low dispersion, as well as the good vortex preservation property. For the sake of illustration, several two-dimensional problems are considered that typically present a challenge for conventional Eulerian numerical schemes. The problems include the preservation of steady vortex in a box domain, acoustic wave scattering by a vortex field and the dynamics and acoustics of counter-rotating vortices pairs. For these problems, several computational solutions are presented and discussed, including those obtained with the CABARET scheme developed by the authors. Analytical and reference solutions are provided where applicable. All test problems considered are promoted as the benchmark problems for new Computational Fluid Dynamics codes that are to be used in application for hydrodynamics and acoustics of vortex resolving simulations.

7. References

Bogey, C. and Bailly, C., "A family of low dispersive and low dissipative explicit schemes for flow and noise computations", J. Comput. Physics, 194 (2004), pp. 194-214.

Bogey C. and Bailly C., "Influence of nozzle-exit boundary-layer conditions on the flow and acoustic fields of initially laminar jets", J. Fluid Mech., Vol.25, 2010, pp507-540.

Boris, J.P., Book, D.L., and Hain, K., "Flux-corrected transport: Generalization of the method", J. Comput. Phys, 31,(1975), 335-350.

Bricteux, L., Schram C., Duponcheel M., Winckelmans, G., "Jet flow aeroacoustics at Re=93000: comparison between experimental results and numerical predictions", AIAA-2011-2792, 17th AIAA/CEAS Aeroacoustics Conference (32nd AIAA Aeroacoustics Conference), 6-8 June, Portland, Oregon, 2011

Cockburn B and Shu CW, "Runge–Kutta Discontinuous Galerkin Methods for Convection-Dominated Problems", Journal of Scientific Computing, 16(3): 173-261, Sept 2001.

Colonius T., Lele S.K., and Moin P., "The scattering of sound waves by a vortex: numerical simulations and analytical solutions", J Fluid Mech (1994), 260, pp 271-298.

Colonius T and Lele SK., "Computational aeroacoustics: progress on nonlinear problems of sound generation." Progress in Aerospace sciences, 2004, 40, pp. 345-416.

Dritschel, D.G., Polvani, L.M. and Mohebalhojeh, A.R.: The contour-advective semi-Lagrangian algorithm for the shallow water equations. Mon. Wea. Rev. 127(7), pp. 1551–1565 (1999).

Eldridge, J.D., "The dynamics and acoustics of viscous two-dimensional leapfrogging vortices", J. Sound Vib., 301 (2007) 74–92.

Fritsch G. and Giles M., "Second-order effects of unsteadiness on the performance of turbomachines", ASME Paper 92-GT-389; 1992.

Georges, T.M. "Acoustic ray paths through a model vortex with a viscous core", J.Acoust.Soc. of America, Vol. 51, No. 1 (Part 2) pp. 206-209 (1972).

Goloviznin V.M. and Samarskii, A.A. "Difference approximation of convective transport with spatial splitting of time derivative", Mathematical Modelling, Vol. 10, No 1, pp. 86-100.

Goloviznin V.M. and Samarskii, A.A., "Some properties of the CABARET scheme", Mathematical Modelling, Vol. 10, No 1, 1998, pp. 101–116.

Goloviznin, V.M. "Balanced characteristic method for systems of hyperbolic conservation laws", Doklady. Mathematics, 2005, vol. 72, no1, pp. 619-62313.

Harten, A., Engqist, B., Osher, S., and Chakravarthy, S. "Uniformly High Order Accurate Essentially Non-Oscillatiry Schemes III", J. Comput. Phys, 71, (1987), pp. 231-303.

Hicks, W.M. "On the mutual threading of vortex rings", Proceedings of the Royal Society of London A 10 (1922) 111–131.

Hirsh, C., "Numerical computation of internal and external flows", vol. 2, John Wiley & Sons, 1998.

Howe, M.S., "The generation of sound by aerodynamic sources in an inhomogeneous steady flow", J. Fluid Mech., Vol 67, No. 3, 1975, pp. 597-610.

Inoue, O "Sound generation by the leapfrogging between two coaxial vortex rings", Physics of Fluids 14 (9) (2002) 3361–3364.

Iserles, A. "Generalized Leapfrog Methods", IMA Journal of Numerical Analysis, 6 (1986), 3, 381-392.

Karabasov, S.A., Berloff, P.S. and Goloviznin, V.M. "CABARET in the Ocean Gyres", J. Ocean Model., 30 (2009), pp. 155–168.

Karabasov, S. A. and Goloviznin, V.M.. "A New Efficient High-Resolution Method for Non-Linear problems in Aeroacoustics", AIAA Journal, 2007, vol. 45, no. 12, pp. 2861 – 2871.

Karabasov, S.A. and Goloviznin, V.M. "Compact Accurately Boundary Adjusting high-REsolution Technique for Fluid Dynamics", J. Comput.Phys., 228(2009), pp. 7426–7451.

Kim S., "High-order upwind leapfrog methods for multidimensional acoustic equations", Int J. Numer. Mech. Fluids, 44 (2004), pp. 505-523.

Kinsler, L.E., Frey, A.R., Coppensand, A.B., Sanders, J.V. "Fundamentals of acoustics", Wiley and Sons Inc. (2000).

Kolgan, V.P. "Numerical schemes for discontinuous problems of gas dynamics based on minimization of the solution gradient", Uch. Zap.TsAGI, 1972, v.3, N6, pp.68-77

Kreichnan, R.H., "The scattering of sound in a turbulent medium". J. Acoust. Soc. Am. 25, 1953, pp. 1096-1104.

Lele, S.K., "Compact finite-difference scheme with spectral-like resolution", J.Comput. Physics, 103 (1992), 16-42.

Liu, X.D., Osher, S., and Chan, T., "Weighted essentially non-oscillatory schemes", J.Comp. Phys, 115 (1994), 200-212.

Margolin, L.G. and Shashkov, M. 2004. "Remapping, recovery and repair on a staggered grid", Comput. Methods Appl. Mech. Engng, 193, pp. 4139 – 4155.

Michelassi, V., Wissink, J., Rodi, W. "Direct numerical simulation, large eddy simulation and unsteady Reynolds-averaged Navier-Stokes simulations of periodic unsteady flow in a low-pressure turbine cascade: a comparison. Proc. IMechI, Part A: Journal of Power and Energy, 2003; 217(4), pp. 403-411.

Roe P.L. "Characteristic based schemes for the Euler equations", Annual. Rev. Fluid. Mech. 1986. V.18. pp.337-365.

Roe, P.L. "Linear bicharacteristic schemes without dissipation", SISC, 19, 1998, pp. 1405-1427.

Tam, C.K.W. and Webb. J.C., "Dispersion-relation-preserving finite difference schemes for computational acoustics", J.Comput. Physics, 107 (1993), 262-281

Thompson, K.W. "Time dependent boundary conditions for hyperbolic systems, II." Journal of Computational Physics, 89, pp. 439-461, 1990.

Toro, E.F. "Godunov methods: theory and applications", Kluwer Academic/Plenum Publishers, 2001

Tran, Q.H. and Scheurer, B. "High-Order Monotonicity-Preserving Compact Schemes for Linear Scalar Advection on 2-D Irregular Meshes", J. Comp. Phys. 2002, Vol. 175, Issue 2, pp. 454 - 486.

Tucker, P.G. and Karabasov, S.A. "Unstructured Grid Solution Approach for Eikonal Equation with Acoustics in Mind", International Journal of Aeroacoustics, vol 8 (6), 2009, pp.535-554.

Van Leer B., "Towards the ultimate conservative difference scheme. V. A second-order sequel to Godunov's method", J. Comput. Phys., 32 (1979), pp. 101-136.

Yakovlev P.G., Karabasov, S.A. and Goloviznin, V.M. "On the acoustic super-directivity of jittering vortex systems for the study of jet noise", AIAA-2011-2889, 17th AIAA/CEAS Aeroacoustics Conference (32nd AIAA Aeroacoustics Conference), 6-8 June, Portland, Oregon, 2011.

4

Fuel Jet in Cross Flow – Experimental Study of Spray Characteristics

E. Lubarsky, D. Shcherbik, O. Bibik, Y. Gopala and B. T. Zinn
School of Aerospace Engineering, Georgia Institute of Technology, Atlanta Georgia
USA

1. Introduction

Injection of the liquid fuel across the incoming air flow is widely used in gas turbine engine combustors. Thus it is important to understand the mechanisms that control the breakup of the liquid jet and the resulting penetration and distribution of fuel droplets. This understanding is needed for validation of Computational Fluid dynamics (CFD) codes that will be subsequently incorporated into engine design tools. Additionally, knowledge of these mechanisms is needed for interpretation of observed engine performance characteristics at different velocity/altitude combinations of the flight envelope and development of qualitative approaches for solving problems such as combustion instabilities (Bonnel et al., 1971). This chapter provides an introduction and literature review into the subject of cross-flow fuel injection and describes the fundamental physics involved. Additionally highlighted are experimental technique and recent experimental data describing the variables involved in fuel spray penetration and fuel column disintegration.

In recent years, there has been a great drive to reduce harmful emissions of oxides of Nitrogen oxides (NOx) from aircraft engines. One of the several approaches to achieve low emissions is to avoid hot spots in combustors by creating a lean homogeneous fuel-air mixture just upstream of the combustor inlet. This concept is termed as Lean Premixed Prevaporized (LPP) combustion. Creating such a mixture requires fine atomization and careful placement of fuel to achieve a high degree of mixing. Liquid jet in cross flow, being able to achieve both of these requirements, has gained interest as a likely candidate for spray creation in LPP ducts (Becker & Hassa, 2002). Since the quality of spray formation directly influences the combustion efficiency of engines, it is important to understand the fundamental physics involved in the formation of spray.

As seen in Fig. 1, the field of a spray created by a jet in cross flow can be divided into three modes: 1) Intact liquid column, 2) Ligaments, and 3) Droplets. The liquid column develops hydrodynamic instabilities and breaks up into ligaments and droplets (Marmottant & Villermaux, 2004; Madabushi, 2003; Wu et al., 1997). This process is referred to as primary breakup. The location where the liquid column ceases to exist is known as the column breakup point (CBP) or the fracture point. The ligaments breakup further into smaller droplets and this process is called secondary breakup.

The most relevant parameter for drop breakup criterion is the Weber number, $We = \rho_{air} U_{air}^2 D / \sigma_{fuel}$ (in this formula ρ_{air} and U_{air} - density and velocity of the crossing air respectively, D - diameter of the injection orifice and σ_{fuel} is the surface tension of the fuel).

We is the ratio of disruptive aerodynamic force to capillary restoring force. The critical *We* above which a droplet disintegrates is *We*=10 (Hanson et al., 1963). When Weber number is high (*We* >200), another mode of breakup called the shear breakup becomes dominant. During shear breakup, aerodynamic forces exerted by the flow on the surface of the liquid jet or ligaments strip off droplets by shear. Though both modes of breakup contribute to atomization of the liquid jet, the domination of one mechanism over the other is dependent on *We* and on liquid jet momentum flux to air momentum flux ratio, *q*.

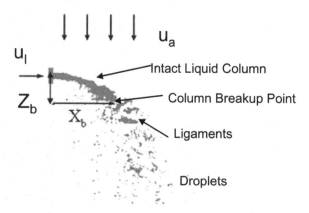

Fig. 1. Schematic of spray created by a liquid jet in cross flow (from Ann et. al., 2006)

Currently two parameters that characterize disintegration of the fuel jet in the cross flow are subjects of great interest among the users of the experimental data. They are (1) column breakup point (CBP) and (2) penetration of spray into the cross flow. The location of CBP is important for the development of computational models for the prediction of spray behavior. Since the aerodynamic drag for the liquid jet is significantly different from that of droplets, it is crucial to know the exact location of jet disintegration into droplets to be able to predict the extent to which the droplets penetrate into the air stream. On the other hand direct measurements of the spray penetration are significant for development of the design tools for use by the engine developers as well as for validation and adjustment of the spray computational models. Various researchers have measured CBP location and spray penetration with reasonable uncertainties. However, these parameters are still not explored extensively because of ambiguities in definition and due to experimental difficulties. A number of experimental studies of column breakup and spray penetration under conditions that simulate those in gas turbine engines were undertaken and are briefly reviewed below.

In the early work on the aerodynamic breakup of liquid droplets in supersonic flows researchers (Ranger & Nichollas, 1969) carried out experiments to find the time required for individual droplets dropped into a supersonic cross flow to breakup to form a trace of mist. They found this time (t_b) to be proportional to the droplet diameter (*d*), inversely proportional to the relative velocity between the droplet and the airflow (u_a), and proportional to the square root of liquid-to-air density ratio (ρ_l / ρ_a). Based on the images taken, they found that the constant of proportionality (t_b/t^*), defined by equation (1) to be 5. Another conclusion of their study was that the effect of the shock wave on the aerodynamic breakup of the droplets was minimal. The main function of the shock wave is to produce the high speed convective flow that is responsible for the disintegration of droplets. This prompted subsequent researchers to use this characteristic time (t*) for droplets in subsonic flows as well by.

$$\frac{t_b}{t^*} = \frac{t_b}{(\rho_l / \rho_a)^{1/2} d \ / u_a} = 5.0 \tag{1}$$

Lower values of t_b/t^*=3.44 were reported later (Wu et al., 1997) for liquid jet disintegration in the cross flow with Weber number in the range of We=71 – 200. The column breakup location for higher We flows could not be determined. They also found that the CBP was located at about eight diameters downstream of the orifice in the direction of airflow for the cases reported.

Other researchers (Sallam et al., 2004) measured column breakup point at We range of 0.5-260. Their studies yielded different value of t_b/t^* = 2.5. However, the uncertainties became high as We of the flow was increased. This can be explained by the fact that the experimental methods that have been employed so far for measuring the CBP position involve the analysis of the spray images obtained by back illumination technique. This method works reasonably well for low We flows in the absence of shear breakup. In the shear breakup regime, that is relevant for the gas turbine applications it becomes very difficult to analyze the spray images and find the location of CBP because of the presence of droplets in high density around the liquid column. This paper demonstrates a method to overcome this shortcoming.

Method used in the current study was first suggested by (Charalompous et al., 2007) who developed a novel technique to locate the CBP for a co-axial air blast atomizer. In this atomizer high density of droplets around the liquid jet column limited optical access to the jet. To overcome this problem, they illuminated the liquid jet column seeded with fluorescent Rhodamine WT dye with a laser beam from the back of the injector. The liquid jet acted as an optical fiber up to the point it breaks up. The jet is visible due to florescence of the dye until the location of the CBP and the light gets scattered beyond that location giving the precise location of the CBP. The current study aims at extending this technique to locate the CBP of liquid jets in cross flow.

Spray penetration into the cross flow have received significant attention by the experimentalists hence placement of fuel in a combustor is significant for its design. In 1990s researchers (Chen et al., 1993, Wu et al., 1997) have carried out experiments at different momentum flux ratios of water jets and developed a correlation of the dependence of the upper surface trajectory of jets in a cross flow with liquid to air momentum flux ratio. Later (Stenzler et al., 2003) a Mie scattering images were used to find the effect of momentum flux ratio, Weber number and liquid viscosity on jet penetration. As in other previous studies, they found that increasing momentum flux ratio increased penetration. Increasing the Weber number decreased the average droplet size and since smaller droplets decelerate faster, the overall penetration of the spray decreased. However, many of these correlations are applicable to specific operating conditions, injector geometries and measurement techniques.

It was also found (Tamaki et al., 1998, 2001) that the occurrence of cavitation inside the nozzle significantly influences the breakup of the liquid jet into droplets. The collapse of cavity bubbles increased the turbulence of the liquid jet accelerating its breakup into droplets. Additional researchers (Ahn et al., 2006) explored the effect of cavitations and hydraulic flip of the orifice internal flow on the spray properties created by a jet in cross flow. They found that while spray trajectories followed the previously obtained correlations (Wu et al., 1997) in absence of cavitations and hydraulic flip, the presence of these phenomena resulted in significant disagreements between the observed trajectories and the ones reported (Wu et al., 1997). Consequently, they concluded that the design of the injector has a significant effect on the spray trajectories.

Practically all previous studies of fuel spray attempted to describe its penetration trajectory into the cross-flow of air in the form of equation that typically incorporate momentum flux ratio of the liquid jet to air flow, $q = \rho_{fuel} U_{fuel}^2 / \rho_{air} U_{air}^2$, Weber number and certain function that describe shape of the outer edge of the spray. Usually, these equations incorporate a number of empiric coefficients that were obtained by processing experimental data. In spite of availability of dozens of correlations their practical use remains problematic because they all provide different results. Figure 2 shows result of application of different correlations to one spray with $q=20$ and $We=1000$.

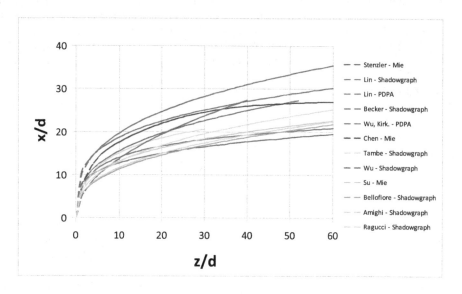

Fig. 2. Comparison of the spray penetration trajectories (x and z – coordinates in the direction of fuel injection and crossing air flow respectively, d - is diameter of the injection orifice)

It can be observed that the spray penetration trajectories differ from each other to an extent of 100%. Among factors that causes such a big difference the following ones seems to be the most important:

• Design of the injector and its position in the cross flow (i.e. l/d, shape and quality of the internal fuel path, presence or absence of the spray well or cavity between the injection orifice and the channel e.t.c).

- Factors that vary flow conditions in the experiment inconspicuously for the researcher such as temperature of the crossing air flow which may change the temperature of the injector and thus surface tension and viscosity of the injected fuel.
- Turbulence of the core and boundary layer characteristics of the crossing air flow that may significantly influence spray penetration but rarely mentioned by researches.
- Imaging technique that was used for many years for capturing spray trajectories was static photography that typically captured superposition of sprays on one image due to the fact that time constant of such oscillatory phenomena as liquid jet disintegration in the cross flow is by several orders lower than expose rate of any available camera used in most of experiments.

The objective of this study was to investigate the spray trajectories and determine locations of the column break up points (CBP) formed by the Jet-A fuel injected from the injectors of different geometries into a cross flow of air while the above mentioned influencing factors will be isolated. For this purpose:

- Both injectors used in the study that had the same diameter of the orifice and a different shape of the internal path were manufactured using the same equipment and technology. They were installed with orifices openings flush with the air channel wall (i.e. with no spray well, or cavity).
- Crossing air flow was of the room temperature. Its turbulence level in the core was ~4%. Thickness of the boundary layer was ~3mm.
- High speed imaging technique (~24,000fps) with spray illumination by the short laser flashes of 30ns duration was used to capture instantaneous images of the spray several times during its movement from maximum to minimum position. That allowed statistically relevant processing of the images and thus extracting information about the averaged spray trajectories and their RMS values.

Sprays penetration into the cross flow were investigated using Jet-A fuel for a wide range of momentum flux ratios between $q=5$ and $q=100$. Velocity of the air flow was varied to attain Weber numbers in the range of $We=400$ to $We=1600$. Air pressure and temperature in the test channel were P=5 atm and T~300K respectively. Column breakups were investigated also at higher air temperature of 550K (in addition to T=300K) and by using water injection in addition to jet fuel experiments in attempt to achieve wider range of non-dimensional parameters.

2. Experimental setup

Figure 3 shows a schematic of the experimental setup used to study the injection of a liquid jet from a flat surface into the cross flow of air at elevated pressure. This setup had a plenum chamber, a rectangular air supply channel, a test section with injector under investigation and a pressurized chamber with four 38mm (1.5 inch) thick windows for optical access to the spray.

Plenum chamber was 203.2 mm in diameter and 457.2mm long. Two perforated screens were installed at the entrance and at the exit of the plenum to achieve necessary level of turbulence and flow uniformity in the test section. The rectangular supply channel was 62.3mm (2.45 inch) by 43.2mm (1.7 inch) in cross-section and was 304.8mm long. It was equipped with a "bell-mouth" air intake which was connected to the bottom of the plenum chamber to smoothen the air flow. On the other end of the channel four aerodynamically shaped plates were attached to the channel creating a test section with a cross-section 31.75 x 25.4mm (1.25 x 1.00 inch).

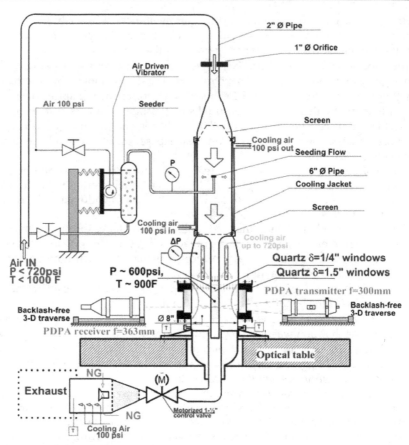

Fig. 3. Schematic of the test facility

This test section has ~50mm (2.00 inch) long, 6mm (1/4 inch) thick windows on three sides for optical access to the spray zone. The fuel injectors were installed on the centerline of the plate 10mm downstream of transparent section. The whole system was fixed to a massive optical table while optical tools were installed on a traversing mechanisms, which provides precise movement (minimal step is 0.0254mm) in three mutually orthogonal directions using step motors and electronic drivers controlled using a computer. In the current study, 1mm increments of movement were typically used for characterizing the spray. Maximum possible flow conditions in the test sections were P=4.2MPa (600 psi) and T=755K (900F) which correspond to supercritical flow conditions for the Jet-A fuel. These flow conditions were achieved by supplying preheated air flow from the controllable high pressure air supply at P < 5.0Mpa (720 psi) and T < 800K (1000ºF) into the plenum, where it then enters the 1.25" × 1.00" test section.

Velocity in the test section was controlled by the motorized control valve in the exhaust line (see Figure 3). Cooling of the test channel, test section as well as inner and outer windows in case of the preheated air use was achieved by pressurizing of the pressure vessel with the high pressure air flow (P<5.0MPa, T~295K). This cooling air was eventually mixed with the high temperature air from the test section in the exhaust path. Pressure of this cooling air was ~1.4KPa (2 psi) higher than in the test section to keep temperature in its surrounding below 100°C. Mixture of the air passing the test section, injected Jet-A fuel and cooling air left the rig through the exhaust line, passing through the control valve, flow straightener and afterburner where fuel was burned in the pilot flame of natural gas to prevent fuel from entering the atmosphere.

Flow conditions in the test section were monitored using 3mm (1/8inch) diameter Pitot tube and thermocouple, which were located within the 2.45" × 1.70" test channel (see Figure 4). An additional pressure transducer and thermocouple were installed just downstream of the test section. Differential pressure sensor measured pressure drop along test section to support flow velocity measurements by the Pitot tube. Axes of the coordinate system used in this study were designated as shown on the Figure 5. X was direction of fuel injection. Y – Lateral spread of the spray and Z – Direction of the air flow.

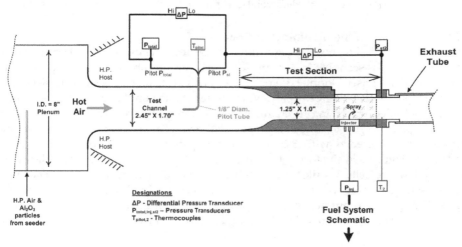

Fig. 4. Instrumentation of the test section

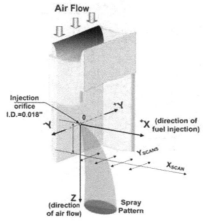

Fig. 5. Coordinate system for spray characterization

3. Results and discussion

This section consists of several parts including

- Characteristics of the incoming air flow;
- Characteristics of the tested fuel injectors which include:
  - Hydraulic characteristics
  - Images of the fuel jet exiting from both injectors in the absence and in the presence of the crossing air flow
  - Droplet sizes
- Locating of the jet breakup position
- Results of the spray penetration measurements obtained by processing of images obtained at different Weber numbers and different momentum ratios
- Development of the empirical correlations for spray penetration into the cross flow

3.1 Characteristics of the incoming air flow

Velocity profiles of the incoming air flow in the test channel were measured in three representative cross-sections in the presence and in the absence of spray using three dimensional (3-D) Laser Doppler Velocimetry (LDV) system. This system consisted of two transceivers oriented 90 degrees apart, which were installed on the rail connected to the 3-D remotely controlled traversing mechanism. This system optically accessed test section from the orifice plate (X=0) to the coordinate X<25mm. To obtain velocity measurements incoming air was seeded with 3-5mkm alumina particles. Results of measurements are presented on Figure 6 in the form of the mean and RMS velocity profiles. It is clear that the mean and RMS velocity profiles are of trapeze-shape form typical for turbulence flow in tubes. Presence and absence of spray did not produce any significant differences in velocity profiles. No significant differences in the profiles were indicated while measured across the test channel 5mm upstream ($z/d\sim 10$) and 20mm downstream ($z/d\sim40$) of the point of injection.

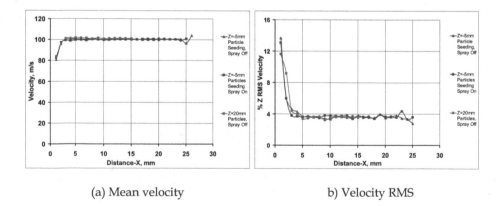

(a) Mean velocity   b) Velocity RMS

Fig. 6. Characterization of the crossing air velocity field in the test section

3.2 Characteristics of injectors

The main difference between the investigated injectors was shape of the surface between the plenum and the injection orifice.

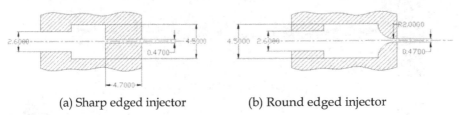

(a) Sharp edged injector   (b) Round edged injector

Fig. 7. Schematics of the tested injectors

One injector had sharp edge as shown on the Fig 7-a and the other one had smooth transition path from the plenum to the orifice (i.e., round edge, see Fig 7-b). Their hydraulic characteristics presented on the Fig. 8 reflect this difference in the injector's internal shape.

Specifically, discharge coefficient $C_d = \dfrac{\dot{m}_{fuel}}{A_{inj.}\sqrt{2\rho_{fuel}\Delta P_{inj.}}}$ of the sharp edge orifice was

relatively constant $C_d\sim0.75$ in the tested range of Re_D numbers while the discharge coefficient of the round edge orifice is $C_d\sim0.96$ at the Reynolds numbers exceeding $Re_D=10,000$ $(\Delta P_{inj.}>60\text{psi})$ which is relevant to the current study.

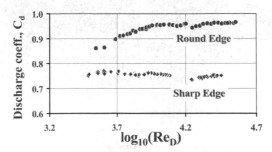

Fig. 8. Hydraulic characteristics of the tested injectors $(Re_D=\rho_{fuel}D_{inj.}U_{fuel}/\mu_{fuel})$

Effect of injector geometry on jet disintegration was first demonstrated without cross flow of air. Images of the fuel jets injected from both injectors into the atmosphere are presented on the Fig. 9. It is clearly seen that the jet coming out of the sharp edged orifice disintegrated forming spray structures, ligaments and droplets (see Figure 9-a) while jet injected from the round edge orifice was relatively smooth and intact (Figure 9-b).

A closer look on these fuel jets without cross flow in a near field (see Figure. 10) reveals that the jet injected from the sharp edge orifice expands and disintegrates while the jet from the round edge orifice shows the development of the hydrodynamic instabilities (see Figures 10-a and 10-b respectively). This observation suggests that internal turbulence created by the sharp edge at the entrance of the cylindrical orifice $(L/D\sim10)$ dramatically change jet boundaries and may lead to the differences in spray creation especially when the mechanism of the jet disintegration in the cross flow at elevated Weber numbers $(We>200)$ is "shearing". In fact images of the fuel jets shown on the Figure 11 clearly indicate that significant scale difference in liquid border structure on the outer edge of the jet remain while jets are injected into the cross flow. Size of the outer border structures on the jet exiting from the round edge orifice (Figure 11-b) is at least ten times smaller and more organized than on the jet exiting from the sharp edged orifice (Figure 11-a).

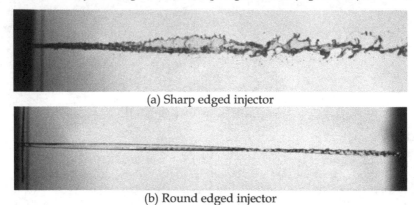

(a) Sharp edged injector

(b) Round edged injector

Fig. 9. Images of the fuel jet injected into the atmosphere (no cross flow) from injectors

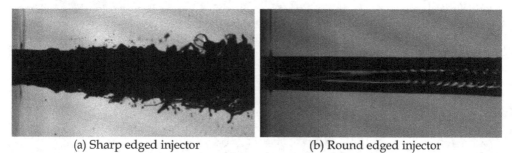

(a) Sharp edged injector   (b) Round edged injector

Fig. 10. Zoom in the liquid jets injected into the atmosphere (no cross flow) from injectors

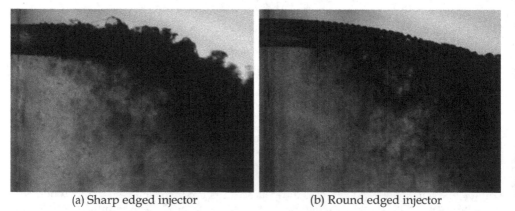

(a) Sharp edged injector   (b) Round edged injector

Fig. 11. Images of the fuel jet injected into the cross-flow of air at We=1000, momentum flux ratio $q=20$ and $Re=14,700$.

The above mentioned difference in the outer border structure of the jet can potentially influence size of the created droplets. In fact, sharp edged injector used in the current study produces larger droplets as indicated on the counter plots of the Sauter Mean Diameter $SMD = D_{32} = \sum D_i^3 n_i / \sum D_i^2 n_i$, with D_i – diameter of the individual droplet) presented for both tested orifices (sharp and round edged) on the Figure 12 (-a and –b respectively). Measurements were undertaken using PDPA in the representative cross-section of the spray located 60 orifice diameters downstream of point of injection ($z/d=60$) where spray was fully developed at the same flow conditions ($We=1000$ and $q\sim20$) for both orifices. Comparison of the SMD along the center line in the same plane ($z/d=60$) presented in the Figure 13 reveals ~10% larger droplets on the periphery of the spray produced by the sharp edge orifice.

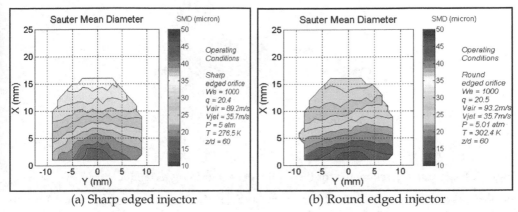

(a) Sharp edged injector   (b) Round edged injector

Fig. 12. Sauter Mean Diameter (SMD) in the cross plane of the spray at z/d=60 for tested injectors

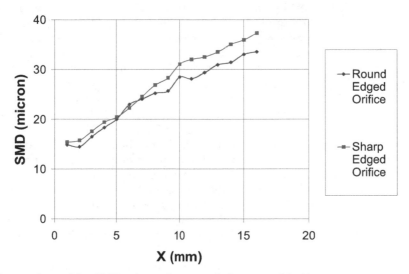

Fig. 13. Comparison of the SMDs along the central plane at z/d=60

3.3 Locating of the jet breakup position

Liquid column breakups were investigated using the same pair of the injectors (sharp and round edge) shown on the Figure 7. For this purpose injectors were modified to allow installation of the fiber optic connector coaxially with the injector orifice to provide capabilities for application of the light guiding technique. Measurements were conducted at the room and elevated temperature of the crossing air flow (T=300K and 555K respectively). Two liquids (Jet-A and water) were used to extend range of possible correlation of the jet location versus non-dimensional parameter.

Figure 14 schematically shows liquid jet light guiding technique that was used for locating the column breakup point (CBP) by letting the liquid jet act as an optical fiber and transmit light through it.

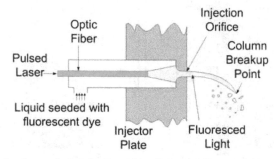

Fig. 14. Experimental schematic for the liquid jet light guiding technique

Pulsed laser light was introduced from the back of the injector using an optical fiber to illuminate the liquid jet. The laser light propagates through the liquid in the injector and reaches the liquid jet column. Light coming out of the orifice undergoes total internal reflection and is guided by the liquid jet like in optical fibers. This effect is based on the fact that the critical angle for total internal reflection for the interface between the Jet-A and air is 43°. In other words, if the liquid jet column bends by over 43⁰ abruptly, a ray of light entering the liquid jet parallel to the injector will also be refracted out of the liquid jet column in addition to being reflected. No such abrupt bends were observed in this study. This ensures that the attenuation of light intensity in the liquid jet column due to refraction is not significant enough to completely terminate the light propagating through the jet.

Slightly different jet illumination techniques were used in this study for the Jet–A and water. When the liquid used was Jet A, Metalaser Technology MTS-20 pulsed Copper Vapor laser with tunable pulse frequency (in the range of 5 kHz – 8 kHz) and a power of about 5mJ per pulse was used for illuminating the liquid jet. When water was used as the liquid for creating the spray, a Nd:YAG laser with a frequency of 10 Hz and a power of about 50mJ per pulse was used for illumination. To make the entire mass of the liquid through which light is passing visible both liquids were seeded with a fluorescent dye. The dyes used were Pyrromethene 567 with Jet A and Fluorescein with water. Both these dyes absorb the laser light and fluoresce in the yellow region. An optical filter was used to cut off the scattered light. The farthest visible point from the center of the orifice in the image is considered to be the CBP.

Figure 15-a shows a typical image of a jet in cross flow obtained by employing the liquid jet light guiding technique. This raw image was eventually inverted into a binary field shown in Figure 15-b by application of the threshold that was set to the intensity of the image which corresponds to the sharp fall in intensity of the liquid jet. The edge of this binary field was tracked to obtain the complete boundary of the liquid jet (see Figure 15-c). The farthest point on this boundary from the center of the orifice is defined as the CBP in this study. This CBP position was averaged over 150 images. Figure 15-d shows the averaged image of the liquid jet obtained using this technique with crosses indicating individual CBPs and circle indicating the average CBP location for the investigated operating conditions.

Figures 16-a and -b show the coordinates of the mean location of the CBP in the direction of fuel injection (X) and airflow (Z) downstream of the orifice respectively. Data of all four experimental series demonstrate the same effect of the CBP approximation to the orifice with the growth of momentum flux ratio (q). Two competing factors control position of the CBP: (1) Increase of the liquid jet velocity with the growth of q and (2) acceleration of the jet disintegration with the growth of the liquid velocity and thus its internal turbulence. This competition is clearly indicated by the maximum on the graph, which shows X/d coordinate of CBP on the Figure 16-a. This effect is much stronger for the sharp edged orifice at higher temperature of the crossing air flow. This fact supports hypothesis of the influence of internal turbulence of liquid jet upon the location of CBP because of possibility of cavitation at increased temperature of the injector internal surfaces caused by the high temperature of the crossing air.

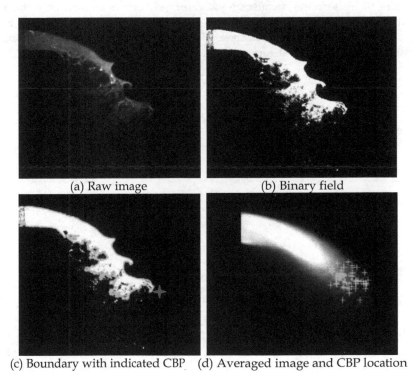

(a) Raw image   (b) Binary field

(c) Boundary with indicated CBP   (d) Averaged image and CBP location

Fig. 15. Methodology for locating the column breakup point (CBP)

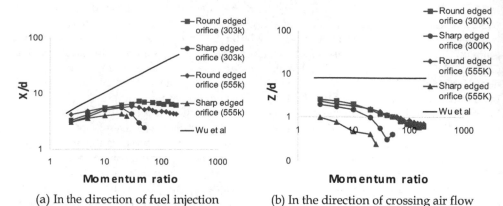

(a) In the direction of fuel injection   (b) In the direction of crossing air flow

Fig. 16. Location of the column breakup point (CBP)

Figure 17 shows position of the CBP as a function of Weber (We) number. In fact CBP location was determined to be at about 1-4 diameters downstream of the orifice. This distance is reduced with increase of We similar to the dependence upon the momentum flux ratio in Figure 16. This occurs because an increase of We causes an increase of the fuel flow rate and thus velocity of liquid which in turn enlarge the scale of structures (see Figure 11) in the jet boundary. Presumably these larger structures accelerate process of jet disintegration by aerodynamic shearing.

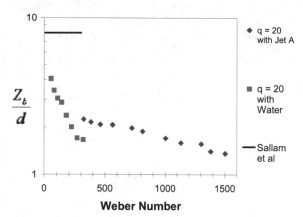

Fig. 17. Typical dependence of the CBP location upon the Weber number for the round edge orifice

It is worth to note that distances at which fuel jet disintegrates in this study are much shorter compared to prior studies (Wu et al., 1997; Sallam et al., 2004) that reported the CBP to lie at a distance of 8 diameters downstream of the orifice for most of the investigated cases. This discrepancy can be attributed to the difference of operating conditions and measuring techniques used for the CBP locating.

Finally, the entire set of CBP obtained in this study for various values of airflow velocities (66 – 140 m/s) and velocity of the liquid jet (19 – 40 m/s) for two liquids (Jet-A and water) at two different cross flow air temperatures was summarized in the form of non-dimensional breakup time (t_{cb}, defined in equation 1), which was calculated from the experimental data with the assumption that velocity of the jet in the X direction does not change until the column breaks up. t_{cb} was obtained by dividing the X distance of the column breakup point from the orifice by the jet exit velocity. Dependence of the t_{cb} upon the liquid jet Reynolds number (Re_j) is shown in the Figure 18. Non-dimensional breakup time (t_{cb}) is chosen as a parameter that is commonly used in computational models of spray formation (Wu et al., 1995). Choice of the Re number is self explained by the fact that only one injector diameter was used in the current study and any variations in the Weber number (We) and momentum flux ratio (q) led to strong variation of velocity of the liquid jet (19 – 40 m/s) and thus of the Re number. This correlation is described by Equation 2 and as shown on the Figure 17 to be valid in the Re_j range of 2,700 – 45,000.

$$\frac{t_{cb}}{t^*} = 9.98 - 0.908\ln(\mathrm{Re}_j) \,. \tag{2}$$

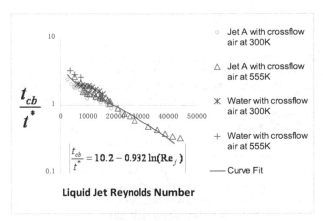

Fig. 18. Non-dimensional breakup time dependence upon the Reynolds number of liquid jet

3.4 Results of the spray penetration measurements

Measurements of spray penetration were obtained using NAC GX-1 high speed camera that captured shadowgraph high definition images of the spray at the rate of 24,000fps at a resolution ~8.5 pixel/mm with a record length of about 20,000 frames. Illumination of the spray was achieved by the copper-vapor laser flashes (30ns) synchronized with the shutter openings. Laser light was introduced into the test section through the 1mm diam. quartz fiber from the laser. Collimator lens and diffusing glass plate created a uniform light beam that illuminated spray from one side through the window in the pressure vessel. Camera that was installed on the other side of the pressure vessel captured shadowgraph images of the spray.

Each of several thousands images (see example on the Figure 19-a) that compose a high speed movie of the fluctuating spray was processed individually in order to characterize the outer border of the spray pattern. For this purpose the following procedure was applied:

- Each image was corrected by subtraction of the averaged background. Images of the background were captured before any fuel was injected at each flow condition and then averaged for the experimental series to be processed.
- Dynamic range of each image was adjusted to eliminate possible influence of laser pulse intensity fluctuations (i.e. to avoid affecting the overall brightness of the image).
- Threshold was applied to all images in the series to equalize pixel intensity value in the spray region to unity and background region pixels to zero. The result of this conversion to a binary field is shown on the Figure 19-b. Line that divided white and black zones on the image represented outer border of the spray.

In the final stage of processing, standard algorithms for calculating mean and maximum values and RMS were applied to the spray border lines.

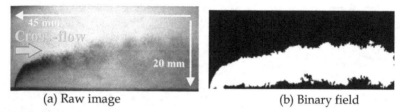

| (a) Raw image | (b) Binary field |

Fig. 19. Procedure for characterization of the outer border of the spray

All together 58 high speed movies of the spray were captured at different flow conditions that are divided into two series. In the first one (so called *We-sweep*) fuel to air momentum flux ratio was kept constant equal to *q*= 20 while Weber number was changed from movie to movie. Spray movies at *We=400, 600, 800, 1000, 1200, 1400,* and *1600* were captured.

In the other series of experiments (so called *q-sweep*) Weber number was kept constant (*We*=1000) while momentum flux ratio was varied from movie to movie. In the *q-sweep* momentum ratios of *q*=5, 10, 20, 40, 60, 80, 100 were examined. *We*-sweep *and q*-sweep were performed for both sharp and round edged injectors.

Typical results of the *We*-sweep are presented on the Fig. 20 in the form of the mean positions of the spray outer boarders at different Weber numbers (see Figure 20-a) and their RMS values (Figure 20-b). It is clearly seen that the position of the spray outer edge and its RMS are practically independent of *We* number. RMS value increases almost linearly with axial position downstream the injection point. Similar result (luck of dependence on the Weber number) was obtained in the *We-sweep* performed with the round edged injector. Luck of dependence of the spray outer border on the Weber number allows significant simplification of the correlation function.

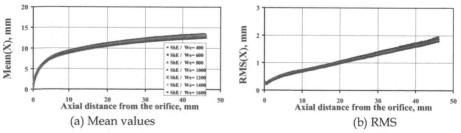

| (a) Mean values | (b) RMS |

Fig. 20. Spray penetration (*X*) into the cross-flow of air at different Weber numbers (*We=400 …1600*) for sharp edge injector

Series of curves each representing the mean position of the spray outer border at a certain momentum flux ratio (*q-sweep*) are shown on the Fig. 21 for the sharp- and round–edged orifices. Graphs reveal strong dependence of the spray border upon the momentum flux ratio. Both series of curves follow the same trend. At the same time they indicate greater spray penetration into the cross flow (~12%) for the sharp edge orifice comparing to the round edged orifice.

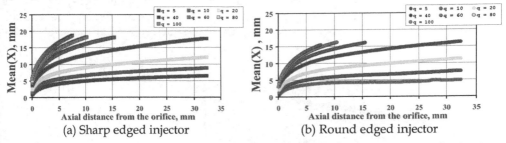

(a) Sharp edged injector   (b) Round edged injector

Fig. 21. Mean spray penetration into the cross flow of air at different momentum flux ratios (q=5...100). Note: We=1000=$const$.

This difference can be attributed to the larger droplets size created by the sharp edge orifice shown on the Figures 12 and 13 and to the difference in the fuel velocity profiles reflected by the difference in flow coefficients C_d of the two tested injectors (see curves on the Fig. 8). Both factors are working towards higher spray penetration. In spite of the fact that the average fuel velocity discharged from the sharp edge orifice is lower than from the round edge orifice because of hydraulic losses, velocity in the center of the jet may be higher and at least some droplets will have higher momentum exclusively because of velocity difference. It is worth to note that the spray border curves obtained for both orifices converge significantly while being normalized by the C_d, (i.e., by the maximum velocity) and by the diameter (D_{32}) of droplets.

Curves on the Fig. 22 were obtained by normalizing the jet penetration into the cross flow by square root of the momentum flux ratio value, q. All the curves obtained in a wide range of q=5...100 and previously shown on the Figure 21 collapsed here in one line. This fact provides a good opportunity for the approximations of the spray penetration X using self explained physical dependence $X{\sim}sqrt(q) \sim U_l$.

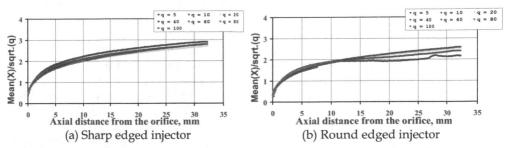

(a) Sharp edged injector   (b) Round edged injector

Fig. 22. Normalized values of the mean spray penetration into the cross flow of air at different momentum flux ratios (q)

Measurements of the spray border obtained in the current study using high speed imaging technique were compared with the spray border data obtained using Phase Doppler method. For this purpose the data rate measured with the PDPA is used as a metric to locate the edge of the spray. The edge of the spray is assumed to be around a region showing 10% of the maximum data rate as shown in Fig. 23-b. Figure 23-a demonstrates a good agreement between the spray trajectories obtained using statistically relevant high speed imaging technique and borders of the spray measured by the processing of the PDPA data rate. It is clearly seen that the maximum spray penetration determined as $X^*=X_{mean}+ 2.8RMS$ is equal to the border determined at the level of 10% threshold of the PDPA data rate curve maximum.

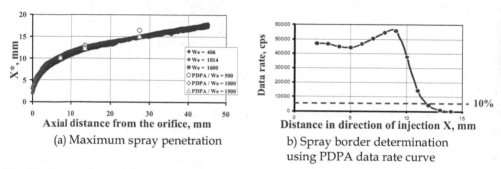

(a) Maximum spray penetration

b) Spray border determination using PDPA data rate curve

Fig. 23. Comparison of the maximum spray penetration (i,e $X^*=X_{mean}+ 2.8RMS$) at $q=20$ measured by the high speed (HS) imaging technique and by the PDPA

3.5 Development of the empirical correlations for spray penetration into the cross flow

Literature sources suggest correlations for the spray outer border $x/d=f(z/d)$ in several different forms that definitely include power function of the momentum flux ratio q^n. Correlations may or may not include power function of Weber number. Shape of the spray pattern is typically described using logarithmic or power function. In spite of the fact that the accuracy of correlation can be improved by increasing number of empiric constants, current study seeks to simplify correlations. This was achieved by using self explained proportionality of droplets penetration into the cross flow to their velocity at the point of discharge (i.e. $x/d \sim U_r \sim q^{0.5}$) and reducing number of the empiric constants by one (i,e $q^n = q^{0.5}$). This significant simplification was proved experimentally on both tested injectors in a wide range of momentum ratios between $q=5$ and $q=100$.

Another simplification of correlation function was attained by limitation of the Weber number range between $We=400$ and $We=1600$. This in turn limited number of possible mechanisms of the jet disintegration to only one mode of liquid jet breakup; i.e., shear breakup excluding column break up. Independence of spray penetration upon the Weber number in the investigated range allowed an exclusion of the Weber number from correlations.

As a result spray penetration for both injectors was correlated using only one empiric coefficient (a_1) that depends only upon the shape of the injector internal surface by the following formula:

$$\frac{x}{d} = a_1 \sqrt{q} \left(\frac{1}{(1 + a_2 \frac{z}{d})} + \ln(1 + a_2 \frac{z}{d}) \right)$$ (3)

The other coefficient (a_2) only shaped the spray border described by the logarithmic function and was independent of the injector design. Thus average and maximum spray penetrations were correlated using coefficients a_1 and a_2 presented in the table 1.

Penetration→	Average		Maximum	
Injector Type↓	a_1	a_2	a_1	a_2
Sharp Edge	1.2181		1.9866	
		1.8806		0.7403
Round Edge	1.0724		1.8641	

Table 1. Empirical correlation coefficients for the average and maximum spray penetration into the cross flow.

Comparison of the experimentally measured and correlated spray penetrations **X** are presented on the Fig. 24 for the average and maximum penetration of the spray created by the sharp edged injector.

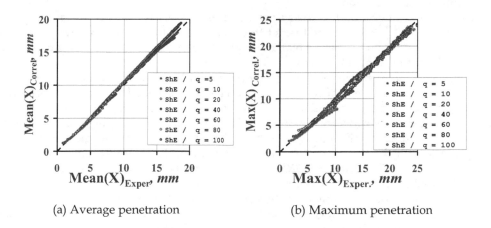

(a) Average penetration   (b) Maximum penetration

Fig. 24. Comparison between the correlated and experimentally measured values of spray penetration X

4. Conclusions

1. Outer borders of the Jet-A spray trajectories created as a result of fuel jet disintegration in the cross flow of cold air at elevated pressure of 5 atm were measured by application the high speed imaging technique that allowed obtaining series of instantaneous images of the fluctuating spray. Locations of the liquid column breakup points (CBP) were determined using the light guiding technique that make mass of liquid illuminated from inside fluoresce till the moment jet losses its continuity.

2. Crossing air flow had core turbulence ~4% and thickness of the boundary layer near the rectangular channel walls ~3mm.

3. Both injectors used in the study had the same diameter of the orifice *d=0.47mm* and a different shape of the internal path (i.e., sharp and round edge orifice) were manufactured using the same equipment and technology. They were installed with orifices openings flush with the channel wall.

4. Application of light guiding technique significantly improved accuracy of the jet in cross flow column breakup point (CBP) determination especially at elevated Weber number (We>200) when traditional shadowgraph methods are not effective because of presence of droplets in high density around the liquid column.

5. CBP was found to be strongly dependent upon velocity of the jet and internal turbulence of liquid inside the orifice. Jet injected from the sharp edge orifice disintegrates earlier compared to the round edge orifice. Dependence of the CBP location upon temperature of injector is much stronger in the sharp edge orifice compared to the round edge orifice.

6. CBP locations were well correlated while converted to the non-dimensional form of characteristic time against the liquid Reynolds number. In fact, CBP location determined in this study were found to be 1-4 diameters of the jet downstream from the injection orifice which is much closer than it was reported in the previous studies (z/d~8).

7. Spray trajectories were found to be independent upon Weber number in the investigated range between *We=400* and *We=1600* due to only shear breakup mode of liquid jet disintegration.

8. Spray penetration into the cross flow was found to be proportional to square root of momentum flux ratio of the fuel jet to crossing air in the investigated range between *q=5* and *q=100* due to self explained dependence of droplet penetration upon the jet velocity at the point of injection.

9. Spray created by the sharp edge injector penetrated 12% further into the cross flow than from the round edge orifice. This observation was attributed to a larger droplet size created by sharp injector and, possibly by the higher velocities of some droplets.

10. Good agreement between the spray trajectories obtained using high speed imaging technique used in the current study and borders of the spray measured by the processing of the PDPA data. It was found that that the maximum spray penetration determined as $X_{max}=X_{mean}+ 2.8RMS$ is equal to the border determined at the level of 10% threshold of the PDPA data rate maximum.

11. Simple correlations for the spray trajectories were obtained using only two empirical coefficients. One of them corresponded to the shape of the injector internal path and the other one only adjusted shape of the logarithmic function that determined average or maximum penetration of the spray and was independent of the injector design.

5. Acknowledgment

The authors would like to thank General Electric–Aviation (Dr. Nayan Patel, contract monitor) for the fund that allowed conduct this study and for additional technical guidance.

6. References

Ahn, K., Kim, J., & Yoon, Y. (2006). Effects of Orifice Internal Flow on Transverse Injection into Subsonic CrossFlows: Cavitation and Hydraulic Flip. *Atomization and Sprays*, Vol. 16, pp.15-34

Becker, J. & Hassa, C. (2002). Breakup and Atomization of a Kerosene Jet in Cross flow at Elevated Pressure. *Atomization and Sprays*, Vol. 11, pp. 49-67

Bonnell, J. M., Marchall, R. L., & Riecke, G. T. (1971) Combustion Instability in Turbojet and Turbofan Augmentors", AIAA 71-698, *Proceedings of 7th AIAA/SAE Propulsion Joint Specialist Conference Exhibit*, Salt Lake City, UT 1971.

Charalampous, G., Hardalupas, Y., & Taylor, A. M. K. P. (2007). A Novel technique for measurements of the intact liquid jet core in a coaxial air-blast atomizer, AIAA 2007-1337, *Proceedings of 45th Aerospace Science Meeting & Exhibit*, Reno, NV 2007.

Chen, T. H., Smith, C. R., & Schommer, D. G. (1993). Multi-Zone Behavior of Transverse Liquid Jet in High-Speed Flow. *AIAA Paper*, 93-0453, *Proceedings of 31st Aerospace Science Meeting & Exhibit*, Reno, NV, 1993

Hanson, A. R., Domich, E. G., & Adams, H. S. (1963) Shock tube investigation of the breakup of drops by air blasts. *Physics of Fluids* Vol. 6, pp.1070-1080

Madabushi, R. (2003). A Model for Numerical Simulation of Breakup of a Liquid Jet in Crossflow, *Atomization and Sprays*, Vol 13, pp. 413-424

Marmottant, P. & Villermaux, E. (2004). On spray formation. *Journal of Fluid Mechanics*, Vol. 498, pp. 73-111

Ranger, A. A. & Nicholls, J. A. (1969). The Aerodynamic Shattering of Liquid Drops. *AIAA Journal*, Vol. 7, No. 2, pp. 285-290.

Sallam, K.A., Dai, Z., & Faeth, G.M. (2002). Liquid breakup at the surface of turbulent round liquid jets in still gases. *International Journal of Multiphase Flow* Vol. 28.

Sallam, K. A., Aalburg, C., & Faeth, G. M. (2004). Breakup of Round Non-Turbulent Liquid Jets in Gaseous Cross Flow. *AIAA Journal*, Vol. 42, No. 12, December 2004.

Stenzler, J. N., Lee, J. G., and Santavicca, D. A., "Penetration of Liquid Jets in a Crossflow", AIAA 2003-1327, *41st Aerospace Science Meeting & Exhibit*, , Reno, NV 2003.

Tamaki, N., Shimizu, M., Nishida, K., and Hiroyasu, H., "Effects of cavitation and internal flow on atomization of liquid jet," *Atomization and Sprays*, Vol. 8, No. 2, 1998, pp. 179-197.

Tamaki, N., Shimizu, M., and Hiroyasu, H., "Enhancement of the atomization of a liquid jet by cavitation in a nozzle hole," *Atomization and Sprays*, Vol. 11, No. 2, 2001, pp. 125-137.

Wu, P.-K., Miranda, R. F., & Faeth, G. M. (1995). "Effects of Initial Flow Conditions on Primary Breakup of NonTurbulent and Turbulent Round Liquid Jets," *Atomization and Sprays*, Vol. 5, No. 2, pp. 175 – 196.

Wu, P.-K., Kirkendall, K. A., Fuller, R. P., & Nejad, A. S. (1997), Breakup Processes of
    Liquid Jets in Subsonic Cross-flows. *Journal of Propulsion and Power*, Vol. 13, No. 1,
    pp. 64 -73.

5

Internal Flows Driven by Wall-Normal Injection

Joseph Majdalani and Tony Saad
University of Tennessee Space Institute
USA

1. Introduction

The internal motion through porous chambers generated by wall-normal injection has received considerable attention in the second half of the twentieth century. This may be attributed to its relevance to a large number of phenomenological applications. In actuality, the motion of fluids driven by either wall injection or suction can be used to describe a variety of practical problems that encompass a wide range of industries and research areas. To name a few, these include: paper manufacturing (Taylor, 1956), ablation or sweat cooling (Peng & Yuan, 1965; Yuan & Finkelstein, 1958), boundary layer control (Acrivos, 1962; Libby, 1962; Libby & Pierucci, 1964), peristaltic pumping (Fung & Yih, 1968; Uchida & Aoki, 1977), gaseous diffusion or filtration, isotope separation (Berman, 1953; 1958a;b), irrigation, and the mean flow modeling of both solid (Culick, 1966; Zhou & Majdalani, 2002) and hybrid rockets (Majdalani, 2007a).

Wall injected flows are initiated by the injection or suction of a fluid across the boundaries of a ducted region having an arbitrary shape and cross-sectional area. This is illustrated in Figure 1 for the special cases of porous channels and tubes. In general, one is required to solve a reduced-order form of the equations of motion for a bounded fluid in order to retrieve a meaningful solution (Terrill & Thomas, 1969). For a general three dimensional setting, this effort leads to a formidable task that is often intractable. However, when simplifying assumptions are invoked, as in the case of an incompressible stream in a channel or tube with uniform injection or suction, Berman (1953) has shown that the Navier-Stokes equations can be reduced to a fourth order nonlinear ODE that may be susceptible to both analytical and numerical treatment. Berman's approach is based on a spatial similarity that transforms the Navier-Stokes equations to a more manageable ODE by assuming that the transverse velocity component v is axially invariant; this immediately translates into a streamfunction that varies linearly in the streamwise direction, i.e. $\psi(x,y) = xF(y)$ (Berman, 1953; White, 2005). Then by considering the limiting case of a small suction Reynolds number, Re $\sim \varepsilon$, Berman employs a regular perturbation series in Re to obtain an approximate expansion for the mean flow function $F(y)$. Berman's Reynolds number, Re $= U_w a/\nu$, is based on the injection speed at the wall, U_w, and the channel half height, a. As for the case of large suction, Berman (1953) first remarks that the limit of the reduced ODE cannot be used to obtain a solution owing to the reduction in order of the governing equation. Later, Sellars (1955) and Terrill (1964) invoke a procedure that permits the extraction of a closed-form analytical approximation for the large Re case by implementing a coordinate transformation that takes into account the spatial relocation of the boundary layer to the sidewall region.

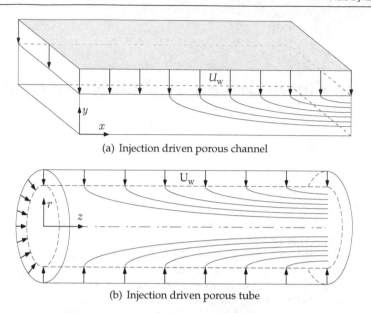

(a) Injection driven porous channel

(b) Injection driven porous tube

Fig. 1. Schematics of porous channels and tubes in which motion is sustained through wall-normal injection.

It is widely believed that Berman (1953) was among the earliest to examine the problem of laminar viscous flow bounded by porous surfaces (see Dauenhauer & Majdalani, 2003; Zhou & Majdalani, 2002). Although his first similarity transformation only applied to a planar configuration with wall suction, it has set forth the foundation for a number of follow-up investigations that relied on either analytical or numerical techniques to explore a variety of geometric configurations with either injection, suction, or both (Proudman, 1960).

Chronologically, these start with Sellars (1955) who extended Berman's solution to very large suction Reynolds numbers. He accomplished this by relaxing the no-slip boundary condition that became immaterial under this limiting condition. At the outset, he extracted a leading order approximation that corresponded to uniform axial motion, i.e. $F(y) = y$. Sellars integrated the ensuing equation based on his leading order approximation. His model thus uncovered the outer solution of this problem when viewed from a boundary layer perspective. Sellars' identification of a thin boundary layer at the wall for the large suction case would later prove crucial in subsequent developments of this problem.

Of particular interest to this chapter is a classic article by Taylor (1956) in which he derived an inviscid rotational solution for both planar and axisymmetric channel flow configurations, in addition to cones and wedges. The absence of viscosity in his model led to approximations that were consistent with Berman's leading order solution of the Navier-Stokes equations expressed at large injection Reynolds numbers, i.e. $F(y) = \sin(\frac{1}{2}\pi y)$. The most peculiar characteristic of Taylor's mean flow profile stood in its ability to satisfy the no-slip boundary condition at the sidewall despite its inviscid nature. This could be attributed to its wall-normal injection that disallowed any axial velocity contribution along the porous boundary.

Returning to the viscous flow problem in a porous channel, Yuan (1956) may have been the first to develop a solution for moderate to large Reynolds numbers and either suction or injection. His solution asymptotically reproduced Taylor's in the limit of a large injection

Reynolds number. However, Yuan's model suffered from a singularity that appeared in the third derivative of the mean flow function $F(y)$ taken at the centerline. This of course signaled the presence of a thin boundary layer that necessitated special treatment. The corresponding boundary layer would later be captured by Terrill (1965) who also described an insightful technique to solve this problem numerically.

In the interim, Berman (1958a) published his second work in which he extended the original planar problem to various geometric settings. This included the familiar case of a straight axisymmetric tube with permeable walls. Almost concurrently, White et al. (1958) advanced a series approximation to the porous channel problem for all ranges of the Reynolds number. However, White and co-workers employed a power series expansion that was centered around Re = 0. They also supplied a numerical solution to this problem. Despite the accuracy of their technique, their power series depended on two arbitrary constants that could only be determined numerically through a trial and error procedure. According to Terrill (1964), their method could be viewed as suitable for intermediate values of Re ($15 \leq \mathrm{Re} \leq 35$). Otherwise, a transformation of the governing equation could be more effective at achieving direct numerical integration. Due to the penalty involved in evaluating the analytical constants of the attendant power series, this particular approach would be later abandoned. Nonetheless, it remained somewhat unique in its ability to provide a single analytical approximation that applied over the entire range of Re, a feat that standard perturbation methods failed to accomplish.

Along similar lines, Terrill (Terrill, 1964; 1965) compiled a comprehensive and detailed résumé of the perturbation solutions of this problem over all ranges of the Reynolds number. Therein, he derived and discussed several limiting cases such as Re = 0, $|\mathrm{Re}| \ll 1$, $\mathrm{Re} \to +\infty$, $\mathrm{Re} \to -\infty$, and compared the various solutions with numerical simulations based on Runge–Kutta integration. For the numerical integration scheme, he introduced a transformation that would lead to a direct numerical solution with no need for predictor-corrector steps or shooting. On the flip side, his technique did not allow the pre-selection of the Reynolds number but rather the post-determination of Re at the conclusion of the numerical procedure. Before leaving this topic, we also note the work of Eckert et al. (1957) who, as far as the authors could verify, were the first to present a numerical solution for the laminar viscous motion in a porous channel.

As far as stability is concerned, the variety of analytical models considered for the planar case appeared to be both unique and stable (Terrill & Thomas, 1969; White, 2005). However, Robinson (1976) reported that dual solutions could exist for large suction while Zaturska et al. (1988) furnished a detailed stability analysis that rigorously showed that (at least) three types of solutions could co-exist. Even more intricate structures would arise in the case of axisymmetric flow in a porous tube. In this context, Terrill & Thomas (1969) have shown that, at least, dual solutions existed for the entire range of injection and suction Reynolds numbers while no steady solutions could be identified for $2.3 < \mathrm{Re} < 9.1$. At the time of this writing, the issue of stability of wall-injected flows remains an open area of investigation especially among applied mathematicians and fluid dynamicists.

1.1 Relevance to propulsion systems

In propulsive applications involving solid and hybrid rocket motors, modeling the mean flow proves to be important for a variety of reasons (Culick, 2006). The instantaneous flow field plays a key role in describing acoustic instability, particle-mean flow interactions, erosive burning, nozzle erosion, and thrust performance. The traditional modus operandi is to decompose the instantaneous motion into a steady average flow and an amalgam of unsteady wave contributions (Chedevergne et al., 2007; Culick, 2006; Majdalani, 2009). In this context, the mean flow represents the bulk motion of the gases and can be approximated by the steady-state solution for a porous tube or channel with wall-normal injection. As for the unsteady field, it refers to any perturbed disturbance that propagates within the chamber. Typical fluctuations are attributed to acoustic, vorticity, entropy, and hydrodynamic instability waves (Chu & Kovásznay, 1958). The importance of the mean flow is therefore evident due to the tight coupling between the steady and unsteady motions.

Although the earliest studies of solid rocket motor (SRM) stability treated the motors as porous enclosures, they failed to consider a suitable mean flow field. For example, the first theoretical study that explored the acoustic instability of rockets may be attributed to Grad (1949) (see Culick, 2006, for greater detail). However, Grad assumed that the mean flow could be ignored as in the case of a stagnant medium, thus limiting his analysis to that of aeroacoustic instability in a cylindrical chamber with no mean flow motion.

Nearly a decade later, the work of McClure and coworkers would prove instrumental in the understanding of rocket motor stability, especially in the development of the energy balance framework. However, principal efforts in this direction have focused on the thin region near the injecting surface (Hart & McClure, 1965; Hart et al., 1960; Hart & Cantrell, 1963; Hart & McClure, 1959; McClure et al., 1960). In fact, McClure et al. (1963) may have been the first to employ a mean flow approximation in their analysis of the aeroacoustic field in SRMs. Their model of choice corresponded to the irrotational motion of an ideal gas in a porous cylinder or between two parallel porous plates. It hence constituted a substantial improvement over the stagnation flow model and, for the first time, succeeded in identifying the intimate coupling between the mean flow and the unsteady wave motion.

It was not until Culick (1966) that a robust representation of the mean flow in circular port motors would be introduced. Despite its inviscid nature, Culick's model was rotational and could satisfy the no-slip requirement at the sidewall. The profile itself coincided with that obtained by Taylor (1956) a decade earlier, albeit in an entirely different application (i.e. paper manufacturing). Culick (1966) derived his solution in the context of a propulsive application that quickly proved to be quintessential to several combustion instability studies, particle-mean-flow interactions, turbulence characterization, and other related investigations of solid propellant rocket motors. It is usually referred to as the Taylor–Culick profile and remains one of the most cited models in rocket motor analysis. For example, Chedevergne et al. (2006), Abu-Irshaid et al. (2007), Griffond et al. (2000), Beddini (1986) and Flandro & Majdalani (2003) made extensive use of the Taylor–Culick model as a basis for their instability work.

1.2 Beyond Culick's solution

Going beyond the Taylor-Culick solution, Majdalani and coworkers have explored a variety of avenues that extended the classic model by providing higher order approximations that could take into account additional factors that are omitted in the inviscid formulation. These include the effects of viscosity, grain taper, wall regression, compressibility, and headwall injection. For example, the sensitivity of the mean flow to viscosity is discussed by Majdalani & Akiki (2010) whereas the effects of tapering of internal bores are addressed by Saad et al. (2006) for the rectangular port slab geometry and by Sams et al. (2007) for the internal burning cylinder with circular cross-section. Other improvements include the work of Kurdyumov (2006) who extended the Taylor–Culick solution to chambers with irregular cross-sections, such as those with a star-shaped perforation. Furthermore, Tsangaris et al. (2007) generalized

Terrill's treatment of the porous tube to include unsteady injection or suction at the sidewall. In the same vein, Erdogan & Imrak (2008) presented a laminar solution for the flow in a porous tube. Their solution was obtained by expanding the velocity field as a series of modified Bessel functions of order n. As for the problem involving wall regression, it was tackled by Dauenhauer & Majdalani (2003), Zhou & Majdalani (2002), and Majdalani & Zhou (2003) for the slab with regressing sidewall, and by Goto & Uchida (1990) and Majdalani et al. (2002) for the internal burning cylinder with expanding walls (see also Majdalani et al., 2009, for an error-free form).

The next noteworthy improvement in this area consists of the compressible Taylor–Culick profile that was first presented in multiple dimensions by Majdalani (2007b). His solution faithfully retained the essential ingredients of Culick's model, yet fully incorporated the effects of compressibility. This was accomplished through the use of a Rayleigh-Janzen expansion jointly with the vorticity-streamfunction approach for a compressible fluid. In asymptotic theory, the Rayleigh-Janzen expansion refers to a regular perturbation expansion in even powers of the Mach number that is ideally suited for the treatment of high speed flows (see Janzen, 1913; Rayleigh, 1916). A similar and equally impactful treatment of the planar configuration was subsequently presented by Maicke & Majdalani (2008) for the compressible Taylor flow analogue. Both analyses give rise to velocity fields that exhibit steep streamline curvatures that are consistent with numerical simulations of the compressible Navier-Stokes equations.

As we move closer to the central topic of this chapter, we consider recent work in which the Taylor–Culick solution is reconstructed for the case of solid rocket motors with headwall injection or hybrid motors with a large headwall-to-sidewall velocity ratio (Majdalani, 2007a). The corresponding problem is analyzed in both axisymmetric and planar configurations by Majdalani & Saad (2007b) and Saad & Majdalani (2009b), respectively. This will be the topic of Section 2 where the solutions for the Taylor–Culick flow with arbitrary headwall injection are derived and compared to steady state, second order accurate inviscid computations. In subsequent work, Majdalani & Saad (2007a) and Saad & Majdalani (2010) manage to introduce a variational procedure based on Lagrangian multipliers to identify solutions of the Taylor–Culick type with varying kinetic energies. As it will be seen in Section 3, these will help to uncover a wide array of motions ranging from purely irrotational to highly rotational fields. The same approach is later applied to slab rocket motors (Saad & Majdalani, 2008a) and to swirl-driven cyclonic chambers with either single (Saad & Majdalani, 2008b) or multiple mantles (Saad & Majdalani, 2009a). In what follows, the main emphasis will be placed on the motion driven by wall-normal injection in a porous, axisymmetric tube.

2. Rotational models with headwall injection

2.1 Arbitrary injection

In this section, we present a model for the mean flow in simulated solid or hybrid rocket motors with headwall injection. Our approach is based on a technique introduced by Majdalani (2007a) and Majdalani & Saad (2007b). The ability to account for arbitrary headwall injection will extend the Taylor-Culick approximation to a wider range of problems. For example, it will enable us to handle both solid and hybrid rocket motors in a unified analysis, the difference being in the relative magnitudes of the headwall-to-sidewall injection speeds. Our approach will be based on the vorticity-streamfunction formulation in which the vorticity transport equation will be used to obtain a functional relation between the streamfunction and the vorticity. The solution will then be retrieved from the vorticity equation. In the

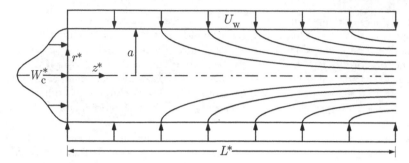

Fig. 2. Schematic of an idealized solid rocket motor with sidewall injection.

process, a multitude of injection profiles will be extracted using superposition. Despite the nonlinearity of the vorticity transport equation near the headwall, it will be shown that the solution becomes progressively more linear in the downstream direction, a factor that permits the use of superposition. Incidentally, the linearity of the vorticity-streamfunction relation used in these studies has been shown by Kurdyumov (2008) to hold true away from the headwall. Finally, the resulting approximations will be tested using three representative injection profiles for which comparisons with finite volume CFD simulations of the Euler equations will be performed.

2.2 Mathematical idealization

A rocket motor can be idealized as a cylindrical chamber of porous length L^* and radius a with both a reactive headwall and a nozzleless aft end as shown in Figure 2. The radial and axial velocities are represented by u^* and w^*, respectively, while r^* and z^* stand for the radial and axial coordinates used to describe the solution from the headwall to the typical nozzle attachment point at the chamber outlet. At the headwall, a fluid stream (which may denote an oxidizer or gaseous propellant mixture) is injected into the chamber at a prescribed velocity $w_0^*(r^*)$. This could be given by

$$w_0^*(r^*) = w^*(r^*, z^* = 0) = \begin{cases} W_c^* = \text{const} & \text{uniform} \\ W_c^* \cos(\tfrac{1}{2}\pi r^{*2}/a^2) & \text{cosine} \\ W_c^*[1 - (r^*/a)^m] & \text{laminar and turbulent} \\ W_c^*(1 - r^*/a)^{1/m} & \text{turbulent} \end{cases} \tag{1}$$

where $W_c^* = w^*(0,0)$ is the centerline speed at the headwall (a constant), m is some integer, and the asterisk denotes a dimensional variable. The incoming stream merges with the cross flow generated by uniform mass addition along the porous sidewall. Naturally, the sidewall injection velocity $U_w = -u^*(a, z^*)$ is commensurate with propellant or fuel regression rates. In hybrids, U_w can be appreciably smaller than W_c^* due to slow fuel pyrolysis; in SRM analysis, these two values can be identical.

2.2.1 Normalization

It is useful to normalize all recurring variables and operators. This can be done by following Majdalani & Saad (2007b) and setting

$$r = \frac{r^*}{a}; \; z = \frac{z^*}{a}; \; \nabla = a\nabla^*; \; p = \frac{p^*}{\rho U_w^2}; \; \psi = \frac{\psi^*}{a^2 U_w};$$

$$u = \frac{u^*}{U_w}; \; w = \frac{w^*}{U_w}; \; \Omega = \frac{\Omega^* a}{U_w}; \; W_c = \frac{W_c^*}{U_w}; \; L = \frac{L^*}{a} \tag{2}$$

where starred variables denote dimensional quantities. Note that this normalization applies to all subsequent developments.

2.2.2 Euler-based formulation

A non-reactive motion may be assumed, prompted by the thin reactive zone above the grain surface. Following Culick (1966), the flow can be taken to be steady, inviscid, incompressible, rotational, and axisymmetric. It should be noted that Majdalani (2007b) and Maicke & Majdalani (2008) have provided compressible Taylor–Culick solutions under isentropic flow conditions. These confirm the suitability of the present model for a variety of applications in which the effects of compressibility are small. Chu et al. (2003) and Vyas et al. (2003) have also demonstrated that the flow field above the thin flame zone may be treated as non-reactive. At the outset, the normalized Euler equations with no swirl can be written as

$$\frac{1}{r}\frac{\partial(ru)}{\partial r} + \frac{\partial w}{\partial z} = 0 \tag{3a}$$

$$u\frac{\partial u}{\partial r} + w\frac{\partial u}{\partial z} = -\frac{\partial p}{\partial r} \tag{3b}$$

$$u\frac{\partial w}{\partial r} + w\frac{\partial w}{\partial z} = -\frac{\partial p}{\partial z} \tag{3c}$$

or, in vector form

$$\nabla \cdot \mathbf{u} = 0 \tag{4a}$$

$$\mathbf{u} \cdot \nabla \mathbf{u} = -\nabla p \tag{4b}$$

One may now invoke the dyadic vector identity $\mathbf{u} \cdot \nabla \mathbf{u} \equiv \nabla(\frac{1}{2}\mathbf{u} \cdot \mathbf{u}) - \mathbf{u} \times \nabla \times \mathbf{u}$. Then, by taking the curl of the resulting expression into (4b), one obtains the vorticity transport equation for steady, inviscid motion

$$\nabla \times (\mathbf{u} \times \mathbf{\Omega}) = 0 \tag{5}$$

where

$$\mathbf{\Omega} = \nabla \times \mathbf{u} \tag{6}$$

Finally, four boundary conditions can be prescribed by writing

$$\begin{cases} u(0,z) = 0 & \text{no flow across centerline} \\ w(1,z) = 0 & \text{no slip at sidewall} \\ u(1,z) = -1 & \text{constant radial inflow at sidewall} \\ w(r,0) = w_0(r) & \text{axial inflow at headwall} \end{cases} \tag{7}$$

where the headwall injection profile may take any of the following plausible forms

$$w_0(r) = \begin{cases} W_c = \text{const} \\ W_c \cos(\frac{1}{2}\pi r^2) \\ W_c(1 - r^m) \end{cases} \tag{8}$$

Here m is the power-law exponent that may be taken as 2 for laminar and 7 or 8 for turbulent-like behavior.

2.3 Vorticity-streamfunction formulation

Continuity is fulfilled by the Stokes streamfunction in cylindrical coordinates when written as

$$u = -\frac{1}{r}\frac{\partial \psi}{\partial z}; \quad w = \frac{1}{r}\frac{\partial \psi}{\partial r} \tag{9}$$

Having a single nonzero component in the azimuthal direction, the vorticity reduces to

$$\Omega = \Omega_\theta \mathbf{e}_\theta \equiv \Omega \mathbf{e}_\theta \tag{10}$$

Its substitution into the vorticity transport equation (5) yields

$$\frac{\partial \psi}{\partial r}\frac{\partial}{\partial z}\left(\frac{\Omega}{r}\right) - \frac{\partial \psi}{\partial z}\frac{\partial}{\partial r}\left(\frac{\Omega}{r}\right) = 0 \quad \text{or} \quad \frac{(\Omega/r)_z}{(\Omega/r)_r} = \frac{\psi_z}{\psi_r} \tag{11}$$

where the subscripts denote differentiation with respect to r or z, respectively. Equation (11) may be satisfied by taking $\Omega = rF(\psi)$ since

$$\frac{(\Omega/r)_z}{(\Omega/r)_r} = \frac{[F(\psi)]_z}{[F(\psi)]_r} = \frac{F_\psi \psi_z}{F_\psi \psi_r} = \frac{\psi_z}{\psi_r} \tag{12}$$

So we follow Culick (1966) and set $\Omega = C^2 r\psi$. Despite the non-uniqueness of this relation, it enables us to secure (5). At this point, straightforward substitution into the vorticity equation (6) renders immediately the second-order PDE associated with the Taylor–Culick problem,

$$\frac{\partial^2 \psi}{\partial z^2} + \frac{\partial^2 \psi}{\partial r^2} - \frac{1}{r}\frac{\partial \psi}{\partial r} + C^2 r^2 \psi = 0 \tag{13}$$

with the particular set of constraints,

$$\lim_{r\to 0} \frac{1}{r}\frac{\partial \psi(r,z)}{\partial z} = 0 \tag{14a}$$

$$\frac{\partial \psi(1,z)}{\partial r} = 0 \tag{14b}$$

$$\frac{\partial \psi(1,z)}{\partial z} = 1 \tag{14c}$$

$$\frac{1}{r}\frac{\partial \psi(r,0)}{\partial r} = w_0(r) \tag{14d}$$

By virtue of L'Hôpital's rule, removing the singularity in (14a) requires that both

$$\frac{\partial \psi(0,z)}{\partial z} = 0 \tag{15a}$$

$$\frac{\partial^2 \psi(0,z)}{\partial r \partial z} = 0 \tag{15b}$$

Being linear, (13) is solvable by separation of variables; it yields

$$\psi(r,z) = (\bar{\alpha}z + \bar{\beta})[A\cos(\tfrac{1}{2}Cr^2) + B\sin(\tfrac{1}{2}Cr^2)] \tag{16}$$

This expression satisfies (15b) identically. Henceforth, (14a) may be superseded by (15a). We then proceed to implement the problem's constraints so that a solution may be realized.

2.4 Solution by eigenfunction expansion

The application of the boundary conditions must be carefully carried out, preferably in the order in which they appear. Starting with (15a), we obtain:

$$\frac{\partial \psi(0,z)}{\partial z} = \bar{a} A \cos(\tfrac{1}{2}Cr^2) + \bar{a} B \sin(\tfrac{1}{2}Cr^2)\Big|_{r=0} = 0 \tag{17}$$

or $A = 0$. Without loss of generality, we set $B = 1$ and rewrite (14b) as

$$\frac{\partial \psi(1,z)}{\partial r} = rC(\bar{a}z + \bar{\beta})\cos(\tfrac{1}{2}Cr^2)\Big|_{r=1} = 0; \quad \forall z \in \mathbb{R}_0^+ \tag{18}$$

and so $\cos(\tfrac{1}{2}C) = 0$. This is satisfied by

$$C = C_n = (2n+1)\pi; \ \forall n \in \mathbb{N}_0 \tag{19}$$

Using $C_n = (2n+1)\pi$, we obtain an infinite series solution to (13). This process introduces an error term in (5) that will be examined in Section 2.6. In the interim, we take

$$\psi_n(r,z) = (\alpha_n z + \beta_n)\sin[(n+\tfrac{1}{2})\pi r^2] \tag{20}$$

For convenience, we introduce $\chi_n \equiv \tfrac{1}{2}(2n+1)\pi r^2$ so that the total streamfunction may be compacted into

$$\psi(r,z) = \sum_{n=0}^{\infty}(\alpha_n z + \beta_n)\sin\chi_n \tag{21}$$

At this juncture, we apply the sidewall injection condition (14c) to produce

$$\frac{\partial \psi(1,z)}{\partial z} = \sum_{n=0}^{\infty} \alpha_n \sin[(n+\tfrac{1}{2})\pi] = 1 \quad \text{or} \quad \sum_{n=0}^{\infty}(-1)^n \alpha_n = 1 \tag{22}$$

This keystone equality encapsulates several possible outcomes depending on the behavior of α_n. One such case corresponds to Taylor's family of solutions for which

$$\alpha_0 = 1 \quad \text{and} \quad \alpha_n = 0; \quad \forall n \neq 0 \tag{23}$$

Accordingly, by setting $\beta_n = 0$, we recover Culick's original solution

$$\psi(r,z) = z\sin(\tfrac{1}{2}\pi r^2) \tag{24}$$

Other forms of α_n will be discussed in Section 3. At present, we let $\alpha_0 = 1$ and reduce (21) into

$$\psi(r,z) = z\sin(\tfrac{1}{2}\pi r^2) + \sum_{n=0}^{\infty} \beta_n \sin\chi_n \tag{25}$$

Lastly, the headwall condition (14d) may be fulfilled through the use of orthogonality. Starting with

$$\frac{1}{r}\frac{\partial \psi(r,0)}{\partial r} = \pi \sum_{n=0}^{\infty}(2n+1)\beta_n \cos\chi_n = w_0(r) \tag{26}$$

one can take advantage of the orthogonality of the cosine function to secure

$$\beta_n \int_0^1 (2n+1)\cos^2\chi_n \, r \, dr = \frac{1}{\pi}\int_0^1 w_0(r)\cos\chi_n \, r \, dr \tag{27}$$

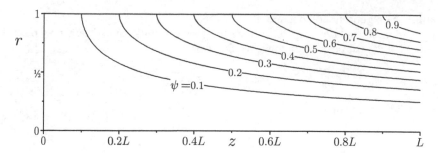

Fig. 3. Streamline patterns corresponding to the classic Taylor–Culick profile with no headwall injection.

or

$$\beta_n = \frac{4}{(2n+1)\pi} \int_0^1 w_0(r) \cos \chi_n \, r \, dr \tag{28}$$

With β_n in hand, the streamfunction is fully determined, namely,

$$\psi(r,z) = z \sin(\tfrac{1}{2}\pi r^2) + \sum_{n=0}^{\infty} \left[\frac{4}{(2n+1)\pi} \int_0^1 w_0(r) \cos \chi_n \, r \, dr \right] \sin \chi_n \tag{29}$$

The radial and axial velocities follow and these may be expressed as

$$u(r) = -r^{-1} \sin(\tfrac{1}{2}\pi r^2);$$

$$w(r,z) = \pi z \cos(\tfrac{1}{2}\pi r^2) + \pi \sum_{n=0}^{\infty} (2n+1)\beta_n \cos \chi_n \tag{30}$$

Interestingly, the radial velocity remains independent of the headwall injection sequence, β_n. Finally, the vorticity may be deduced from

$$\Omega(r,z) = \pi^2 rz \sin(\tfrac{1}{2}\pi r^2) + \pi^2 r \sum_{n=0}^{\infty} (2n+1)^2 \beta_n \sin \chi_n \tag{31}$$

This extended form of the Taylor–Culick profile represents a solution for an arbitrary headwall injection pattern $w_0(r)$ that may be prescribed by the proper specification of β_n through (28). By way of confirmation, the classical Taylor–Culick solution with inert headwall may be readily recovered by setting $\beta_n = 0$ everywhere. The streamline patterns associated with this historical benchmark are illustrated in Figure 3.

2.5 Axisymmetric headwall injection profiles
The framework may be tested using a variable headwall injection profile. To be consistent with the underlying flow assumptions, we employ an axisymmetric function to specify the injection pattern at $z = 0$, namely,

$$w_0(r) = \begin{cases} W_c = \text{const} & \text{uniform} \\ W_c \cos(\tfrac{1}{2}\pi r^2) & \text{half cosine} \\ W_c(1 - r^2) & \text{parabolic} \end{cases} \tag{32}$$

These are prescribed by classic profiles used by Berman (1953) (half cosine), Poiseuille (White, 2005), and others (uniform flow).

2.5.1 Uniform injection
In this case, the headwall injection sequence β_n collapses into

$$\beta_n = \frac{4(-1)^n W_c}{\pi^2 (2n+1)^2} \tag{33}$$

whence

$$\psi(r,z) = z \sin(\tfrac{1}{2}\pi r^2) + \frac{4W_c}{\pi^2} \sum_{n=0}^{\infty} \frac{(-1)^n}{(2n+1)^2} \sin \chi_n \tag{34}$$

The axial velocity and vorticity may be easily determined to be

$$w(r,z) = \pi z \cos(\tfrac{1}{2}\pi r^2) + \frac{4W_c}{\pi} \sum_{n=0}^{\infty} \frac{(-1)^n}{(2n+1)} \cos \chi_n \tag{35}$$

$$\Omega(r,z) = \pi^2 rz \sin(\tfrac{1}{2}\pi r^2) \tag{36}$$

The character of (34) is illustrated in Figure 4. Using $W_c = U_w = 1$, a balance between sidewall and headwall injection causes the streamline originating at the corner ($r = 1, z = 0$) to bisect the flow field at an angle of $\pi/4$ as shown in Figure 4(b). By concentrating on a thin region near the sidewall in Figure 4(c), it may be seen that the solution conforms to the stated boundary conditions. It is also evident that $w_0(r) = W_c = 1$ corresponds to a simulated solid propellant grain that is burning evenly along its headwall and sidewall boundaries.

2.5.2 Similarity-conforming cosine injection
For the cosine injection profile, we use (28) to obtain

$$\beta_n = \begin{cases} \dfrac{W_c}{\pi} \equiv W_h; & n = 0 \\ 0; & \text{otherwise} \end{cases} \tag{37}$$

Using (9), the streamfunction becomes

$$\psi(r,z) = (z + W_h) \cos(\tfrac{1}{2}\pi r^2) \tag{38}$$

The streamlines associated with the cosine headwall injection case are depicted in Figure 5. Their axial velocity and vorticity correspond to

$$w(r,z) = \pi(z + W_h) \cos(\tfrac{1}{2}\pi r^2);$$
$$\Omega(r,z) = \pi^2 r(z + W_h) \sin(\tfrac{1}{2}\pi r^2) \tag{39}$$

It should be noted that while the solutions derived for most injection profiles are approximate, the one corresponding to the similarity-conforming Berman injection will prove to be exact. This behavior will be discussed in Section 2.6.

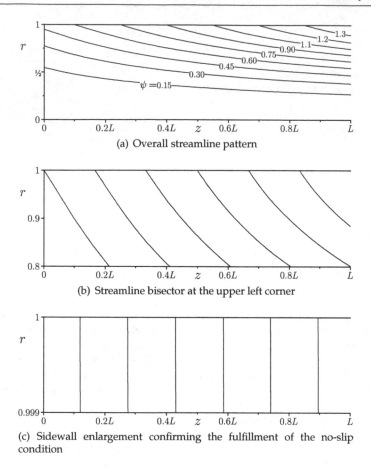

(a) Overall streamline pattern

(b) Streamline bisector at the upper left corner

(c) Sidewall enlargement confirming the fulfillment of the no-slip condition

Fig. 4. Streamlines corresponding to uniform headwall injection with $W_c = 1$.

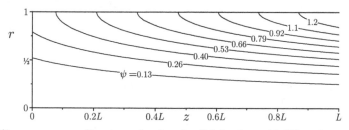

Fig. 5. Streamlines corresponding to cosine headwall injection with $W_c = 1$.

2.5.3 Parabolic injection

For the parabolic, laminar-like profile, one may substitute $w_0(r) = W_c(1 - r^2)$ into (28) and retrieve

$$\beta_n = \frac{8W_c}{(2n + 1)^3 \pi^3} \tag{40}$$

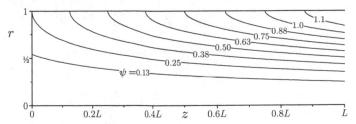

Fig. 6. Streamlines corresponding to parabolic headwall injection with $W_c = 1$.

Consequently, the streamfunction, axial velocity, and vorticity may be deduced one-by-one:

$$\psi(r,z) = z\sin(\tfrac{1}{2}\pi r^2) + \frac{8W_c}{\pi^3}\sum_{n=0}^{\infty}\frac{1}{(2n+1)^3}\sin\chi_n \tag{41}$$

$$w(r,z) = \pi z\cos(\tfrac{1}{2}\pi r^2) + \frac{8W_c}{\pi^2}\sum_{n=0}^{\infty}\frac{1}{(2n+1)^2}\cos\chi_n \tag{42}$$

$$\Omega(r,z) = \pi^2 rz\sin(\tfrac{1}{2}\pi r^2) + \frac{8W_c}{\pi}r\sum_{n=0}^{\infty}\frac{1}{(2n+1)}\sin\chi_n \tag{43}$$

The streamlines corresponding to this case are shown in Figure 6.

2.6 Nonlinear residual error

To test the accuracy of the solutions presented heretofore, we substitute (25) into (5). Terms that do not entirely cancel are hereafter referred to as the residual error $Q(r,z)$. It is straightforward to see that Q may be calculated from

$$Q(r,z) = \|\nabla \times \mathbf{u} \times \mathbf{\Omega}\| = -\frac{\partial}{\partial r}(u\Omega) - \frac{\partial}{\partial z}(w\Omega) \tag{44}$$

In terms of the streamfunction and the vorticity, we have

$$Q(r,z) = -\frac{\Omega}{r^2}\frac{\partial\psi}{\partial z} + \frac{1}{r}\frac{\partial\psi}{\partial z}\frac{\partial\Omega}{\partial r} - \frac{1}{r}\frac{\partial\psi}{\partial r}\frac{\partial\Omega}{\partial z} \tag{45}$$

For each eigensolution given by (29), the vorticity transport equation is fulfilled with zero residual. Using $\Omega = \Omega_n = C_n^2 r\psi_n$, it is clear that (45) becomes

$$Q_n = -\frac{C_n^2\psi_n}{r}\frac{\partial\psi_n}{\partial z} + \frac{1}{r}\frac{\partial\psi_n}{\partial z}\frac{\partial}{\partial r}(C_n^2 r\psi_n) - \frac{1}{r}\frac{\partial\psi_n}{\partial r}\frac{\partial}{\partial z}(C_n^2 r\psi_n) \tag{46}$$

$$= -\frac{C_n^2\psi_n}{r}\frac{\partial\psi_n}{\partial z} + \frac{1}{r}\frac{\partial\psi_n}{\partial z}C_n^2\psi_n + C_n^2\frac{\partial\psi_n}{\partial z}\frac{\partial\psi_n}{\partial r} - C_n^2\frac{\partial\psi_n}{\partial r}\frac{\partial\psi_n}{\partial z} = 0$$

It may hence be seen that the summation of (46) over all eigenmodes will be identically zero if a hypothetical case may be considered for which all eigensolutions coexist independently. In practice, however, the eigensolutions must be taken collectively, and so coupling between eigenmodes must be allowed. The total vorticity and streamfunction must be determined and substituted into the vorticity transport equation. Insertion into (45) requires evaluating

$$Q = -\frac{1}{r^2}\sum_{n=0}^{\infty}\Omega_n\sum_{n=0}^{\infty}\frac{\partial\psi_n}{\partial z} + \frac{1}{r}\sum_{n=0}^{\infty}\frac{\partial\psi_n}{\partial z}\sum_{n=0}^{\infty}\frac{\partial\Omega_n}{\partial r} - \frac{1}{r}\sum_{n=0}^{\infty}\frac{\partial\psi_n}{\partial r}\sum_{n=0}^{\infty}\frac{\partial\Omega_n}{\partial z} \tag{47}$$

where
$$\begin{cases} \psi_n = (\alpha_n z + \beta_n)\sin\chi_n; & \dfrac{\partial\psi_n}{\partial z} = \alpha_n \sin\chi_n \\[2mm] \dfrac{\partial\psi_n}{\partial r} = rC_n(\alpha_n z + \beta_n)\cos\chi_n; & \dfrac{\partial\Omega_n}{\partial z} = rC_n^2\alpha_n \sin\chi_n \end{cases}$$
(48)

Furthermore, for the Taylor–Culick class of solutions, $\alpha_0 = 1$ and $\alpha_n = 0,\ \forall\, n \neq 0$. This leaves us with

$$\frac{\partial\psi_n}{\partial z} = \frac{\partial\psi_0}{\partial z} = \sin(\tfrac{1}{2}\pi r^2); \quad \frac{\partial\Omega_n}{\partial z} = \frac{\partial\Omega_0}{\partial z} = C_0^2 r\frac{\partial\psi_0}{\partial z}$$
(49)

Note that the axial derivatives are solely due to the zeroth eigenmode. This reduces (47) into

$$Q = \frac{\partial\psi_0}{\partial z}\left(-\frac{1}{r^2}\sum_{n=0}^{\infty}\Omega_n + \frac{1}{r}\sum_{n=0}^{\infty}\frac{\partial\Omega_n}{\partial r} - C_0^2\sum_{n=0}^{\infty}\frac{\partial\psi_n}{\partial r}\right)$$
(50)

Finally, noting that

$$\frac{\partial\Omega_n}{\partial r} = C_n^2\psi_n + C_n^2 r\frac{\partial\psi_n}{\partial r}$$
(51)

we retrieve

$$Q(r) = \frac{\partial\psi_0}{\partial z}\sum_{n=0}^{\infty}\left(C_n^2 - C_0^2\right)\frac{\partial\psi_n}{\partial r} = \sin(\tfrac{1}{2}\pi r^2)\, r\sum_{n=1}^{\infty}C_n\beta_n(C_n^2 - C_0^2)\cos\chi_n$$
(52)

Equation (52) represents the net residual of the vorticity transport equation due to nonlinear coupling. It is not necessarily zero except for inert ($\beta_n = 0,\ \forall n$) or sinusoidal headwall injection profiles ($\beta_n = 0,\ \forall n \geq 1$). To further explore the behavior of the residual error, we expand (52) into

$$Q(r) = 4\pi^3 r\sin(\tfrac{1}{2}\pi r^2)\sum_{n=1}^{\infty}D_n\cos\left[\tfrac{1}{2}(2n+1)\pi r^2\right]$$
(53)

where

$$D_n \equiv \frac{C_n}{4\pi^3}\beta_n(C_n^2 - C_0^2) \equiv n(n+1)(2n+1)\beta_n$$
(54)

Clearly, the residual error vanishes at $r = (0,1)$ and is otherwise controlled by the behavior of D_n. This sequence represents the deviation from the exact solution corresponding to the cosine profile for which $C_n^2 - C_0^2 = 0$. In the case of no headwall injection, $\beta_n = D_n = 0$, thus leading to an exact representation. As $D_n \to 0$, the solutions become more accurate. Generally, $\beta_n \neq 0$ and so D_n will only vanish when $C_n^2 = C_0^2$. To illustrate this character, we consider two examples, namely, those corresponding to parabolic and uniform injection. For parabolic injection, we find a quickly converging sequence, specifically

$$D_{n,\text{parabolic}} \sim \frac{n(n+1)}{(2n+1)^2}\xrightarrow[n\to\infty]{}\frac{1}{4}$$
(55)

In this case, the residual is sufficiently small, albeit non-vanishing, because of the first few terms in (55). However, for the uniform flow, we get alternating infinity, namely

$$D_{n,\text{uniform}} \sim (-1)^n\frac{n(n+1)}{(2n+1)}\xrightarrow[n\to\infty]{}\pm\infty$$
(56)

In this case, the residual is undefined because the alternating sequence of increasing terms in (56) diverges. This may be corroborated by the nature of the uniform profile known for its sharp discontinuity at the sidewall.

In all cases for which the residual converges, the error vanishes along the centerline and at the chamber sidewall. This grants our model the character of a rational approximation. Moreover, because the residual remains independent of z, the error that is entailed decreases as we move away from the headwall. This improvement in the streamwise direction makes the approximation more suitable for modeling elongated chambers, such as SRMs. Its behavior near the headwall is consistent with the Taylor–Culick model that is known for its subtle discontinuity at $z = 0$. In all cases considered, the core flow approximations become increasingly more accurate away from the headwall, a condition that is compatible with the parallel flow assumption used in many stability investigations of solid and hybrid rocket flow fields. A similar conclusion is reached by Kurdyumov (2008) whose work confirms the nonlinearity of the vorticity-streamfunction relation in the vicinity of the headwall and its progressive linearity with successive increases in z.

2.7 Pressure evaluation

The steady momentum equation (4b) may be readily solved for the pressure distribution. One may start with $\mathbf{u} \cdot \nabla \mathbf{u} = -\nabla p$ and integrate in two spatial directions to retrieve

$$p = p_0 - \tfrac{1}{2}\mathbf{u} \cdot \mathbf{u} - \int u \frac{\partial w}{\partial r} \, dz \tag{57}$$

where $p_0 = p(0,0)$ represents the centerline pressure at the headwall. To ensure a viable expression for the pressure, the total differential of p must be exact, or

$$\frac{\partial^2 p}{\partial r \partial z} = \frac{\partial^2 p}{\partial z \partial r} \tag{58}$$

This identity stands in fulfillment of Clairaut's theorem (Clairaut, 1739; 1740). In terms of the velocity field, (58) yields

$$u \frac{\partial^2 w}{\partial r^2} + w \frac{\partial^2 w}{\partial r \partial z} - \frac{u}{r} \frac{\partial w}{\partial r} = 0 \tag{59}$$

In short, (57) will produce an analytical expression for the pressure only when (59) is valid. For the classic Taylor–Culick solution, (59) is identically satisfied and the pressure can be integrated into

$$p(r,z) = p_0 - \tfrac{1}{2}\pi^2 z^2 - \tfrac{1}{2} r^{-2} \sin^2(\tfrac{1}{2}\pi r^2) \tag{60}$$

For the cosine profile, we have

$$\mathbf{u} = -r^{-1} \sin(\tfrac{1}{2}\pi r^2)\mathbf{e}_r + \pi(z + W_h) \cos(\tfrac{1}{2}\pi r^2)\mathbf{e}_z \tag{61}$$

and so (59) is fully secured. Integration of Euler's equation renders

$$p(r,z) = p_0 - \tfrac{1}{2}\pi^2 z^2 - W_c \pi z - \tfrac{1}{2} r^{-2} \sin^2(\tfrac{1}{2}\pi r^2) \tag{62}$$

For uniform or parabolic injection, axial velocities may be determined only approximately through (35) and (42); as such, the integrability constraint (59) is no longer satisfied. However, along the centerline, the constraint remains valid. At $r = 0$, the radial velocity shared by both injection profiles vanishes in view of

$$u(0,z) = \lim_{r \to 0} r^{-1} \sin(\tfrac{1}{2}\pi r^2) = 0 \tag{63}$$

As for the axial velocities, they become equal viz.

$$w_{\text{uniform}}(0,z) = w_{\text{parabolic}}(0,z) = \pi z + W_c \tag{64}$$

This enables us to integrate (57) and collect

$$p(0,z) = p_0 - \tfrac{1}{2}\pi^2 z^2 - \pi z W_c \tag{65}$$

Interestingly, all injection profiles generate the same expression for the centerline pressure. To overcome the pitfalls of pressure integrability of a non-exact velocity, approximate representations of p may be sought based on a linear expansion that becomes increasingly more accurate as z is increased. This is

$$p(r,z) = \sum_{n=0}^{\infty} p_n(r,z) \tag{66}$$

where p_n is the pressure corresponding to the nth eigenmode in (20). Integration of the pressure in this case is possible because each eigensolution given by $\psi_n(r,z)$ consists of an exact solution of the Euler equations that directly satisfies (59). Using

$$u_n = -\alpha_n r^{-1} \sin \chi_n;$$
$$w_n = (2n+1)\pi(\alpha_n z + \beta_n)\cos \chi_n \tag{67}$$

one can integrate for the pressure to find

$$p_n(r,z) = p_0 - \tfrac{1}{2}(2n+1)^2 \pi^2 \alpha_n^2 z^2 - (2n+1)^2 \pi^2 \alpha_n \beta_n z - \tfrac{1}{2}\alpha_n^2 r^{-2} \sin^2 \chi_n \tag{68}$$

or

$$p(r,z) = \sum_{n=0}^{\infty} p_n = p_0 - \tfrac{1}{2}\pi^2 z^2 - \beta_0 \pi^2 z - \tfrac{1}{2}r^{-2} \sin^2(\tfrac{1}{2}\pi r^2) \tag{69}$$

As shown in Figure 7, this linear approximation stands in better agreement with the numerical data than the result obtained in (65) for the pressure based on the total velocity. This may be connected to the increasing accuracy associated with a linear vorticity-streamfunction assumption and the superposition of eigensolutions with successive increases in z (Kurdyumov, 2008).

2.8 Numerical verification

So far we have introduced an approximate Euler solution for the Taylor–Culick profile with variable headwall injection. By way of confirmation, an inviscid numerical solution is presented for the mean flow using three illustrative headwall injection profiles. Our simulations are carried out using a finite-volume CFD solver. The targeted flow is that corresponding to a rocket motor with an average sidewall Mach number of 0.03 and strictly inviscid conditions. For the sake of comparison, the working fluid is taken to be ambient air. The aspect ratio of the domain is set at $L = 16$. The actual length and radius are

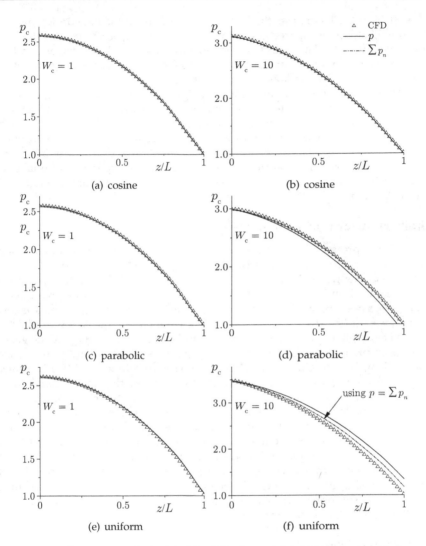

Fig. 7. Comparison between analytical (—) for (65), (— · —) for (69) and numerical simulations ($\triangle$) for the centerline pressure using (a, b) cosine, (c, d) parabolic, and (e, f) uniform injection. Curves are shown for $z/L = 0.1, 0.3, 0.5, 0.7$, and 0.9.

taken at $1.6\,\mathrm{m} \times 0.1\,\mathrm{m}$ and the wall injection velocity is taken at 10 m/s for the simulated SRM. The boundary condition at the sidewall is specified as a velocity inlet to closely mimic the mathematical model where injection is imposed uniformly along the grain surface. The headwall is also defined as a velocity inlet. On the right-hand-side of the domain, a pressure outlet boundary condition is prescribed where the exit pressure is set to be atmospheric as in the case of sea level testing. Although an outflow boundary condition can also be imposed at the downstream section, it is avoided here to avert the possible case of partially developed flow (White, 2005). The difference between an outflow and a pressure outlet boundary condition is that, in the latter case, the exit pressure is fixed at the boundary. The domain is meshed into 589,824 equally spaced control volumes consisting of 3072×192 cells. While the Quadratic Upwind Interpolation for Convection Kinematics (QUICK) scheme is called upon for spatial discretization, the Semi Implicit Method for Pressure Linked Equation (SIMPLE) algorithm is used to resolve the pressure–velocity coupling.

Results for the inviscid simulations are shown in Figures 7–8. These are carried out for $W_c = 1$ and 10 (i.e., solid and hybrid motors); their purpose is to show the streamwise evolution of the axial velocity, vorticity, and centerline pressure at $z/L = 0.1, 0.3, \ldots, 0.9$. It may be seen that the agreement with the computations is excellent except in the case of uniform injection with a large W_c. This may be attributed to the discontinuity that the uniform injection profile experiences at the sidewall. Furthermore, according to (56), we expect the residual error to be large. These limited numerical runs reaffirm the viability of the analytical approximations as simple predictive tools.

3. Generalized Taylor-Culick formulation

In Section 2.1, we presented a mean flow model for solid and hybrid rocket motors that could assimilate a rather arbitrary headwall injection profile based on a specific form of β_n. Initially, the solutions were obtained in series form that depended on two parameters, α_n and β_n, the sidewall and headwall injection sequences. While β_n was prescribed by the headwall injection pattern, the choice of α_n appeared to be flexible provided that the constraint given by (22) remained satisfied. In this section, we follow Majdalani & Saad (2007a) by applying the Lagrangian optimization technique to the total kinetic energy of the generalized Taylor–Culick solution to the extent of producing a variational constraint on α_n (see also Saad & Majdalani, 2010). After some effort, two types of solutions will be identified with increasing or decreasing kinetic energies; of the two families, the Taylor–Culick model will be recovered as a special case. The new approximations will be shown to exhibit velocity profiles with energy dependent curvatures that are reminiscent of turbulent or compressible motions. In practice, steeper profiles have been observed in either experimental or numerical tests, particularly in the presence of intense levels of acoustic energy (Apte & Yang, 2000; 2001; 2002). Interestingly, the energy-based models will range from irrotational to rotational fields with increasing vorticity, thus covering a wide spectrum of admissible motions that observe the problem's physical requirements. A second law analysis will be later used to test the physicality of these solutions and establish the Taylor–Culick motion as an equilibrium state to which all profiles will tend to converge.

3.1 Kinetic energy optimization

As shown in Section 2.1, the sidewall injection sequence must observe a key constraint associated with the wall-normal injection velocity:

$$\sum_{n=0}^{\infty} (-1)^n \alpha_n = 1 \tag{70}$$

Clearly, numerous sequences of α_n exist that can be made to satisfy (70). One of these choices may be arrived at by optimizing the total volumetric kinetic energy in the chamber. The guiding principle is based on the hypothesis that a flow may follow the path of least or most energy expenditure. To test this behavior, we evaluate the local kinetic energy at (r, θ, z) for

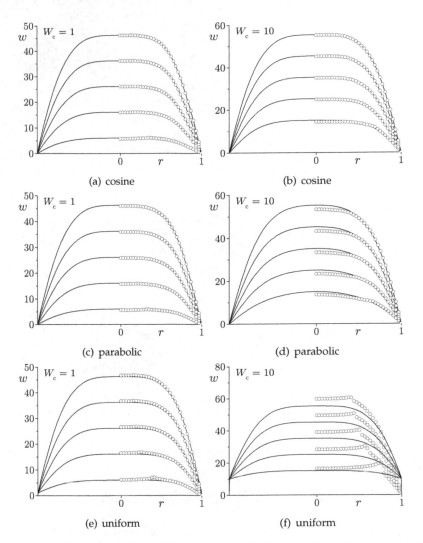

Fig. 8. Comparison between analytical (—) and numerical simulations (○) for the axial velocity using (a, b) cosine, (c, d) parabolic, and (e, f) uniform injection. Curves are shown for $z/L = 0.1, 0.3, 0.5, 0.7$, and 0.9.

each eigensolution using

$$E_n(r, \theta, z) = \tfrac{1}{2}\mathbf{u}_n^2 = \tfrac{1}{2}(u_n^2 + v_n^2 + w_n^2) \tag{71}$$

where each mode is an exact solution that is given by

$$\begin{cases} u_n = -r^{-1}\alpha_n \sin\chi_n; & v_n = 0 \\ w_n = \pi\alpha_n z(2n+1)\cos\chi_n \end{cases} \qquad \chi_n \equiv (n + \tfrac{1}{2})\pi r^2 \tag{72}$$

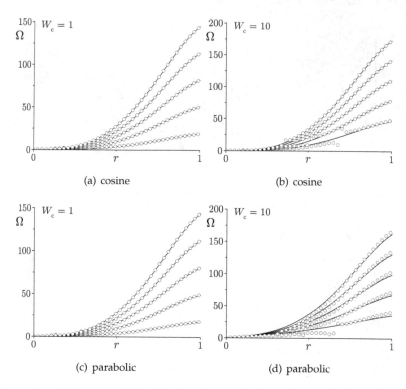

Fig. 9. Comparison between analytical (—) and numerical simulations (○) for the vorticity magnitude using (a, b) cosine and (c, d) parabolic injection. Curves are shown for $z/L = 0.1$, 0.3, 0.5, 0.7, and 0.9.

We now define the cumulative local kinetic energy as the sum of contributions from individual eigensolutions. This can be written as

$$E(r,\theta,z) = \sum_{n=0}^{\infty} E_n(r,\theta,z) = \tfrac{1}{2} \sum_{n=0}^{\infty} \left[\alpha_n^2 r^{-2} \sin^2 \chi_n + \pi^2 \alpha_n^2 z^2 (2n+1)^2 \cos^2 \chi_n \right] \tag{73}$$

Subsequently, the total kinetic energy in a chamber of volume $\mathcal{V}$ may be calculated by integrating the local kinetic energy over the length and chamber cross-section,

$$E_{\mathcal{V}} = \iiint_{\mathcal{V}} E(r,\theta,z) r \, dr \, d\theta \, dz = \pi \sum_{n=0}^{\infty} \int_0^L \int_0^1 \alpha_n^2 \left[\frac{\sin^2 \chi_n}{r^2} + \pi^2 z^2 (2n+1)^2 \cos^2 \chi_n \right] r \, dr \, dz$$

Straightforward evaluation and simplification yield

$$E_{\mathcal{V}} = \tfrac{1}{12} \pi^3 L^3 \sum_{n=0}^{\infty} (\alpha_n^2 a_n + \alpha_n^2 \pi^{-2} L^{-2} d_n); \quad \begin{cases} a_n = (2n+1)^2 \\ d_n = 3\mathrm{Cin}[(2n+1)\pi] \end{cases} \tag{74}$$

where $\mathrm{Cin}(x) \equiv \displaystyle\int_0^x (1 - \cos t) t^{-1} \, dt$ is the Entire Cosine Integral. At this point, one may seek the extremum of (74) subject to the fundamental constraint (70). The latter enables us to introduce the constrained energy function

$$\mathcal{G}(\alpha_0, \alpha_1, \alpha_2, \ldots, \lambda) = E_V + \lambda \left[\sum_{n=0}^{\infty} (-1)^n \alpha_n - 1 \right] \tag{75}$$

where λ is a Lagrangian multiplier. Equation (75) can be maximized or minimized by imposing $\nabla \mathcal{G}(\alpha_0, \alpha_1, \alpha_2, \ldots, \lambda) = 0$. In shorthand notation, we put

$$\nabla \mathcal{G}(\alpha_n, \lambda) = 0; \quad n \in \mathbb{N}_0 \tag{76}$$

Naturally, the constrained energy function may be differentiated with respect to each of its variables to obtain

$$\frac{\partial \mathcal{G}}{\partial \alpha_n} = \frac{1}{6} \pi^3 L^3 \left(\alpha_n a_n + \frac{\alpha_n d_n}{\pi^2 L^2} \right) + (-1)^n \lambda = 0 \tag{77}$$

and

$$\frac{\partial \mathcal{G}}{\partial \lambda} = \sum_{n=0}^{\infty} (-1)^n \alpha_n - 1 = 0 \tag{78}$$

Equation (77) may be used to extract α_n in terms of λ such that

$$\alpha_n = -\frac{6(-1)^n \lambda}{\pi^3 L^3 (a_n + \pi^{-2} L^{-2} d_n)} \tag{79}$$

Then, through substitution into (78), one retrieves

$$\lambda = -\frac{\pi^3 L^3}{6 \sum_{n=0}^{\infty} (a_n + \pi^{-2} L^{-2} d_n)^{-1}} \tag{80}$$

Finally, when λ is inserted into (79), a general solution for α_n emerges, specifically

$$\alpha_n = \frac{(-1)^n}{(a_n + \pi^{-2} L^{-2} d_n) N}; \quad N = \sum_{i=0}^{\infty} \frac{1}{a_i + \pi^{-2} L^{-2} d_i} \tag{81}$$

Clearly, (81) satisfies the fundamental constraint which, by inspection, returns

$$\sum_{n=0}^{\infty} (-1)^n \alpha_n = \frac{1}{N} \sum_{n=0}^{\infty} \frac{1}{(a_n + \pi^{-2} L^{-2} d_n)} = \frac{N}{N} = 1 \tag{82}$$

Some values of α_n are posted in Table 1 at four different aspect ratios corresponding to $L = 1, 5, 10$, and 100. With this expression at hand, the total energy E_V is completely determined.

3.2 Critical length

Equation (74) can be normalized by L^3 and simplified into an energy density form. This can be accomplished by setting

$$\mathcal{E} = E_V / L^3 \tag{83}$$

By plotting $\mathcal{E}$ versus L in Figure 10, it can be seen that $\mathcal{E}$ approaches a constant asymptotic value of $\mathcal{E}_\infty = 2\pi/3$. Granted this behavior, a critical aspect ratio L_{cr} may be defined beyond

n	$L=1$	$L=5$	$L=10$	$L=100$
0	0.7524	0.8095	0.8115	0.8121
1	-0.1146	-0.0914	-0.0905	-0.0902
2	0.0434	0.0329	0.0326	0.0324
3	-0.0225	-0.0168	-0.0166	-0.0165
4	0.0137	0.0101	0.0100	0.0100
5	-0.0092	-0.0068	-0.0067	-0.0067

Table 1. Convergence of the sidewall injection sequence α_n for $L = 1, 5, 10$, and 100.

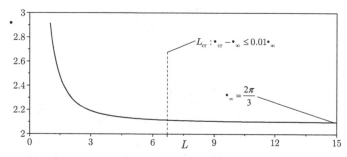

Fig. 10. Kinetic energy density variation with L. Note that for $L > 6.7$, the energy density will be within 1% of its final asymptotic value $\mathscr{E}_\infty$.

which the energy density will vary by less than one percent from its asymptotic value $\mathscr{E}_\infty$. We therefore set

$$\mathscr{E}_{cr} - \mathscr{E}_\infty \leq 0.01\, \mathscr{E}_\infty \tag{84}$$

For a chamber of length $L \geq L_{cr}$, one may evaluate the limiting behavior of (81) by taking $L \to \infty$. For SRMs with inert headwalls, the critical length is found to be 6.7. In practice, most SRMs are designed with an aspect ratio that exceeds 20 and so the assumption of a large L may be safely employed in describing their flow fields. With this simplification, the expression for α_n collapses into

$$\lim_{L \to \infty} \alpha_n = (-1)^n \left(a_n \sum_{i=0}^{\infty} \frac{1}{a_i} \right)^{-1} = \frac{8(-1)^n}{\pi^2 (2n+1)^2} \tag{85}$$

Note that (85) identically satisfies the sidewall constraint viz.

$$\sum_{n=0}^{\infty} (-1)^n \alpha_n = \frac{8}{\pi^2} \sum_{n=0}^{\infty} \frac{1}{(2n+1)^2} = 1 \tag{86}$$

The large-L approximation of α_n quickly converges as illustrated in Table 2.

3.3 Least kinetic energy solution
While the use of Lagrangian multipliers enables us to identify the problem's extremum, straightforward substitution of (81) into (74) allows us to compare the energy content of the present approximation to that of Taylor–Culick's. We find that the extremum obtained through Lagrangian optimization corresponds to the solution with least kinetic energy. Given

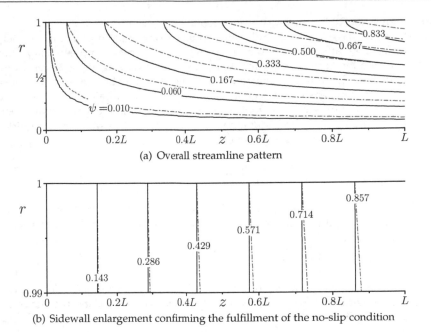

(a) Overall streamline pattern

(b) Sidewall enlargement confirming the fulfillment of the no-slip condition

Fig. 11. Streamlines corresponding to the minimum kinetic energy solution (87) for an inert headwall. Solid lines: Taylor-Culick; broken lines: minimum energy solution.

an inert headwall, the minimum energy approximation reduces to

$$\psi(r,z) = \frac{8}{\pi^2} z \sum_{n=0}^{\infty} \frac{(-1)^n}{(2n+1)^2} \sin \chi_n \mapsto r^2 z \qquad (87)$$

The right–oriented mapping arrow '$\mapsto$' in (87) is used to indicate that the compacted expression is valid inside the domain, $0 \le r < 1$, thus excluding the sidewall. We also remark that, in evaluating (87), the large L approximation is used. The corresponding streamfunction, velocity, and vorticity are catalogued in Table 3. The streamlines are shown in Figure 11(a) using solid lines to denote the Taylor–Culick benchmark, and broken lines to describe the minimum energy solution.

m	α_m	$\sum_{n=0}^{m}(-1)^n \alpha_n$
0	0.8105	0.8105
1	-0.0900	0.9006
2	0.0324	0.9330
3	-0.0165	0.9495
4	0.0100	0.9596
5	-0.0066	0.9663
∞	0.0000	1.0000

Table 2. Convergence of the sidewall injection sequence α_n when $L \to \infty$.

3.4 Type I solutions with increasing energy levels

At this juncture, we have identified only one profile bearing the minimum kinetic energy that the flow can possibly afford. It may be hypothesized that two complementary families of solutions exist with the unique characteristics of exhibiting varying energy levels from which the Taylor-Culick model may be recovered. To this end, it may be useful to seek mean flow solutions with either increasing or decreasing energies. It would also be instructive to rank the Taylor-Culick solution according to its energy content within the set of possible solutions. In the interest of simplicity, we consider long chambers and make use of (85) as a guide. Based on the form obtained through Lagrangian optimization, we note that

$$\alpha_n = \frac{8(-1)^n}{\pi^2(2n+1)^2} \sim \frac{(-1)^n A_2}{(2n+1)^2} \tag{88}$$

where $A_2 = 8/\pi^2$ can be deduced from the radial inflow requirement given by (70). Its subscript is connected with the power of $(2n+1)$ in the denominator. To generalize, we posit the generic Type I form

$$\alpha_n^-(q) = \frac{(-1)^n A_q}{(2n+1)^q}; \quad q \geq 2 \tag{89}$$

where the exponent q will be referred to as the *kinetic energy power index*. This is due to its strong connection with the kinetic energy density as it will be shown shortly. The constant A_q can be used to make (89) consistent with (70). This enables us to retrieve

$$\sum_{n=0}^{\infty} (-1)^n \frac{(-1)^n A_q}{(2n+1)^q} = 1 \tag{90}$$

or

$$A_q = \frac{1}{\sum_{n=0}^{\infty}(2n+1)^{-q}} = \frac{1}{\zeta(q)(1-2^{-q})}; \quad \zeta(q) = \sum_{k=1}^{\infty} k^{-q} \tag{91}$$

where $\zeta(s)$ is Riemann's zeta function. Clearly, the case corresponding to $q = 2$ reproduces the state of least energy expenditure. Furthermore, the $q \geq 2$ condition is needed to ensure

$w(r,0)$	$\psi^-(r,z)$	$w^-(r,z)$
0	$\psi_{\text{ref}}^- \equiv \frac{8}{\pi^2} z \sum_{n=0}^{\infty} \frac{(-1)^n}{(2n+1)^2} \sin\chi_n \mapsto r^2 z$	$w_{\text{ref}}^- \equiv \frac{8}{\pi} z \sum_{n=0}^{\infty} \frac{(-1)^n}{(2n+1)} \cos\chi_n \mapsto 2z$
W_c	$\psi_{\text{ref}}^- + \frac{4W_c}{\pi^2} \sum_{n=0}^{\infty} \frac{(-1)^n}{(2n+1)^2} \sin\chi_n$	$w_{\text{ref}}^- + \frac{4W_c}{\pi} \sum_{n=0}^{\infty} \frac{(-1)^n}{(2n+1)} \cos\chi_n$
$W_c \cos(\frac{1}{2}\pi r^2)$	$\psi_{\text{ref}}^- + \frac{W_c}{\pi} \sin(\frac{1}{2}\pi r^2)$	$w_{\text{ref}}^- + W_c \cos(\frac{1}{2}\pi r^2)$
$W_c(1-r^2)$	$\psi_{\text{ref}}^- + \frac{8W_c}{\pi^3} \sum_{n=0}^{\infty} \frac{\sin\chi_n}{(2n+1)^3}$	$w_{\text{ref}}^- + \frac{8W_c}{\pi^2} \sum_{n=0}^{\infty} \frac{\cos\chi_n}{(2n+1)^2}$

Table 3. Summary of least kinetic energy solutions.

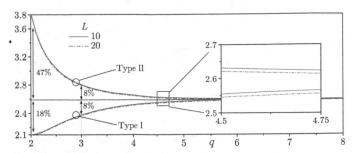

Fig. 12. Variation of the kinetic energy density with the energy power index for Type I (lower branch) and Type II (upper branch) solutions. These are shown at two aspect ratios, $L = 10$ (—) and 20 ($- \cdot -$).

series convergence down to the vorticity. Backward substitution allows us to extract the final form of α_n, namely,

$$\alpha_n^-(q) = \frac{(-1)^n (2n+1)^{-q}}{\sum_{k=0}^{\infty}(2k+1)^{-q}} = \frac{(-1)^n (2n+1)^{-q}}{\zeta(q)(1-2^{-q})}; \quad q \geq 2 \quad \text{(Type I)} \tag{92}$$

To understand the effect of the energy power index q on the kinetic energy density, we use (92) and (74) to plot $\mathscr{E}$ versus q at two aspect ratios. This plot corresponds to the lower branch of Figure 12 for both $L = 10$ and 20. Interestingly, as $q \to \infty$, Taylor–Culick's classic solution is recovered. In fact, using (92), it can be rigorously shown that

$$\lim_{q \to \infty} \alpha_n^-(q) = \begin{cases} 1; & n = 0 \\ 0; & \text{otherwise} \end{cases} \tag{93}$$

This result identically reproduces Taylor–Culick's expression. All of the Type I solutions derived from (92) possess kinetic energies that are lower than Taylor–Culick's; this explains the negative sign in the superscript of α_n^-. They can be bracketed between (87) and $\psi(r,z) = z\sin(\frac{1}{2}\pi r^2)$. In practice, profiles with $q \geq 5$ will be indiscernible from Taylor–Culick's as their energies will then differ by less than one percent. The most distinct solutions will correspond to $q = 2$, 3, and 4 with energies that are 81.1, 91.7, and 97.3 percent of Taylor–Culick's, respectively.

3.5 Type II solutions with decreasing energy levels
To capture solutions with energies that exceed that of Taylor-Culick's, a modified form of α_n is needed. We begin by introducing

$$\alpha_n^+(q) = \frac{B_q}{(2n+1)^q}; \quad q \geq 2 \tag{94}$$

The key difference here stands in the exclusion of the $(-1)^n$ multiplier that appears in (89). The remaining steps are similar. Substitution into (70) unravels

$$\sum_{n=0}^{\infty} \frac{(-1)^n B_q}{(2n+1)^q} = 1 \tag{95}$$

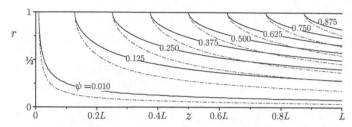

Fig. 13. Comparison of the Taylor–Culick streamlines (—) and the Type II energy-maximized solution ($q = 2$) with stretched streamline curvature (— · —). Results are shown for an inert headwall.

or

$$B_q = \frac{1}{\sum_{n=0}^{\infty}(-1)^n(2n+1)^{-q}} = \frac{4^q}{\zeta(q,\frac{1}{4}) - \zeta(q,\frac{3}{4})}; \tag{96}$$

where $\zeta(q,a)$ is the generalized Riemann zeta function given here as

$$\zeta(q,a) = \sum_{k=0}^{\infty}(k+a)^{-q}; \quad \forall a \in \mathbb{R} \tag{97}$$

Equation (94) yields the general structure of the Type II complementary family of solutions

$$\alpha_n^+(q) = \frac{(2n+1)^{-q}}{\sum_{k=0}^{\infty}(-1)^k(2k+1)^{-q}}$$

$$= \frac{4^q(2n+1)^{-q}}{\zeta(q,\frac{1}{4}) - \zeta(q,\frac{3}{4})}; \quad q \geq 2 \quad \text{(Type II)} \tag{98}$$

Note that the Type II solutions emerging from (98) dispose of kinetic energies that are higher than Taylor-Culick's. The variation of the solution with respect to q is embodied in the upper branch of Figure 12. According to this form of α_n^+, Taylor-Culick's model is recoverable asymptotically by taking the limit as $q \to \infty$. Here too, most of the solutions exhibit energies that fall within one percent of Taylor-Culick's. The most interesting solutions are those corresponding to $q = 2$, 3, and 4 with energies that are 47.0, 8.08, and 2.4 percent larger than Taylor-Culick's. When the energy level is fixed at $q = 2$, a simplification follows for the Type II representation. Catalan's constant emerges in (98), namely,

$$\mathscr{C} = \sum_{k=0}^{\infty}(-1)^k(2k+1)^{-2} \simeq 0.915966 \tag{99}$$

The Type II solution that carries the most energy at $q = 2$ is plotted in Figure 13 and listed in Table 4. In Figure 13, the Type II approximation is seen to overshoot the Taylor-Culick streamline curvature. In view of the two types of solutions with energies that either lag or surpass that of Taylor-Culick's, one may perceive the $q \to \infty$ case as a saddle point to which other possible forms will quickly converge when their energies are shifted. Later in Section 3.12, we will use the entropy maximization principle to establish the Taylor–Culick model as a local equilibrium solution to which all other profiles will be attracted to.

$w(r,0)$	$\psi^+(r,z)$	$w^+(r,z)$
0	$\psi^+_{\text{ref}} \equiv \dfrac{z}{\mathscr{C}} \displaystyle\sum_{n=0}^{\infty} \dfrac{\sin \chi_n}{(2n+1)^2}$	$w^+_{\text{ref}} \equiv \dfrac{\pi}{\mathscr{C}} z \displaystyle\sum_{n=0}^{\infty} \dfrac{\cos \chi_n}{(2n+1)}$
W_c	$\psi^+_{\text{ref}} + \dfrac{4W_c}{\pi^2} \displaystyle\sum_{n=0}^{\infty} \dfrac{(-1)^n}{(2n+1)^2} \sin \chi_n$	$w^+_{\text{ref}} + \dfrac{4W_c}{\pi} \displaystyle\sum_{n=0}^{\infty} \dfrac{(-1)^n}{(2n+1)} \cos \chi_n$
$W_c \cos(\tfrac{1}{2}\pi r^2)$	$\psi^+_{\text{ref}} + \dfrac{W_c}{\pi} \sin(\tfrac{1}{2}\pi r^2)$	$w^+_{\text{ref}} + W_c \cos(\tfrac{1}{2}\pi r^2)$
$W_c(1-r^2)$	$\psi^+_{\text{ref}} + \dfrac{8W_c}{\pi^3} \displaystyle\sum_{n=0}^{\infty} \dfrac{\sin \chi_n}{(2n+1)^3}$	$w^+_{\text{ref}} + \dfrac{8W_c}{\pi^2} \displaystyle\sum_{n=0}^{\infty} \dfrac{\cos \chi_n}{(2n+1)^2}$

Table 4. Summary of solutions with most kinetic energy for various headwall injection patterns. Here, $\chi_n \equiv \tfrac{1}{2}(2n+1)\pi r^2$.

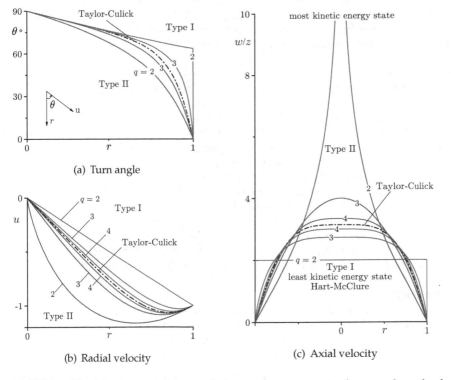

(a) Turn angle

(b) Radial velocity

(c) Axial velocity

Fig. 14. Effect of the kinetic energy power index on the two types of energy-based solutions. Results are shown for (a) turn angle, (b) radial velocity, and (c) axial velocity.

3.6 Behavior of the velocity and vorticity fields

The sensitivity of the solution to the energy power index q is illustrated in Figure 14 where both components of the velocity are displayed in addition to the streamline turn angle θ.

3.6.1 Turn angle

This angle represents the slope of the local velocity measured from the radial injection direction. Making use of axial similarity, θ may be expressed as

$$\theta(r) = \frac{180}{\pi} \tan^{-1} \left(-\frac{1}{z} \frac{w}{u} \right) \qquad (100)$$

The turn angle is shown in Figure 14(a) where, irrespective of q, the flow enters radially at the sidewall with $\theta(1) = 0$. This feature confirms that the flow enters the chamber perpendicularly to the surface in fulfilment of the no-slip requirement. Conversely, for all cases considered, the establishment of strictly parallel motion along the centerline is reflected in $\theta(0) = 90$. Crossing the region between the wall and the centerline, the $q = 2$ Type I case is accompanied by the sharpest change in the turn angle from 0 to 90 degrees. However, as we shift toward the state of most kinetic energy, the smoothing process causes the turn angle to change more gradually. This may be explained by the relative magnitudes of the radial and axial velocities. Specifically, for the Type I, $q = 2$ case, the axial velocity remains practically constant at any chamber cross-section, whereas the radial velocity magnitude increases with r. As we cross into the Type II region, the flow starts turning in the vicinity of the sidewall and progresses smoothly as the centerline is approached.

3.6.2 Radial velocity

The radial velocity is illustrated in Figure 14(b) for representative energy power indices. Starting with the Type II region, the $q = 2$ solution is seen to exhibit a maximum radial velocity overshoot of 16.5 percent relative to the sidewall injection speed. This overshoot reaches its peak at $r = 0.66$ and is required to compensate for the decreasing circumferential area $(2\pi r L)$ normal to the injected stream. Recalling that the Taylor-Culick radial velocity exhibits a 7 percent overshoot at $r = 0.861$, the maximum overshoot calculated here is more than twice as large; it also occurs at a greater distance from the sidewall. Overall, the Type II solutions exhibit smoother curvatures as q is increased. In contrast, by examining the case of least kinetic energy in Figure 14(b), no radial overshoot is observed. Instead, the radial velocity displays its lowest absolute value by diminishing linearly from 1 at the wall to 0 at the centerline. This linear variation is accompanied by an essentially uniform axial velocity depicted for the Type I $q = 2$ case in Figure 14(c). At the outset, the locus of the overshoot varies between $0.66 < r < 1$ as one moves from the Type II, $q = 2$ to the Type I, $q = 2$ case.

3.6.3 Axial velocity

In Figure 14(c), it is clear that the Type I axial velocities are initially blunt, with the flattest curve being the one corresponding to the top-hat profile at $q = 2$. As q is increased, all curves evolve into a sinusoid that approaches the Taylor–Culick model for $q = 5$ and above. Furthermore, as we cross into the Type II region, the centerline velocity continues to increase with increasing energy levels. Due to mass conservation, $Q = 2\pi \int_0^1 wr \, dr = 2\pi z$, and so the centerline speed at each power index is compelled to vary with its corresponding shape to preserve Q. The lowest centerline speed will thus accompany the spatially uniform

$w(r,0)$	$\Omega^-(r,z)$	$\Omega^+(r,z)$
0	0	$\Omega_{\text{ref}}^+ \equiv \dfrac{\pi^2}{2\mathscr{C}} rz\,\csc(\tfrac{1}{2}\pi r^2)$
W_c	0	Ω_{ref}^+
$W_c\cos(\tfrac{1}{2}\pi r^2)$	$\pi W_c r\sin(\tfrac{1}{2}\pi r^2)$	$\Omega_{\text{ref}}^+ + \pi W_c r\sin(\tfrac{1}{2}\pi r^2)$
$W_c(1-r^2)$	$2W_c r$	$\Omega_{\text{ref}}^+ + 2W_c r$

Table 5. Vorticity for least or most kinetic energy solutions.

distribution whereas the highest speed will emerge in the narrowest and most elongated profile connected with the state of most kinetic energy. Interestingly, although this profile slowly diverges at the centerline, it observes mass conservation. This may be explained by the fact that $\lim\limits_{r\to 0} rw^+(r,z) = 0$.

3.6.4 Vorticity
Having fully determined the velocity field, its vorticity companion may be determined from

$$\Omega = \Omega_\theta = \pi^2 r \sum_{n=0}^{\infty} (2n+1)^2 \alpha_n z \sin \chi_n \tag{101}$$

This expression is evaluated for the least and most kinetic energy forms ($q = 2$) and provided in Table 5.

3.6.5 Irrotational motion
For the least kinetic energy solution (Type I, $q = 2$), the linear variation that accompanies the radial velocity as well as the uniformity of the axial velocity are characteristics of an irrotational motion. The vorticity in this case vanishes and the corresponding velocity field collapses into $\mathbf{u} = -r\mathbf{e}_r + 2z\mathbf{e}_z$. This potential analogue of the Taylor–Culick velocity has been historically used by McClure et al. (1963) and Hart & McClure (1959) in modeling the internal flow in SRMs. It is recovered here as an extreme state with the lowest kinetic energy.

3.7 Pressure evaluation
One may approximate the pressure by taking

$$p(r,z) = \sum_{n=0}^{\infty} p_n(r,z) \tag{102}$$

By substituting u_n and w_n from (72) into (4b), the pressure eigenmodes may be integrated. One gets

$$p_n = p_0 - \tfrac{1}{2}(2n+1)^2 \pi^2 z^2 \alpha_n^2 - \tfrac{1}{2}\frac{\alpha_n^2}{r^2}\sin^2 \chi_n \tag{103}$$

The total pressure is then determined by summing over all eigensolutions

$$p(r,z) = p_0 - \tfrac{1}{2}\pi^2 z^2 \sum_{n=0}^{\infty} (2n+1)^2 \alpha_n^2 - \tfrac{1}{2}\sum_{n=0}^{\infty} \frac{\alpha_n^2}{r^2}\sin^2 \chi_n \tag{104}$$

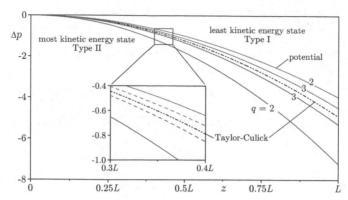

Fig. 15. Centerline pressure drop for the two types of energy-driven solutions.

where $p_0 = p(0,0)$. Interestingly, the pressure drop along the centerline collapses into

$$\Delta p = p(0,z) - p_0 = -\tfrac{1}{2}\pi^2 z^2 \sum_{n=0}^{\infty} (2n+1)^2 \alpha_n^2 \tag{105}$$

Equation (105) is plotted in Figure 15 for $q = 2, 3,$ and ∞. Unsurprisingly, the largest pressure excursion is seen to accompany the Type II state with most kinetic energy while the smallest pressure loss is accrued in the least kinetic energy expression, specifically, in the $q = 2$ potential case.

3.8 Asymptotic limits of the kinetic energy density

When the large L approximation is employed with $q = 2$, the Type II kinetic energy density $\mathscr{E}^+$ approaches a constant value of $\mathscr{E}_\infty^+(2) = \pi^5/(96\mathscr{C}^2) \approx 3.79944$. Note that the asymptotic value for Taylor–Culick's (i.e. when both L and q approach infinity), $\mathscr{E}_\infty^\infty \equiv \pi^3/12 \approx 2.5838$, is recovered as $q \to \infty$. In general, when $L \to \infty$, the limit of the kinetic energy density can be written as

$$\mathscr{E}_\infty = \tfrac{1}{12}\pi^3 \sum_{n=0}^{\infty}(2n+1)^2\alpha_n^2 = \mathscr{E}_\infty^\infty \sum_{n=0}^{\infty}(2n+1)^2\alpha_n^2 \tag{106}$$

For the Type I solutions, substitution of (106) yields a closed-form expression,

$$\mathscr{E}_\infty^-(q) = \mathscr{E}_\infty^\infty \left[\sum_{k=0}^{\infty}\frac{1}{(2k+1)^q}\right]^{-2}\sum_{n=0}^{\infty}(2n+1)^{2-2q} = \mathscr{E}_\infty^\infty \frac{4^q-4}{(2^q-1)^2}\frac{\zeta(2q-2)}{\zeta(q)^2} \tag{107}$$

In like manner, for the Type II solutions, (106) leads to

$$\mathscr{E}_\infty^+(q) = \mathscr{E}_\infty^\infty \left[\sum_{k=0}^{\infty}\frac{(-1)^k}{(2k+1)^q}\right]^{-2}\sum_{n=0}^{\infty}(2n+1)^{2-2q} = \mathscr{E}_\infty^\infty \frac{4^q(4^q-4)\zeta(2q-2)}{[\zeta(q,\frac{1}{4})-\zeta(q,\frac{3}{4})]^2} \tag{108}$$

As shown in Figure 16 both types approach $\mathscr{E}_\infty^\infty$ either from below or above, depending on q. The Taylor–Culick limit of 2.5838 is practically reached by both Type I and Type II solutions with differences of less than 0.287 and 0.265 percent at $q = 6$. The maximum range occurs at $q = 2$ while the total allowable excursion in energy that the mean flow can undergo may be estimated at $[\mathscr{E}_\infty^+(2) - \mathscr{E}_\infty^-(2)]/\mathscr{E}_\infty^\infty = 0.66$. From an academic standpoint, the Type I family of

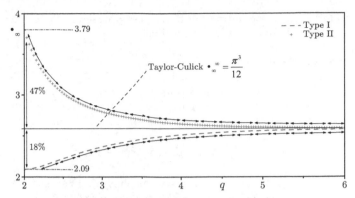

Fig. 16. Asymptotic behavior of the kinetic energy density for both Type I (- - -) and Type II (+ + +) solutions.

solutions bridges the gap between an essentially potential flow at $q = 2$ and a fully rotational field at $q \to \infty$, thus yielding intermediate formulations with energies that vary across the range $[0.81 - 1]\mathscr{E}_\infty^\infty$.

3.9 Convergence properties

Using the absolute convergence and ratio tests, the series representations can be individually shown to be unconditionally convergent for $q \geq 2$. The most subtle solutions to examine correspond to the Type II inert headwall case with maximum kinetic energy. The attendant velocity and vorticity forms require special attention. For the sake of illustration, we consider the Type II streamfunction, specifically

$$\psi(r,z) = z \sum_{n=0}^{\infty} \frac{B_q}{(2n+1)^q} \sin \chi_n; \quad \chi_n \equiv \tfrac{1}{2}(2n+1)\pi r^2 \tag{109}$$

The absolute convergence test may be applied to show that

$$\sum_{n=0}^{\infty} \left| \frac{1}{(2n+1)^q} \sin \chi_n \right| \leq \sum_{n=0}^{\infty} \frac{1}{(2n+1)^q} \tag{110}$$

where the right-hand-side converges for $q > 1$. In evaluating quantities that require one or more differentiations (such as the vorticity), we find it useful to substitute, whenever possible, the closed-form analytical representations of the series in question. The equivalent finite expressions enable us to overcome the pitfalls of term-by-term differentiation which, for some infinite series, can lead to spurious results. The Type II axial velocity for the inert headwall configuration presents such an example at $q = 2$. This series can be collapsed into a combination of inverse hyperbolic tangent functions by writing

$$w^+ = \sum_{n=0}^{\infty} \frac{B_2}{(2n+1)} \cos \chi_n = \frac{1}{2\mathscr{C}} \left[\tanh^{-1}\left(e^{i\frac{1}{2}\pi r^2} \right) + \tanh^{-1}\left(e^{-i\frac{1}{2}\pi r^2} \right) \right] \tag{111}$$

While term-by-term differentiation of the infinite series representation of w^+ diverges, the derivative of the closed-form equivalent yields the correct outcome of

$$\Omega^+ = -\frac{\pi}{2\mathscr{C}} r \csc(\tfrac{1}{2}\pi r^2) \tag{112}$$

As it may be expected, the corresponding solution is accompanied by finite kinetic energy and mass flowrate despite its singularity at the centerline.

3.10 Arbitrary headwall injection

For T-burners, solid rocket motors with reactive fore-ends, and hybrid rocket chambers with injector faceplates, a model that accounts for headwall injection is required. For these problems, our analysis may be repeated assuming an injecting headwall with an axisymmetrically varying profile defined by (8). The streamfunction becomes

$$\psi(r,z) = \sum_{n=0}^{\infty} (\alpha_n z + \beta_n) \sin[\tfrac{1}{2}(2n+1)\pi r^2] \tag{113}$$

In the resulting expressions, β_n does not vanish. As shown by Majdalani & Saad (2007b) and detailed in Section 2.1, orthogonality may be applied to obtain β_n for an axisymmetric headwall injection pattern. Application of Lagrangian optimization in conjunction with the large L approximation yield identical results for α_n as those obtained in (92) and (98). The streamfunction, axial velocity, and vorticity for several injection profiles are available through Tables 3, 4, and 5 where the least and most kinetic energy solutions are identified.

3.11 Numerical verification

Our analytical expansions may be verified by solving (13) using Runge-Kutta integration. We begin by introducing the transformation $\psi = zf(r)$ through which (13) may be reduced to a second order ODE

$$F''(r) - \frac{1}{r}F'(r) + C^2 r^2 F(r) = 0 \tag{114}$$

In order to numerically capture the different variational solutions, the boundary conditions of (14) have to be carefully selected. Because our solutions are in series form, we first decompose $F(r)$ into its eigenmode components by taking

$$F(r) = \sum_{n=0}^{\infty} F_n(r) \tag{115}$$

and so (114) becomes

$$F_n''(r) - \frac{1}{r}F_n'(r) + C_n^2 r^2 F_n(r) = 0; \quad n = 0, 1, \cdots, \infty \tag{116}$$

where n corresponds to the eigenmode associated with $C_n = (2n+1)\pi$. Finally, the boundary conditions may be written as

$$\begin{cases} F_n(0) = 0; & F(0) = \sum\limits_{n=0}^{\infty} F_n(0) = 0 \\ F_n(1) = (-1)^n \alpha_n; & F(1) = \sum\limits_{n=0}^{\infty} F_n(1) = \sum\limits_{n=0}^{\infty} (-1)^n \alpha_n = 1 \end{cases} \tag{117}$$

Using 120 terms to reconstruct the series expansions, both numerical and analytical solutions for $F(r)$ and $F'(r)$ are displayed in Figures 17(a) and 17(b), respectively. This comparison is held at representative values of the kinetic energy power index corresponding to $q = 2, 3$, and ∞. It is gratifying that, irrespective of q and n, the variational solutions are faithfully simulated by the numerical data to the extent that visual differences between full circles (numerical) and solid lines (analytical) are masked.

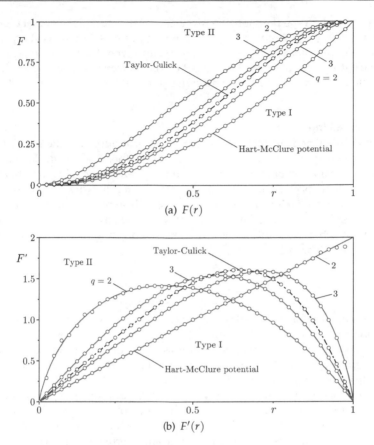

(a) $F(r)$

(b) $F'(r)$

Fig. 17. Comparison between analytical (—) and numerical (○) solutions for (a) $F(r)$, and (b) $F'(r)$ for Type I (blue) and Type II (red) solutions. Plots are shown for $q = 2, 3$ and ∞. Here, $\psi(r, z) = zF(r)$.

3.12 Unphysicality of the Type II family of solutions

To explore the physicality of our variational solutions, a second law analysis is helpful. To better understand the mechanisms responsible for the system to opt for one energy state over another, or one type of solution over another, the principle of entropy maximization may be referred to. This principle states that a system will tend to maximize entropy at equilibrium and may hence be applied to our problem by considering the different energy solutions as different states of the same system. As shown by Saad & Majdalani (2010), the second law analysis reveals that the volumetric entropy of the Type I family grows with successive increases in q but depreciates in the Type II case. So given an initial profile, the system may evolve according to one of two scenarios that are described below.

3.12.1 Type I branching

If the system is initialized on the Type I branch, it will evolve toward the Taylor–Culick solution to the extent of maximizing its total entropy. While entropy could be further increased by branching out to the Type II region, it may be argued that such development is not possible for two reasons. Firstly, the character of the two types of solutions is sharply dissimilar, especially in the expressions for α_n^- and α_n^+. Secondly, given that the Taylor–Culick solution maximizes the entropy for the Type I branch, it can be viewed as a local equilibrium state. As such, there is no necessity for the system to switch branches once it reaches the Taylor–Culick state.

3.12.2 Type II branching

If the system is initialized on the Type II branch, it will approach the solution with most vorticity (i.e. Type II, $q = 2$). Although this may be a mathematically viable outcome, it may not be physically realizable because it would be practically impossible to initialize a system with such a high level of vorticity without the aid of external work. The most natural flow evolution corresponds to an irrotational system originally at rest in which vorticity generation is initiated at the sidewall during the injection process. The ensuing motion will subsequently progress until it reaches the stable Taylor–Culick equilibrium state wherein it can settle with no further tendency to branch out.

4. Conclusions

For four decades and counting, the motion of incompressible fluids through porous tubes with wall-normal injection (or suction) has been extensively used in the propulsion and flow separation industries. In this chapter, the focus has been on the inviscid form of the Taylor-Culick family of incompressible solutions. The originality of the analysis stands, perhaps, in the incorporation of variable headwall injection using a linear series expansion that may be attributed to Majdalani & Saad (2007b).

The extended Taylor-Culick framework has profound implications as it permits the imposition of realistic conditions that may be associated with solid or hybrid propellant rockets with reactive fore ends or injecting faceplates. The procedure that we follow starts with Euler's steady equations, and ends with an approximation that is exact only at the sidewall, the centerline, or when using similarity-conforming inlet velocities. For similarity-nonconforming profiles, our approach becomes increasingly more accurate as the distance from the headwall is increased; this property makes our model well suited to describe the bulk motion in simulated solid and hybrid rockets where the blowing speed is assumed to be uniformly distributed along the grain surface. The justification for using a linear summation of eigensolutions and the reason for its increased accuracy in elongated chambers may be connected, in part, to the quasi-linear behavior of the vorticity transport equation for large z. Such behavior is corroborated by the residual error analysis that we carried out in Section 2.6. Furthermore, as carefully shown by Kurdyumov (2008), the vorticity-streamfunction relation appears to be strongly nonlinear in the direct vicinity of the headwall, yet becomes increasingly more linear with successive increases in z.

Another advantage of the present formulation may be ascribed to its quasi-viscous character, being observant of the no-slip requirement at the sidewall. Based on numerical simulations conducted under both inviscid and turbulent flow conditions, the closed-form expressions that we obtain appear to provide reasonable approximations for several headwall injection patterns associated with conventional laminar and turbulent flow profiles. Everywhere, our comparisons are performed for the dual cases of small and large headwall injection in an effort to mimic the internal flow character in either SRM or hybrid rocket motors. Overall, we find that the flow field evolves to the self-similar Taylor–Culick sinusoid far downstream irrespective of the headwall injection pattern. Nonetheless, the details of headwall injection remain important in hybrid motors, short chambers, and T-burners where the foregoing approximations may be applied. In hybrid rockets, our models seem to capture the streamtube motion quite effectively.

The other chief contribution of this chapter is the discussion of variational solutions that may be connected with the Taylor-Culick problem. Based on the Lagrangian optimization of the total volumetric energy in the chamber, we are able to identify two families of solutions with dissimilar energy signatures. These are accompanied by lower or higher kinetic energies that vary, from one end of the spectrum to the other, by up to 66 percent of their mean value. After identifying that $\alpha_n^- \sim (-1)^n (2n+1)^{-2}$ yields the profile with least kinetic energy, a sequence of Type I solutions is unraveled in ascending order, $\alpha_n^- \sim (-1)^n (2n+1)^{-q}$; $q > 2$, up to Taylor-Culick's. The latter is asymptotically recovered in the limit of $q \to \infty$, a case that corresponds to an equilibrium state with maximum entropy. In practice, most solutions become indiscernible from Taylor-Culick's for $q \geq 5$. Indeed, those obtained with $q = 2, 3$, and 4 exhibit energies that are 18.9, 8.28, and 2.73 percent lower than their remaining counterparts. The least kinetic energy solution with $q = 2$ returns the classic, irrotational Hart-McClure profile. It can thus be seen that the application of the Lagrangian optimization principle to this problem leads to the potential form that historically preceded the Taylor-Culick motion. It can also be inferred that the Type I solutions not only bridge the gap between a plain potential representation of this problem and a rotational formulation, but also recover a continuous spectrum of approximations that stand in between. When the same analysis is repeated using $\alpha_n^+ \sim (2n+1)^{-q}$; $q \geq 2$, a complementary family of Type II solutions is identified with descending energy levels. These are shown to be purely academic, although they represent a class of exact solutions to the modified Helmholtz equation. Their most notable profiles correspond to $q = 2, 3$, and 4 with energies that are 47.0, 8.08, and 2.40 percent higher than Taylor-Culick's. Their entropies are also higher than that associated with the equilibrium state. Despite their dissimilar forms, both Type I and II solutions converge to the Taylor-Culick representation when their energies are incremented or reduced. Yet before using the new variational solutions to approximate the mean flow profile in porous tubes or the bulk gaseous motion in simulated rocket motors, it should be borne in mind that no direct connection exists between the energy steepened states and turbulence. For this reason, it is hoped that additional numerical and experimental investigations are pursued to test their physicality and the particular configurations in which they are prone to appear. As for the uniqueness of the Taylor-Culick equilibrium state, it may be confirmed from the entropy maximization principle and the Lagrangian-based solutions where, for a given set of boundary conditions, the equilibrium state may be asymptotically restored as $q \to \infty$ irrespective of the form of $\alpha_n \sim (-1)^n (p\,n + m)^{-q}$, provided that the Lagrangian constraint $\sum (-1)^n \alpha_n = 1$ is faithfully secured.

Lastly, we note that the collection of variational solutions that admit variable headwall injection increase our repertoire of Euler-based approximations that may be used to model the incompressible motion in porous tubes. For the porous channel flow analogue, the planar solutions are presented by Saad & Majdalani (2008a; 2009b). As for tapered grain configuration, the reader may consult with Saad et al. (2006) or Sams et al. (2007).

5. Acknowledgments

This material is based on work supported partly by the National Science Foundation, and partly by the University of Tennessee Space Institute and the H. H. Arnold Chair of Excellence in Advanced Propulsion.

6. References

Abu-Irshaid, E. M., Majdalani, J. & Casalis, G. (2007). Hydrodynamic stability of rockets with headwall injection, *Physics of Fluids* 19(2): 024101–11.
  URL: *http://dx.doi.org/10.1063/1.2434797*

Acrivos, A. (1962). The asymptotic form of the laminar boundary-layer mass-transfer rate for large interfacial velocities, *Journal of Fluid Mechanics* 12(3): 337–357.
  URL: *http://dx.doi.org/10.1017/S0022112062000257*

Apte, S. & Yang, V. (2000). Effect of acoustic oscillation on flow development in a simulated nozzleless rocket motor, *Solid Propellant Chemistry, Combustion, and Motor Interior Ballistics*, Vol. 185 of *Progress in Astronautics and Aeronautics*, AIAA Progress in Astronautics and Aeronautics, Washington, DC, pp. 791–822.

Apte, S. & Yang, V. (2001). Unsteady flow evolution in porous chamber with surface mass injection, part 1: free oscillation, *AIAA Journal* 39(8): 1577–1630.
  URL: *http://dx.doi.org/10.2514/2.1483*

Apte, S. & Yang, V. (2002). Unsteady flow evolution in porous chamber with surface mass injection, part 2: acoustic excitation, *AIAA Journal* 40(2): 244–253.
  URL: *http://dx.doi.org/10.2514/2.1666*

Beddini, R. A. (1986). Injection-induced flows in porous-walled ducts, *AIAA Journal* 24(11): 1766–1773.
  URL: *http://dx.doi.org/10.2514/3.9522*

Berman, A. S. (1953). Laminar flow in channels with porous walls, *Journal of Applied Physics* 24(9): 1232–1235.
  URL: *http://dx.doi.org/10.1063/1.1721476*

Berman, A. S. (1958a). Effects of porous boundaries on the flow of fluids in systems with various geometries, *Proceedings of the Second United Nations International Conference on the Peaceful Uses of Atomic Energy* 4: 351–358.

Berman, A. S. (1958b). Laminar flow in an annulus with porous walls, *Journal of Applied Physics* 29(1): 71–75.
  URL: *http://dx.doi.org/10.1063/1.1722948*

Chedevergne, F., Casalis, G. & Féraille, T. (2006). Biglobal linear stability analysis of the flow induced by wall injection, *Physics of Fluids* 18(1): 014103–14.
  URL: *http://dx.doi.org/10.1063/1.2160524*

Chedevergne, F., Casalis, G. & Majdalani, J. (2007). DNS investigation of the Taylor-Culick flow stability, *43rd AIAA/ASME/SAE/ASEE Joint Propulsion Conference*, AIAA Paper 2007-5796, Cincinnati, OH.

Chu, B.-T. & Kovásznay, L. S. G. (1958). Non-linear interactions in a viscous heat-conducting compressible gas, *Journal of Fluid Mechanics* 3(5): 494–514.
  URL: *http://dx.doi.org/10.1017/S0022112058000148*

Chu, W. W., Yang, V. & Majdalani, J. (2003). Premixed flame response to acoustic waves in a porous-walled chamber with surface mass injection, *Combustion and Flame*

133(3): 359–370.
URL: *http://dx.doi.org/10.1016/S0010-2180(03)00018-X*

Clairaut, A.-C. (1739). Recherches générales sur le calcul intégral, *Histoire de l'Académie Royale des Science avec les Mémoires de Mathématique et de Physiques* 1: 425–436.
URL: *http://www.clairaut.com/n29avril1733.html*

Clairaut, A.-C. (1740). Sur l'intégration ou la construction des équations différentielles du premier ordre, *Histoire de l'Académie Royale des Science avec les Mémoires de Mathématique et de Physiques* 1: 293–323.
URL: *http://www.clairaut.com/n29avril1733.html*

Culick, F. E. C. (1966). Rotational axisymmetric mean flow and damping of acoustic waves in a solid propellant rocket, *AIAA Journal* 4(8): 1462–1464.
URL: *http://dx.doi.org/10.2514/3.3709*

Culick, F. E. C. (2006). Unsteady motions in combustion chambers for propulsion systems, *Agardograph*, Advisory Group for Aerospace Research and Development.

Dauenhauer, E. C. & Majdalani, J. (2003). Exact self-similarity solution of the Navier–Stokes equations for a porous channel with orthogonally moving walls, *Physics of Fluids* 15(6): 1485–1495.
URL: *http://dx.doi.org/10.1063/1.1567719*

Eckert, E. R. G., Donoughe, P. L. & Moore, B. J. (1957). Velocity and friction characteristics of laminar viscous boundary-layer and channel flow over surfaces with ejection or suction, *Technical report*, NACA Technical Note 4102.
URL: *http://aerade.cranfield.ac.uk/ara/1957/naca-tn-4102.pdf*

Erdogan, M. E. & Imrak, C. E. (2008). On the flow in a uniformly porous pipe, *International Journal of Non-Linear Mechanics* 43(4): 292–301.
URL: *http://dx.doi.org/10.1016/j.ijnonlinmec.2007.12.006*

Flandro, G. A. & Majdalani, J. (2003). Aeroacoustic instability in rockets, *AIAA Journal* 41(3): 485–497.
URL: *http://dx.doi.org/10.2514/2.1971*

Fung, Y. C. & Yih, C. S. (1968). Peristaltic transport, *Journal of Applied Mechanics* 35(4): 669–675.

Goto, M. & Uchida, S. (1990). Unsteady flows in a semi–infinite expanding pipe with injection through wall, *Journal of the Japan Society for Aeronautical and Space Sciences* 38: 434.

Grad, H. (1949). Resonance burning in rocket motors, *Communications on Pure and Applied Mathematics* 2(1): 79–102.
URL: *http://dx.doi.org/10.1002/cpa.3160020105*

Griffond, J., Casalis, G. & Pineau, J.-P. (2000). Spatial instability of flow in a semiinfinite cylinder with fluid injection through its porous walls, *European Journal of Mechanics B/Fluids* 19(1): 69–87.
URL: *http://dx.doi.org/10.1016/S0997-7546(00)00105-9*

Hart, R. & McClure, F. (1965). Theory of acoustic instability in solid-propellant rocket combustion, *Tenth Symposium (International) on Combustion*, Vol. 10, pp. 1047–1065.
URL: *http://dx.doi.org/10.1016/S0082-0784(65)80246-6*

Hart, R. W., Bird, J. F. & McClure, F. T. (1960). *The Influence of Erosive Burning on Acoustic Instability in Solid Propellant Rocket Motors*, Storming Media.

Hart, R. W. & Cantrell, R. H. (1963). Amplification and attenuation of sound by burning propellants, *AIAA Journal* 1(2): 398–404.
URL: *http://dx.doi.org/10.2514/3.1545*

Hart, R. W. & McClure, F. T. (1959). Combustion instability: Acoustic interaction with a burning propellant surface, *The Journal of Chemical Physics* 30(6): 1501–1514.
    URL: *http://dx.doi.org/10.1063/1.1730226*

Janzen, O. (1913). Beitrag Zu Einer Theorie Der Stationaren Stromung Kompressibler Flussigkeiten (Towards a theory of stationary flow of compressible fluids), *Physikalische Zeitschrift* 14: 639 – 643.

Kurdyumov, V. N. (2006). Steady flows in the slender, noncircular, combustion chambers of solid propellant rockets, *AIAA Journal* 44(12): 2979–2986.
    URL: *http://dx.doi.org/10.2514/1.21125*

Kurdyumov, V. N. (2008). Viscous and inviscid flows generated by wall-normal injection into a cylindrical cavity with a headwall, *Physics of Fluids* 20(12): 123602–7.
    URL: *http://dx.doi.org/10.1063/1.3045738*

Libby, P. A. (1962). The homogeneous boundary layer at an axisymmetric stagnation point with large rates of injection, *Journal of the Aerospace Sciences* 29(1): 48–60.

Libby, P. A. & Pierucci, M. (1964). Laminar boundary layer with hydrogen injection including multicomponent diffusion, *AIAA Journal* 2(12): 2118–2126.
    URL: *http://dx.doi.org/10.2514/3.2752*

Maicke, B. A. & Majdalani, J. (2008). On the rotational compressible Taylor flow in injection-driven porous chambers, *Journal of Fluid Mechanics* 603: 391–411.
    URL: *http://dx.doi.org/10.1017/S0022112008001122*

Majdalani, J. (2007a). Analytical models for hybrid rockets, *in* K. Kuo & M. J. Chiaverini (eds), *Fundamentals of Hybrid Rocket Combustion and Propulsion*, Progress in Astronautics and Aeronautics, AIAA Progress in Astronautics and Aeronautics, Washington, DC, chapter Chap. 5, pp. 207–246.

Majdalani, J. (2007b). On steady rotational high speed flows: the compressible Taylor–Culick profile, *Proceedings of the Royal Society A: Mathematical, Physical and Engineering Science* 463(2077): 131–162.
    URL: *http://dx.doi.org/10.1098/rspa.2006.1755*

Majdalani, J. (2009). Multiple asymptotic solutions for axially travelling waves in porous channels, *Journal of Fluid Mechanics* 636(1): 59–89.
    URL: *http://dx.doi.org/10.1017/S0022112009007939*

Majdalani, J. & Akiki, M. (2010). Rotational and quasiviscous cold flow models for axisymmetric hybrid propellant chambers, *Journal of Fluids Engineering* 132(10): 101202–7.
    URL: *http://link.aip.org/link/?JFG/132/101202/1*

Majdalani, J. & Saad, T. (2007a). Energy steepened states of the Taylor-Culick profile, *43rd AIAA/ASME/SAE/ASEE Joint Propulsion Conference and Exhibit*, AIAA Paper 2007-5797, Cincinnati, OH.

Majdalani, J. & Saad, T. (2007b). The Taylor–Culick profile with arbitrary headwall injection, *Physics of Fluids* 19(9): 093601–10.
    URL: *http://dx.doi.org/10.1063/1.2746003*

Majdalani, J., Vyas, A. B. & Flandro, G. A. (2002). Higher mean–flow approximation for a solid rocket motor with radially regressing walls, *AIAA Journal* 40(9): 1780–1788.
    URL: *http://dx.doi.org/10.2514/2.1854*

Majdalani, J., Vyas, A. B. & Flandro, G. A. (2009). Higher Mean-Flow approximation for a solid rocket motor with radially regressing walls - erratum, *AIAA Journal* 47(1): 286–286.
    URL: *http://dx.doi.org/10.2514/1.40061*

Majdalani, J. & Zhou, C. (2003). Moderate-to-large injection and suction driven channel flows with expanding or contracting walls, *Journal of Applied Mathematics and Mechanics* 83(3): 181–196.
URL: *http://dx.doi.org/10.1002/zamm.200310018*

McClure, F. T., Cantrell, R. H. & Hart, R. W. (1963). Interaction between sound and flow: Stability of T-Burners, *AIAA Journal* 1(3): 586–590.
URL: *http://dx.doi.org/10.2514/3.54846*

McClure, F. T., Hart, R. W. & Bird, J. F. (1960). Acoustic resonance in solid propellant rockets, *Journal of Applied Physics* 31(5): 884–896.
URL: *http://dx.doi.org/10.1063/1.1735713*

Peng, Y. & Yuan, S. W. (1965). Laminar pipe flow with mass transfer cooling, *Journal of Heat Transfer* 87(2): 252–258.

Proudman, I. (1960). An example of steady laminar flow at large Reynolds number, *Journal of Fluid Mechanics* 9(4): 593–602.
URL: *http://dx.doi.org/10.1017/S002211206000133X*

Rayleigh, L. (1916). On the flow of compressible fluid past an obstacle, *Philosophical Magazine* 32(1): 1–6.

Robinson, W. A. (1976). The existence of multiple solutions for the laminar flow in a uniformly porous channel with suction at both walls, *Journal of Engineering Mathematics* 10(1): 23–40.
URL: *http://dx.doi.org/10.1007/BF01535424*

Saad, T. & Majdalani, J. (2008a). Energy based mean flow solutions for slab hybrid rocket chambers, *44th AIAA/ASME/SAE/ASEE Joint Propulsion Conference and Exhibit*, AIAA Paper 2008-5021, Hartford, Connecticut.

Saad, T. & Majdalani, J. (2008b). Energy based solutions of the bidirectional vortex, *44th AIAA/ASME/SAE/ASEE Joint Propulsion Conference and Exhibit*, AIAA Paper 2008-4832, Hartford, Connecticut.

Saad, T. & Majdalani, J. (2009a). Energy based solutions of the bidirectional vortex with multiple mantles, *45th AIAA/ASME/SAE/ASEE Joint Propulsion Conference and Exhibit*, AIAA Paper 2009-5305, Denver, Colorado.

Saad, T. & Majdalani, J. (2009b). Rotational flowfields in porous channels with arbitrary headwall injection, *Journal of Propulsion and Power* 25(4): 921–929.
URL: *http://dx.doi.org/10.2514/1.41926*

Saad, T. & Majdalani, J. (2010). On the Lagrangian optimization of wall-injected flows: from the Hart–McClure potential to the Taylor–Culick rotational motion, *Proceedings of the Royal Society A: Mathematical, Physical and Engineering Science* 466(2114): 331–362.
URL: *http://dx.doi.org/10.1098/rspa.2009.0326*

Saad, T., Sams, O. C., IV & Majdalani, J. (2006). Rotational flow in tapered slab rocket motors, *Physics of Fluids* 18(10): 103601.
URL: *http://dx.doi.org/10.1063/1.2354193*

Sams, O. C., Majdalani, J. & Saad, T. (2007). Mean flow approximations for solid rocket motors with tapered walls, *Journal of Propulsion and Power* 23(2): 445–456.
URL: *http://dx.doi.org/10.2514/1.15831*

Sellars, J. R. (1955). Laminar flow in channels with porous walls at high suction Reynolds numbers, *Journal of Applied Physics* 26(4): 489–490.
URL: *http://dx.doi.org/10.1063/1.1722024*

Taylor, G. (1956). Fluid flow in regions bounded by porous surfaces, *Proceedings of the Royal Society A: Mathematical, Physical and Engineering Sciences* 234(1199): 456–475.
URL: *http://dx.doi.org/10.1098/rspa.1956.0050*

Terrill, R. (1964). Laminar flow in a uniformly porous channel, *Aeronautical Quarterly* 15: 299–310.

Terrill, R. M. (1965). Laminar flow in a uniformly porous channel with large injection, *Aeronautical Quarterly* 16: 323–332.

Terrill, R. M. & Thomas, P. W. (1969). On laminar flow through a uniformly porous pipe, *Applied Scientific Research* 21(1): 37–67.
URL: *http://dx.doi.org/10.1007/BF00411596*

Tsangaris, S., Kondaxakis, D. & Vlachakis, N. (2007). Exact solution for flow in a porous pipe with unsteady wall suction and/or injection, *Communications in Nonlinear Science and Numerical Simulation* 12(7): 1181–1189.
URL: *http://dx.doi.org/10.1016/j.cnsns.2005.12.009*

Uchida, S. & Aoki, H. (1977). Unsteady flows in a semi-infinite contracting or expanding pipe, *Journal of Fluid Mechanics* 82(2): 371–387.
URL: *http://dx.doi.org/10.1017/S0022112077000718*

Vyas, A. B., Majdalani, J. & Yang, V. (2003). Estimation of the laminar premixed flame temperature and velocity in injection-driven combustion chambers, *Combustion and Flame* 133(3): 371–374.
URL: *http://dx.doi.org/10.1016/S0010-2180(03)00017-8*

White, F. M. (2005). *Viscous fluid flow*, 3 edn, McGraw-Hill, New York.

White, F. M., Barfield, B. F. & Goglia, M. J. (1958). Laminar flow in a uniformly porous channel, *Journal of Applied Mechanics* 25: 613.

Yuan, S. W. (1956). Further investigation of laminar flow in channels with porous walls, *Journal of Applied Physics* 27(3): 267–269.
URL: *http://dx.doi.org/10.1063/1.1722355*

Yuan, S. W. & Finkelstein, A. B. (1958). Heat transfer in laminar pipe flow with uniform coolant injection, *Jet Propulsion* 28(1): 178–181.

Zaturska, M., Drazin, P. & Banks, W. (1988). On the flow of a viscous fluid driven along a channel by suction at porous walls, *Fluid Dynamics Research* 4(3): 151–178.
URL: *http://dx.doi.org/10.1016/0169-5983(88)90021-4*

Zhou, C. & Majdalani, J. (2002). Improved mean flow solution for slab rocket motors with regressing walls, *Journal of Propulsion and Power* 18(3): 703–711.
URL: *http://dx.doi.org/10.2514/2.5987*

6

Influence of Horizontal Temperature Gradients on Convective Instabilities with Geophysical Interest

H. Herrero, M. C. Navarro and F. Pla
Universidad de Castilla- La Mancha
Spain

1. Introduction

Since Bénard's experiments on convection and Rayleigh's theoretical work in the beginning of the twentieth century (1)-(2), many experimental, theoretical and numerical works related to Rayleigh-Bénard convection have been done (3)-(10) and different problems have been posed depending on what is to be modelled. Classically, heat is applied uniformly from below and the conductive solution becomes unstable for a critical vertical gradient beyond a certain threshold.

A setup for natural convection more general than that of uniform heating consists of including a non-zero horizontal temperature gradient which may be either constant or not (11)-(29). In those problems a clear difference is marked by the fact that the fluid is simply contained (11)-(19), where stationary and oscillatory instabilities appear depending on the multiple parameters present in the problem: properties of the fluid, surface tension effects, heat exchange with the atmosphere, aspect ratio, dependence of viscosity with temperature, etc., and the case where the fluid can flow throughout the boundaries (29), where vortical solutions can appear reinforcing the relevance of convective mechanisms for the generation of vertical vortices very similar to those found for some atmospheric phenomena as dust devils or hurricanes (29)-(31).

The case where the fluid is simply contained displays stationary and oscillatory instabilities. This problem has been treated from different points of view: experimental (11)-(18) and theoretical, both with semiexact (20)-(21) and numerical solutions (40)-(28). This case contains applications to mantle convection when the viscosity is large (45; 52) or it depends on temperature (19).

There are not experiments yet for the case where a flow throughout the boundaries is allowed, only observations of atmospheric phenomena (30; 33; 34; 36; 37), and theoretic numerical results (29; 31).

In this work we will review this physical problem, focusing on the latest problems addressed by the authors on this topic, where a non-uniform heating is considered in different geometrical configurations, and we will show the relevant results obtained, some of them in the context of interesting atmospheric and geophysical phenomena (30; 36; 37).

2. Theoretical formulation of the problem

The physical set-up (see figure 1) consists of a horizontal fluid layer in a rectangular domain (19; 45) or a cylindrical annulus (18; 25; 28) between two vertical walls at $r = a$ and $r = a + l$. The depth of the domain is d (z coordinate). At $z = 0$ the imposed temperature gradient takes the value T_{max} at a and the value T_{min} at the outer part $(a + l)$. The upper surface is at temperature $T = T_0$. We define $\Delta T_v = T_{max} - T_0$, $\Delta T_h = T_{max} - T_{min}$ and $\delta = \Delta T_h / \Delta T_v$.

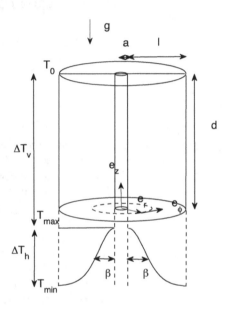

Fig. 1. Physical setup for the cylindrical annulus.

From now on we will consider an annular domain, therefore cylindrical coordinates will be used in the following. The formulation in a rectangular domain and coordinates would be similar. In the governing equations, $\mathbf{u} = (u_r, u_\phi, u_z)$ is the velocity field, T is the temperature, p is the pressure, r is the radial coordinate, and t is the time. They are expressed in dimensionless form after rescaling: $\mathbf{r}' = \mathbf{r}/d$, $t' = \kappa t/d^2$, $\mathbf{u}' = d\mathbf{u}/\kappa$, $p' = d^2 p / (\rho_0 \kappa \nu)$, $\Theta = (T - T_0) / \Delta T_v$. Here $\mathbf{r}$ is the position vector, κ the thermal diffusivity, ν the kinematic viscosity of the liquid, and ρ_0 the mean density at temperature T_0. The domain is $[\bar{a}, \bar{a} + \Gamma] \times [0, 1] \times [0, 2\pi]$ where $\Gamma = l/d$ is the aspect ratio and $\bar{a} = a/d$.

The system evolves according to the mass balance, energy conservation and momentum equations, which in dimensionless form (with primes now omitted) are,

$$\nabla \cdot \mathbf{u} = 0, \tag{1}$$

$$\partial_t \Theta + \mathbf{u} \cdot \nabla \Theta = \nabla^2 \Theta, \tag{2}$$

$$\partial_t \mathbf{u} + (\mathbf{u} \cdot \nabla) \mathbf{u} = Pr \left(-\nabla p + \nabla^2 \mathbf{u} + R \Theta \mathbf{e}_z \right), \tag{3}$$

where the operators and fields are expressed in cylindrical coordinates and the Oberbeck-Boussinesq approximation has been used (25), i.e. density is constant except in the term of gravity, where a linear dependence on temperature is considered. Here $\mathbf{e}_z$ is the unit vector in the z direction. The following dimensionless numbers have been introduced: the Prandtl number $Pr = \nu/\kappa$, and the Rayleigh number $R = g\alpha\triangle T d^3/\kappa\nu$, which represents the effect of buoyancy and in which α is the thermal expansion coefficient and g the gravitational acceleration. In the case of variable viscosity the laplacian operator in Eq. (3) takes the form div $\left(\frac{\nu(\Theta)}{\nu_0}\cdot(\nabla\mathbf{u}+(\nabla\mathbf{u})^t)\right)$, where $\nu(\Theta) = \nu_0 e^{-\eta R\Theta}$.

2.1 Contained fluid

Regarding boundary conditions, several conditions can be considered such as that one where flow through the boundaries is not permitted. For instance at the lateral walls $r = \bar{a}$ and $r = \bar{a} + \Gamma$ the velocity is zero and an insulating wall is considered,

$$u_r = u_\phi = \partial_r u_z = \partial_r\Theta = 0, \quad \text{on } r = \bar{a} \text{ and } r = \bar{a}+\Gamma. \tag{4}$$

On the top surface, the vertical velocity is zero, the normal derivatives of the rest of components of the velocity are zero and the temperature is $T = T_0$, that after rescaling become,

$$\partial_z u_r = \partial_z u_\phi = u_z = \Theta = 0, \quad \text{on } z = 1, \tag{5}$$

and at the bottom

$$u_r = u_\phi = u_z = 0, \quad \text{on } z = 0. \tag{6}$$

For temperature at the bottom we consider a constant horizontal temperature difference, i.e. a linear profile. Here the horizontal temperature gradient appears,

$$\Theta = \theta^1(r) \quad \text{on } z = 0. \tag{7}$$

with $\theta^1(r) = 1 - r\delta/\Gamma$ and a second order polynomial which matches the linear profile such that $\partial_r\theta^1(r) = 0$ on $r = \bar{a}$ and $r = \bar{a}+\Gamma$.

The dimensionless equations and boundary conditions contain five external parameters: R, Γ, Pr, δ, and η.

2.2 Not contained fluid

Regarding boundary conditions, several conditions can be considered like allowing flow through the boundaries. At the lateral inner wall $r = \bar{a}$ the velocity is zero and an insulating wall is considered,

$$u_r = u_\phi = u_z = \partial_r\Theta = 0, \quad \text{on } r = \bar{a}. \tag{8}$$

At $r = \bar{a} + \Gamma$, the lateral outer wall, a constant radial velocity is assumed and an insulating wall is considered,

$$\partial_r u_r = \partial_r u_\phi = \partial_r u_z = \partial_r\Theta = 0, \quad \text{on } r = \bar{a}+\Gamma. \tag{9}$$

On the top surface, the velocity is zero and the temperature is $T = T_0$, that after rescaling become,

$$u_r = u_\phi = u_z = \Theta = 0, \quad \text{on } z = 1, \tag{10}$$

and at the bottom

$$u_r = \partial_z u_\phi = u_z = 0, \quad \text{on } z = 0. \tag{11}$$

For temperature at the bottom we consider a variable horizontal temperature gradient through imposing a Gaussian profile as in Ref. (28),

$$\Theta = 1 - \delta(e^{(\frac{1}{\beta})^2} - e^{(\frac{1}{\beta} - (\frac{r-a}{\Gamma})^2 \frac{1}{\beta})^2}) / (e^{(\frac{1}{\beta})^2} - 1) \quad \text{on } z = 0. \tag{12}$$

The dimensionless equations and boundary conditions contain five external parameters: R, Γ, Pr, δ, and β.

3. Metodology: search for solutions and their linear stability

We look for stationary axisymmetric solutions of the problem, then, the equations to be solved are

$$\nabla^* \cdot \mathbf{u} = 0, \tag{13}$$

$$\mathbf{u} \cdot \nabla^* \Theta = \nabla^{*2} \Theta, \tag{14}$$

$$(\mathbf{u} \cdot \nabla^*) \mathbf{u} = Pr \left(-\nabla^* p + \nabla^{*2} \mathbf{u} + R \Theta \mathbf{e}_z \right), \tag{15}$$

where $\nabla^* = (\partial_r, 0, \partial_z)$, together with the corresponding boundary conditions.

The time independent solution $U^b(r, z)$ to the stationary problem obtained from equations (1)-(3) by eliminating the time dependence, is called basic state. It is a non-conductive state ($\mathbf{u} \neq 0$) as soon as $\delta \neq 0$. The basic state is considered to be axisymmetric and therefore depends only on $r - z$ coordinates (i.e. all ϕ derivatives are zero). The velocity field of the basic flow is restricted to $\mathbf{u} = (u_r, u_\phi = 0, u_z)$.

A linear stability analysis of the stationary solutions is performed. Fixed $(\Gamma, \delta, Pr, \beta)$, the solution $U(r, z, t) = (\mathbf{u}, \Theta, p)(r, z, t)$ of (1)-(3) at given R is expressed as

$$U(r, z, t) = U^b(r, z) + \tilde{U}(r, z)e^{ik\phi + \lambda t}, \tag{16}$$

where $U^b(r, z)$ is the base flow for the given $(R, \Gamma, \delta, Pr, \beta)$ and $\tilde{U}(r, z)$ refers to the perturbation. We have considered Fourier mode expansions in the angular direction, because along it boundary conditions are periodic. Introducing (16) into the full system (1)-(3) and linearizing the resulting system, the following eigenvalue problem in λ is obtained:

$$\nabla^k \cdot \tilde{U} = 0, \tag{17}$$

$$\lambda \tilde{\Theta} + \tilde{U} \cdot \nabla^k \Theta^b + U^b \cdot \nabla^k \tilde{\Theta} = (\nabla^k)^2 \tilde{\Theta}, \tag{18}$$

$$\lambda \tilde{U} + \left(\tilde{U} \cdot \nabla^k \right) U^b + \left(U^b \cdot \nabla^k \right) \tilde{U} = Pr \left(-\nabla^k \tilde{p} + (\nabla^k)^2 \tilde{U} + R \tilde{\Theta} \mathbf{e}_z \right), \tag{19}$$

where $\nabla^k = (\partial_r, ik, \partial_z)$, with the corresponding boundary conditions.

The instability is achieved when the real part of the eigenvalue with maximum real part, $\lambda_{\max}(R)$, changes from a negative value to a positive one as R increases, for a specific wave number k. The critical value of R for which $\lambda_{\max}(R,k) = 0$ is denoted by R_c and the critical wave number, minimum k for which the bifurcation occurs, by k_c.

3.1 Numerical methods

The numerical method is described in detail and tested in Refs. (25; 28). The nonlinearities in the basic state problem are solved with a Newton method. Each step of the Newton method and the linear stability analysis have been numerically solved with a Chebyshev collocation method as explained in Refs. (28; 39; 48). The problem is posed in the primitive variables formulation, and the use of the same order approximations for velocity and pressure in the Chebyshev collocation procedure introduces spurious modes for pressure that are solved by adding convenient boundary conditions (43; 44). In the resulting linear problems any unknown field $\mathbf{x}$ is expanded in Chebyshev polynomials

$$\mathbf{x}^{LN} = \sum_{l=0}^{L-1}\sum_{n=0}^{N-1} a_{ln}^{\mathbf{x}} T_l(r) T_n(z). \tag{20}$$

The corresponding expansions for the four different fields are introduced into the Newton linearized version of equations (13)-(15) and the corresponding boundary conditions and evaluated at the Chebysehv Gauss-Lobatto collocation points (r_j, z_i),

$$r_j = \cos\left(\left(\frac{j-1}{L-1}-1\right)\pi\right), \quad j = 1, ..., L. \tag{21}$$

$$z_i = \cos\left(\left(\frac{i-1}{N-1}-1\right)\pi\right), \quad i = 1, ..., N. \tag{22}$$

Some care is necessary in the evaluation rules at the boundaries as explained in Refs. (28; 48). At each iteration of the Newton method a linear system of the form $AX = B$ is derived, where X is a vector containing $P = 4 \times L \times N$ unknowns and A is a full rank matrix of order $P \times P$. This can be solved with standard routines. The algorithm starts with an approximation to the solution $\mathbf{x}^{0,LN}$ and the iteration procedure is applied until the stop criterion $||\mathbf{x}^{s+1,LN} - \mathbf{x}^{s,LN}|| < 10^{-9}$ is satisfied.

The same discretization is used for the eigenvalue problem (17)-(19) with the corresponding boundary conditions. In this way it is transformed into its discrete form by expanding the perturbations in a truncated series of Chebyshev polynomials (20) as performed for the basic state. The evaluation rules are detailed in Ref. (48). Therefore, the eigenvalue problem in its discrete form is,

$$Cw = \sigma Bw, \tag{23}$$

where w is a vector which contains Q unknowns and C and B are $Q \times Q$ matrices, with $Q = 5 \times L \times N$.

QZ or Arnoldi algorithms are used to solve the eigenvalue problem (42). σ are the eigenvalues and w are coefficients in the Chebyshev basis of the corresponding eigenfunctions.

The discrete eigenvalue problem (23) has a finite number of eigenvalues σ_i. The stability condition must now be imposed upon $\sigma_{\max}$ where $\sigma_{\max} = \max(\mathrm{Re}(\sigma_i))$, bearing in mind that if $\sigma_{max} < 0$ the stationary state is stable while if $\sigma_{max} > 0$ the stationary state becomes unstable. The control or bifurcation parameter is the Rayleigh number R. For fixed values of the parameters, in those cases Γ, Pr, δ, β or η, the critical values are the minimum value of R_c for which there exists a value of k, k_c, such that $\sigma(R_c, k_c) = 0$.

In order to test convergence of the method we include, as an example, the calculation of the critical value of the bifurcation parameter, R_c, and the critical wave number, k_c, for different order expansions in the Chebyshev approximation in the contained fluid case. And we benchmark the method and code to ensure the correctness of the results. Table 1 shows these results for the contained fluid case. When the orders L and N are increased, the critical values tend to a determined value, convergence is very good and for $L = 24$ and $N = 14$ the results are sufficiently accurate, in fact they are exact to the thousandth. The values $L = 24$ and $N = 14$ can be considered in the computations. In a convergence test comparing the critical R_c obtained at different order expansions, the relative difference between the expansions at 26×18 and 24×16 is found to be less than 10^{-4}. There are three significant digits in this calculation. The benchmarking of the method can be done with results in Ref. (48). The critical wave number for $\Gamma = 2.936$, $\eta = 0.0862$ and $\delta = 0$ is $k_c = 0$, so for these values of the parameters the results reported in Ref. (48) are recovered. For this value of the aspect ratio the bifurcation corresponds to a mode 2 in the x direction and the bifurcation takes place at the same value $R_c = 73.5$.

	$N = 12$	$N = 14$	$N = 16$	$N = 18$
$L = 14$	$(2.5, 1203.91)$	$(2.5, 1210.00)$	$(2.5, 1212.73)$	$(2.5, 1208.70)$
$L = 16$	$(2.5, 1220.00)$	$(2.5, 1214.00)$	$(2.5, 1214.92)$	$(2.5, 1214.94)$
$L = 18$	$(2.5, 1220.10)$	$(2.5, 1225.00)$	$(2.5, 1224.92)$	$(2.5, 1224.07)$
$L = 20$	$(2.5, 1220.10)$	$(2.5, 1224.15)$	$(2.5, 1224.90)$	$(2.5, 1224.90)$
$L = 22$	$(2.5, 1220.20)$	$(2.5, 1224.92)$	$(2.5, 1224.92)$	$(2.5, 1224.92)$
$L = 24$	$(2.5, 1220.20)$	$(2.5, 1225.00)$	$(2.5, 1225.00)$	$(2.5, 1224.92)$
$L = 26$	$(2.5, 1220.20)$	$(2.5, 1225.00)$	$(2.5, 1224.92)$	$(2.5, 1224.92)$

Table 1. (k_c, R_c) for different order expansions in L and N in the Chebyshev expansion (20) for a 3D fluid with constant viscosity, $\eta = 0$, aspect ratio $\Gamma = 2.936$ and $\delta = 0.1$.

4. Numerical solutions with geophysical interest

4.1 Contained fluid

In references of small cells the case of large viscosity (or Prandlt number) could be considered as an approximation to mantle convection. The largest value of Prandlt number considered in the experiments is $Pr = 60$ in Ref. (52), in this case boundary layer waves are observed. Numerical results with infinite Pr number are reported in (45). In this case only stationary patterns of rolls perpendicular to the temperature gradient are observed. Also it is of interest the case of variable viscosity dependent on temperature, this case is plenty of references, but all of them consider homogeneous heating without horizontal temperature gradients (35; 36; 47). The only reference in which those gradients are taken into account in a variable viscosity case is (19). Some numerical solutions obtained in the case considered in Ref. (19) are presented in figure 2 at infinite Pr number, aspect ratio $\Gamma = 2.936$, $\eta = 0.0862$, $R = 72.650$ and

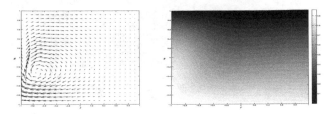

Fig. 2. Basic state for $\Gamma = 2.936$, $\eta = 0.0862$, $\delta = 0.1$ and $R = 72.65$. On the left velocity field **u**. On the right Isotherms of temperature Θ.

$\delta = 0.1$. Figure 2 shows that the structure of the velocity field is more localized close to the zone where the temperature is higher, i.e, at $r = -1$. The presence of the horizontal gradient generates convective basic states, that were conductive without the horizontal gradients. In Ref. (19) it is shown the fluid motion is produced in the region where viscosity is lower.

Regarding the instabilities, in the case of large Γ the influence of the horizontal temperature gradient is considerable, the problem is nearly two-dimensional (2D) in the uniform heating case, but it is three-dimensional (3D) with the horizontal temperature gradient. Figure 3 shows the growing mode or eigenfunction in the case $\Gamma = 2.936$, $\delta = 0$ and $\eta = 0.0862$, the critical wave number in this case is $k_c = 0$, so a 2D structure appears after the bifurcation. Figure 4 shows the growing mode or eigenfunction in the same case as before, but with horizontal gradient $\delta = 0.1$, the critical wave number in this case is $k_c = 1.7$, so a 3D structure appears after the bifurcation. Also we can observe from figures 3 and 4 that the bifurcating pattern is more structured in the $r - z$ plane for $\delta = 0$ and becomes more structured in the $y - z$ plane for $\delta \neq 0$. Hence, a horizontal temperature gradient gives rise to thermal plumes which bifurcate to totally 3D structures.

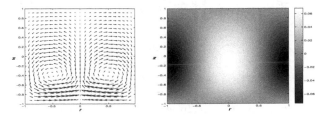

Fig. 3. Growing mode or eigenfunction at the instability threshold for $\Gamma = 2.936$, $\delta = 0$, $\eta = 0.0862$ and $R_c = 72.504$. On the left velocity field **u**. On the right Isotherms of temperature Θ.

4.2 Not contained fluid

This case is plenty of references of direct simulations solving numerically the partial differential equations (31; 34). But under the instability or bifurcation perspective the case in which the fluid can flow through the boundaries is only treated in reference (29). In that paper we show that a vortical structure appears after a stationary bifurcation of a state without angular velocity.

Fig. 4. Growing mode or eigenfunction at the instability threshold for $\Gamma = 2.936$, $\delta = 0.1$, $\eta = 0.0862$ and $R_c = 72.650$. On the left velocity field **u**. On the right Isotherms of temperature Θ.

A numerical solution obtained in the problem considered in Ref. (29) is presented in figure 5. Figure 5 shows the profiles of temperature, pressure and velocity components corresponding to the clockwise vortex for $Pr = 0.7$, $\Gamma = 0.5$ and $\delta = 10$ at $R = 4367$. This vortex appears after a bifurcation of a basic state with zero azimuthal velocity (see Ref. (29)). The main feature of the new steady flow emerging from the convective instability with $k_c = 0$ and $\bar{u}_\phi \neq 0$ is a non-zero azimuthal velocity component. The fluid inside the annulus begins to move in the azimuthal direction, rotating around the inner cylinder.

The linear stability analysis of the vortical structures shows that there is a wide range of parameters for which this state is stable.

The track of a particle in the vortex can be obtained by integrating the evolution of the element of fluid which follows the velocity field,

$$\frac{dr}{dt} = u_r(r,z), \tag{24}$$

$$\frac{d\phi}{dt} = u_\phi(r,z), \tag{25}$$

$$\frac{dz}{dt} = u_z(r,z). \tag{26}$$

In our simulations we observe a spiral upward motion of the particle around the inner cylinder, which implies a transport of mass in the azimuthal direction. Starting from below, the particle goes up, moves towards the inner cylinder and rotates around it. The combination of these movements gives the spiral trajectory shown in figure 6, where the trajectory of a particle in the fluid is presented for $\Gamma = 0.5$, $\delta = 10$ and $R = 4367$ at two different initial conditions. Starting from a point close to the bottom plate but near the inner cylinder, where the effect of u_z is stronger than the effect of u_ϕ, the particle goes up very fast without much turning around the inner cylinder. This can be appreciated in figure 6 a) where the starting point considered is $(r = 0.085, z = 0.05, \phi = 0)$ in $[0.06, 0.56] \times [0, 1] \times [0, 2\pi]$. When the particle reaches the upper part of the structure it describes wider circles around the inner cylinder as u_r becomes positive and u_z is very small at those levels (see figure 5). Figure 6 b) shows the effect of localizing the starting point further from the inner cylinder, e.g. at $(r = 0.31, z = 0.05, \phi = 0)$. In this case, the effect of u_ϕ is stronger and the spiral up motion of the particle starts as soon as the particle begins to move.

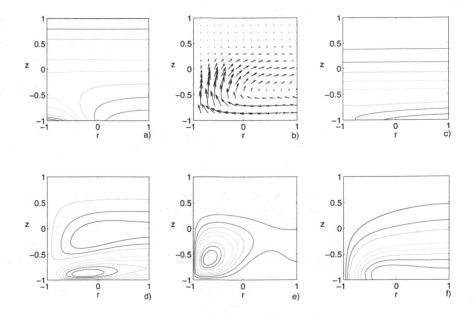

Fig. 5. Clockwise vortex at $\Gamma = 0.5$, $\delta = 10$ and $R = 4367$. a) Isotherms of Θ; b) meridional velocity (u_r, u_z); c) contour plot of the pressure p ; d) contour plot of the radial velocity component u_r; e) contour plot of the vertical velocity component u_z; f) contour plot of the azimuthal velocity component u_ϕ. The contours correspond to equally spaced values within their ranges of [-9:1] for Θ, [-0.02:20.4] $\cdot 10^3$ for p, [-9.3:4.3] for u_r, [-0.6:14.5] for u_z and [-26.1:0] for u_ϕ. The pressure p is determined up to a constant.

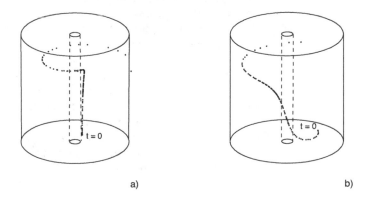

Fig. 6. Track of a particle in the fluid for the stable clockwise vortex. The values of the parameters are $\Gamma = 0.5$, $\delta = 10$ and $R = 4367$. a) Starting point at $(r = 0.085, z = 0.05, \phi = 0)$; b) starting point at $(r = 0.31, z = 0.05, \phi = 0)$.

5. Discusion

5.1 Contained fluid

Regarding the numerical solutions found in this case, the horizontal temperature gradient generates convective states and tends to concentrate motion near the warmer wall. This fact coincides with experiments in Refs. (38; 40; 41) and is consistent with previous numerical results reported in (27; 45). The temperature dependent viscosity localizes motion near the region of lower viscosity, i.e., the bottom plate. This also coincides with experiments in Refs. (36; 46; 49) and numerical results in Ref. (48). It is remarkable that the horizontal temperature gradient favours a threedimensional structure after the bifurcation, while the pattern continues being axisymmetric after the bifurcation in the only vertical gradient case.

5.2 Not contained fluid

As detailed in Ref. (48) we found qualitative similarities between the vortical structures computed numerically and some meteorological phenomena such as dust devils and cyclones. One of the main characteristics of dust devils is a low-pressure region in the center of the dust devil which coincides with the dust devil's warm core (33). This is also observed in our numerical vortices (see figure 5 c). Regarding temperature, in dust devils, the maximum temperature deviation from the environment temperature (i.e. the temperature furthest from the dust devil center) occurs at the lowest levels. This feature is observed in the temperature profile of our vortices.

The experimental measures provided in Ref. (33) show that there is radial inflow at the lower levels of the dust devil and radial outflow in the upper levels. It is also shown that the vertical velocity reaches highest values and then falls off rapidly as the radius is increased. These features are appreciated in the profile of u_r and u_z shown in figures 5 d) and e).

The trajectory of particles around the inner cylinder described in this section appears to be very similar to the trajectory of particles of air (or dust) in a dust devil, characterized by a spiral up motion (33).

Other more complex meteorological phenomena such as cyclones also present these structural characteristics. It is known that the center (eye) of a cyclone is the area of lowest atmospheric pressure in the region, which corresponds to a warm core in some kind of cyclones (e.g. tropical and mesoscale) (31; 34). This coincides with that observed in figures 5 a) and 5 b). Regarding the motion in cyclones, it is observed the inward flow next to the surface, strong upward motion in the eyewall and outflow in a layer near the top of the storm (31; 34). This characteristic is described in the combined effect of the radial and vertical velocity components observed in our vortices as pointed out above (see figures figures 5 c, d) and e)). In cyclones, a counter-clockwise motion (clockwise in the southern hemisphere) is observed around the center of the storm, stronger just above the surface in a ring around the center and sligther as we go up from the surface (31; 34). That coincides with the effect of the azimuthal velocity component observed in the vortices we have computed numerically responsible for the movement of the particles around the inner cylinder.

6. Conclusions

In this work we have reviewed the influence of horizontal temperature gradients on convective instabilities, focusing on results with geophysical interest.

We have distinguished two cases, a first one where the fluid is simply contained in a domain, and a second one where the fluid can flow throughout the boundaries.
In the first case three subcases can be grouped. The case corresponding to small cells and small Pr number displays stationary and oscillatory instabilities depending on the multiple parameters present in the problem: properties of the fluid, surface tension effects, heat exchange with the atmosphere, aspect ratio, dependence of viscosity with temperature, etc. This problem has been treated from experimental, theoretical and numerical points of view. The cases corresponding to small cells and large or infinite Pr number are closer to mantle convection. Boundary layer waves are observed in experiments and 3D stationary patterns of rolls perpendicular to the temperature gradient appear numerically. Finally for the case of infinite Pr number with temperature dependent viscosity, the closest to mantle convection, 3D stationary patterns concentrated in the region of lower viscosity and waves for larger values of the R number appear. Summarizing, horizontal temperature gradients favour the presence of waves and the totally three dimensional patterns.
The problem where the fluid can flow throughout the boundaries has been treated usually as direct numerical simulations. For the first time it has been studied under the perspective of instabilities or bifurcations in Ref. (29). In this reference vortical solutions, very similar to those found for some atmospheric phenomena such as dust devils or hurricanes, appear after a stationary bifurcation. This is a powerfull and simple explanation of those atmospheric phenomena as an instability.

7. Acknowledgements

This work was partially supported by Research Grant MICINN (the Government of Spain) MTM2009-13084 and CCYT (Junta de Comunidades de Castilla-La Mancha) PAI08-0269-1261, which include RDEF funds.

8. References

[1] H. Bénard *Rev. Gén. Sci. Pures Appl.* 11, 1261 (1900).
[2] Lord Rayleigh. On convective currents in a horizontal layer of fluid when the temperature is on the under side. Phil. Mag. 32, pp. 529-46 (1916).
[3] S. Chandrasekhar. *Hydrodynamic and Hydromagnetic Stability.* Dover Publications, New York, 1981.
[4] E. Bodenschatz, J. R. de Bruyn, G. Ahlers, and D. S. Cannell. Transitions between patterns in thermal convection. *Phys. Rev. Lett.* 67, 3078-3081, 1991.
[5] S.W Morris, E. Bodenschatz, D. S. Cannell and G. Ahlers. Spiral defect chaos in large aspect ratio Rayleigh-Bénard convection. *Phys. Rev. Lett.* 71, 2026-2029, 1993.
[6] E.L. Koschmider. Bénard Cells and Taylor Vortices. Cambridge University Press, 1993.
[7] M. Assenheimer, V. Steinberg, "Transition between spiral and target states in Rayleigh-Bénard Convection". *Nature* 367 (6461) 345 (1994).
[8] E. Bodenschatz, W. Pesch and G. Ahlers. Recent developments in Rayleigh-Bénard convection. *Annual Review of Fluid Mechanics* 32, 709-778, 2000.
[9] B. B. Plapp, D. A Egolf, E. Bodenschatz and W. Pesch. Dynamics and selection of giant spirals in Rayleigh-Bénard convection. *Phys. Rev. Lett.* 81, 5334-5337, 1998.

[10] S. Rüdiger and F. Feudel. Pattern formation in Rayleigh Bénard convection in a cylindrical container. *Phys. Rev. E* 62, 4927-4931, 2000.

[11] A.B. Ezersky, A. Garcimartín, J. Burguete, H.L. Mancini and C. Pérz-García, "Hydrothermal waves in Marangoni convection in a cylindrical container." *Phys. Rev. E* 47, 1126 (1993).

[12] M.A. Pelacho and J. Burguete, "Temperature oscillations of hydrothermal waves in thermocapillary-buoyancy convection."*Phys. Rev. E* 59, 835 (1999).

[13] E. Favre, L. Blumenfeld and F. Daviaud, "Instabilities of a liquid layer locally heated on its free surface."*Phys. Fluids* 9, 1473 (1997).

[14] N. Garnier and A. Chiffaudel. "Two dimensional hydrothermal waves in an extended cylindrical vessel." *Eur. Phys. J. B* 19, 87 (2001).

[15] N. Garnier and C. Normand. "Effects of curvature on hydrothermal waves instability of radial thermocapillary flows." *C. R. Acad. Sci., Ser. IV* 2 (8) 1227 (2001).

[16] B.C. Sim, A. Zebib. "Effect of free surface heat loss and rotation on transition to oscillatory thermocapillary convection." *Phys. Fluids* 14 (1), 225 (2002).

[17] B.C. Sim, A. Zebib, D. Schwabe. "Oscillatory thermocapillary convection in open cylindrical annuli. Part 2. Simulations."*J. Fluid Mech.* 491, 259 (2003).

[18] S. Hoyas, A.M. Mancho, H. Herrero, N. Garnier, A. Chiffaudel. "Bénard-Marangoni convection in a differentially heated cylindrical cavity" *Phys. Fluids,* 17, 054104 (2005).

[19] F. Pla and H. Herrero "Effects of non uniform heating in a variable viscosity Rayleigh-Bénard problem". *Theoretical and Computational Fluid Dynamics,* DOI: 10.1007/s00162-010-0189-3, 2010.

[20] M. K. Smith and S. H. Davis, "Instabilities of dynamic thermocapillary layers. 1. Convective instabilities." *J. Fluid Mech.* 132, 119 (1983).

[21] De Saedeleer, C., Garcimartin, A., Chavepeyer, G., Platten, J.K., Lebon, G.: The instability of a liquid layer heated from the side when the upper surface is open to air. Phys. Fluids 8(3), pp. 670-676 (1996).

[22] A. M. Mancho, H. Herrero and J. Burguete, "Primary instabilities in convective cells due to non-uniform heating."*Phys. Rev E* 56, 2916 (1997).

[23] H. Herrero and A. M. Mancho "Influence of aspect ratio in convection due to non-uniform heating." *Phys. Rev E* 57, 7336 (1998).

[24] R.J. Riley and G.P. Neitzel, "Instability of thermocapillary-buoyancy convection in shallow layers. Part 1. Characterization os steady and oscillatory instabilities." *J. Fluid Mech* 359 , 143 (1998).

[25] S. Hoyas, H. Herrero and A.M. Mancho, "Thermal convection in a cylindrical annulus heated laterally." *J. Phys. A: Math and Gen.* 35, 4067 (2002).

[26] S. Hoyas, H. Herrero and A.M. Mancho, "Bifurcation diversity of dynamic thermocapillary liquid layers."*Phys. Rev. E,* 66, 057301 (2002).

[27] S. Hoyas, H. Herrero, A.M. Mancho, "Thermocapillar and thermogravitatory waves in a convection problem". *Theoretical and Computational Fluid Dynamics* 18, 2-4, 309 (2002).

[28] M. C. Navarro, A. M. Mancho and H. Herrero, "Instabilities in buoyant flows under localized heating."*Chaos* 17, 023105 (2007).

[29] M. C. Navarro, H. Herrero, "Vortex generation by a convective instability in a cylindrical annulus non homogeneously heated". *Physica D,* accepted (2011).

[30] N. O. Rennó, M. L. Burkett, M. P. Larkin, "A simple theory for dust devils". *J. Atmos. Sci.* 55 3244 (1998).

[31] K. A. Emanuel. *Divine wind.* Oxford University Press, Oxford, 2005.

[32] L. Battan, Energy of a dust devil, J. Meteor. 15 (1958) 235-237.

[33] P. C. Sinclair, The lower structure of dust devils, J. Atmos. Sci. 30 (1973) 1599-1619.

[34] K. A. Emanuel, Thermodynamic control of hurricane intensity, Nature 401 (1999) 665-669.

[35] Bercovici, D.: The generation of plate tectonics from mantle convection. Earth and Planetary Science Letters 205, pp. 107-121 (2003).

[36] Booker, J.R.: Thermal convection with strongly temperature-dependent viscosity. J. Fluid Mech. 76 (4), pp. 741-754 (1976).

[37] Bunge, H.P., Richards, M.A., Baumgardner, J.R.: Effects of depth-dependent viscosity on the platform of mantle convection. Nature 379, pp. 436-438 (1996).

[38] Burguete, J., Mokolobwiez, N., Daviaud, F., Garnier, N., Chiffaudel, A.: Buoyant-thermocapillary instabilities in extended layers subjected to a horizontal temperature gradient. Phys. Fluids 13, pp. 2773-2787 (2001).

[39] Canuto, C., Hussaini, M.Y., Quarteroni, A., Zang, T.A. Spectral Methods in Fluid Dynamics. Springer, Berlin (1988).

[40] Daviaud, F., Vince, J.M.: Traveling waves in a fluid layer subjected to a horizontal temperature gradient. Phys. Rev. E 48, pp. 4432-4436 (1993).

[41] De Saedeleer, C., Garcimartin, A., Chavepeyer, G., Platten, J.K., Lebon, G.: The instability of a liquid layer heated from the side when the upper surface is open to air. Phys. Fluids 8(3), pp. 670-676 (1996).

[42] Golub, G.F., Van Loan, C.F.: Matrix Computations. The Johns Hopkins University Press. Baltimore and London, (1996).

[43] Herrero, H., Mancho, A.M.: On pressure boundary conditions for thermoconvective problems. Int. J. Numer. Meth. Fluids 39, pp. 391-402 (2002).

[44] Herrero, H., Hoyas, S., Donoso, A., Mancho, A.M., Chacón, J.M., Portugues, R.F., Yeste, B.: Chebyshev Collocation for a Convective Problem in Primitive Variables Formulation. J. of Scientific Computing 18(3), pp. 315-328 (2003).

[45] Mancho, A.M., Herrero, H.: Instabilities in a laterally heated liquid layer. Phys. Fluids 12, pp. 1044-1051 (2000).

[46] Manga, M., Weeraratne, D., Morris, S.J.S.: *Boundary-layer thickness and instabilities in Bénard convection of a liquid with a temperature-dependent viscosity*, Phys. Fluids 13 (3), pp. 802-805 (2001).

[47] Moresi, L.N., Solomatov, V.S.: Numerical investigation of 2D convection with extremely large viscosity variations, Phys. Fluids, 7 (9), pp. 2154-2162 (1995).

[48] Pla, F., Mancho, A.M., Herrero, H.: Bifurcation phenomena in a convection problem with temperature dependent viscosity at low aspect ratio. Physica D, 238, pp. 572-580, (2009).

[49] Richter, F.M., Nataf, H.C., Daly, S.F.,: *Heat transfer and horizontally averaged temperature of convection with large viscosity variations*, J. Fluid Mech. 129, pp. 173-192 (1983).

[50] Trompert, R., Hansen, U.:Mantle convection simulations with rheologies that generate plate-like behaviour, Nature 395 (6703), pp. 686-689 (1998).

[51] Yanagawa, T.K.B., Nakada, M., Yuen, D.A.: A simplified mantle convection model for thermal conductivity stratification. Physics of the Earth and Planetary Interiors 146, pp. 163-177 (2004).

[52] A.B. Ezersky, A. Garcimartín, J. Burguete, H.L. Mancini and C. Pérez García. *Hydrothermal waves in Marangoni convection in a cylindrical container.* Phys. Rev. E 47, pp. 1126-1131, 1993; A.B. Ezersky, A. Garcimartín, H.L. Mancini and C. Pérez García. *ibid.* 48, pp. 4414, 1993.

7

Modelling of Turbulent Premixed and Partially Premixed Combustion*

V. K. Veera[1], M. Masood[1], S. Ruan[1], N. Swaminathan[1] and H. Kolla[2]
[1]Department of Engineering, Cambridge University, Cambridge
[2]Sandia National Laboratory, Livermore, CA
[1]UK
[2]USA

1. Introduction

A significant increase in our energy consumption, from 495 quadrillion Btu in 2007 to 739 quadrillion Btu in 2035 with about 1.4% annual increase, is predicted (US Energy Information Administraion, 2010). This increase is to be met in environmentally friendly means in order to protect our planet. Despite the renewable energy sources are identified to be the fastest growing in the near future, they are expected to meet only one third or less of this energy demand. Also, the renewable generation methods face significant barriers such as economical risks, high capital costs, cost for infrastructure development, low energy conversion efficiency, and low acceptance level from public (US Energy Information Administraion, 2010) at this time. It is likely that improvements will be made on all of these factors in due course. The role of nuclear technology in the energy market will vary from time to time for many cultural and political reasons, and the perceptions of the general public. In the current climate, however, it is clear that the fossil fuels will remain as the dominant source to meet the demand in energy consumption. Hence, optimised design of combustion and power generating systems for improved efficiency and emissions performance are crucial.

The emissions of oxides of nitrogen and sulphur, and poly-aromatic hydrocarbons are known sources of atmospheric pollution from combustion. Their detrimental effects on environment and human health is well known (Sawyer, 2009) and green house gases such as oxides of carbon and some hydrocarbons are also included as pollutants in recent years. The emission of carbon dioxide (CO_2) from fossil, liquid and solid, fuel combustion accounts for nearly 76% of the total emissions from fossil fuel burning and cement production in 2007 (Carbon Dioxide Information Analysis Center, 2007). The global mean CO_2 level in the atmosphere increases each year by about 0.5% suggesting a global mean level of about 420 ppm by 2025 (Anastasi et al., 1990; US Department of Commerce, 2011) Such a forecasted increase has led to stringent emission regulations for combustion systems compelling us to find avenues to improve the environmental friendliness of these systems. Lean premixed combustion is known (Heywood, 1976) to have potentials for effective reduction in emissions and to increase efficiency simultaneously. Significant technological advances are yet to be made for developing fuel lean combustion systems operating over wide range of conditions with

*Draft February 27, 2012, Book Chapter for Fluid Dynamics / Book 1, Ed. Dr. Hyoung Woo Oh Department of Mechanical Engineering, Chungju National University, Chungju, Korea

desirable characteristics. This is because, ignition and stability of this combustion, controlled strongly by turbulence-combustion interaction, are not fully understood yet. Many of the recent studies are focused to improve this understanding as it has been noted in the books edited by Echekki & Mastorakos (2011) and Swaminathan & Bray (2011).

The OEMs (original equipment manufacturers) of gas turbines and internal combustion engines are embracing computational fluid dynamics (CFD) into their design practice to find answers to *what if?* type questions arising at the design stage. This is because CFD provides quicker and more economical solutions compared to "cut metal and try" approach. Thus, having an accurate, reliable and robust combustion modelling becomes indispensable while developing modern combustors or engines for fuel lean operation. In this chapter, we discuss one such modelling method developed recently for lean premixed flames along with its extension to partially premixed combustion. Partial premixing is inevitable in practical systems and introduced deliberately under many circumstances to improve the flame ignitability, stability and safety.

Before embarking on this modelling discussion, challenges in using the standard moment methods for reacting flows, which are routinely used for non-reacting flows, are discussed in the next section along with a brief discussion on three major computational paradigms used to study turbulent flames. Section 1.2 identifies important scales of turbulent flame and discusses a combustion regime diagram. The governing equations for Reynolds averaged Navier Stokes (RANS) simulation of turbulent combustion are discussed in section 2 along with turbulence modelling used in this study. The various modelling approaches for lean premixed combustion are briefly discussed in section 3. The detail of strained flamelet model and its extension to partially premixed flames are presented in section 4. Its implementation in a commercial CFD code is discussed in section 4.2 and the results are discussed in section 5. The final section concludes this chapter with a summary and identifies a couple of topics for further model development.

1.1 Challenges

In the RANS approach, the instantaneous quantities are decomposed into their means and fluctuations. The mean values of density, velocity, etc., in a flow are computed by solving their transport equations along with appropriate modelling hypothesis for correlations of fluctuating quantities. These modelling are discussed briefly in section 2. The simulations of non-reacting flows have become relatively easier task now a days. The presence of combustion however, significantly complicates matters and alternative approaches are to be sought. Combustion of hydrocarbon, even the simplest one methane, with air includes several hundreds of elementary reactions involving several tens of reactive species. If one follows the traditional moment approach by decomposing each scalar concentration into its mean and fluctuation then it is clear that several tens of partial differential equations for the conservation of mean scalar concentration need to be solved. These equations will involve many correlations of fluctuations requiring closure models with a large set of model parameters. More importantly, the highly non-linear reaction rate is difficult to close accurately. To put this issue in a clear perspective, let us consider an elementary chemical reaction $R_1 + R_2 \rightarrow P$, involving two reactants and a product. The law of mass action would give the instantaneous reaction rate for R_1 as $\dot{\omega}_1 = -A\,T^b \rho^2 Y_1 Y_2 \exp\left(-T_a/T\right)$, where A is a pre-exponential factor and T_a is the activation temperature. The temperature is denoted as T and the mass fractions of two reactants are respectively denoted by Y_1 and Y_2. Let us take $b = 0$ for the sake of simplicity. In variable density flows, it is normal to use density-weighted means, defined, for example for mass fraction of reactant R_1, as $\widetilde{Y}_1 = \overline{\rho\, Y_1}/\overline{\rho}$ and its fluctuation y_1''. Substituting this decomposition into the above reaction rate expression, one obtains

$$\dot{\omega}_1 = \overline{\dot{\omega}}_1 + \dot{\omega}_1' = -A\overline{\rho}^2 \left(\widetilde{Y}_1 + y_1''\right)\left(\widetilde{Y}_2 + y_2''\right)\exp\left(-\frac{T_a}{\widetilde{T} + T''}\right). \tag{1}$$

The exponential term can be shown (Williams, 1985) to be

$$\exp\left(\frac{-T_a}{\widetilde{T}(1 + T''/\widetilde{T})}\right) = \exp\left(\frac{-T_a}{\widetilde{T}}\right) \times \left(\sum_{m=0}^{\infty}\frac{(-1)^m}{m!}\left(\frac{T_a}{\widetilde{T}}\right)^m\left[\sum_{n=1}^{\infty}(-1)^n\left(\frac{T''}{\widetilde{T}}\right)^n\right]^m\right). \tag{2}$$

Since $T_a/\widetilde{T}$ is generally large, at least about 20 terms in the above expansion are required to have a convergent series. This is impractical and also $T''/\widetilde{T}$ is seldom smaller than 0.01 in turbulent combustion to neglect higher order terms in the above expansion. There are already some approximations made while writing Eq. (1) and furthermore, while averaging this equation to get $\overline{\dot{\omega}}$ one must not forget that $\overline{\rho \exp(-T_a/T)} \neq \overline{\rho}\,\exp(-T_a/\widetilde{T})$.

It is clear that one needs to solve a large set of coupled partial differential equations with numerous model parameters, which poses a serious question on the accuracy and validity of computed solution using the classical RANS approach which usually tracks the first two moments for each of the quantities involved. This is a well-known problem in turbulent combustion and alternative approaches have been developed in the past (Echekki & Mastorakos, 2011; Libby & Williams, 1994; Swaminathan & Bray, 2011). The first two statistical moments of one or two key scalars are computed instead of solving hundreds of partial differential equations for the statistical moments and correlations of all the reactive scalars. The statistics of the two key scalars, typically a mixture fraction, Z, and reaction progress variable, c, are then used to estimate the thermo-chemical state of the chemically reacting mixture using modelling hypothesis. Many modelling approaches have been proposed in the past and the readers are referred to the books by Libby & Williams (1994), Peters (2000), Echekki & Mastorakos (2011), and Swaminathan & Bray (2011). Approaches relevant for lean premixed and partially premixed flames are briefly reviewed in later parts of this chapter.

The three computational paradigms generally used to study turbulent combustion are (i) direct numerical simulation (DNS), (ii) large eddy simulation (LES) and (iii) RANS. These approaches have their own advantages and disadvantages. For example, the detail and level of information available for analysis decreases from (i) to (iii). Thus DNS is usually used for model testing and validation and it incurs a heavy computational cost because it resolves all the length and time scales (in the continuum sense) involved in the reacting flow. The general background of DNS is discussed elsewhere (Chapter ??) in this book. With the advent of Tera- and Peta-scale computing, it is becoming possible to directly simulate laboratory scale flames with hundreds of chemical reactions. However, direct simulations of practical flames in industry are not to be expected in the near future.

On the other hand, in RANS all the scales of flow and flame are modelled and thus it provides only statistical information and it is possible to include different level of chemical kinetics detail as will be discussed later in this chapter. If the RANS simulations are performed carefully, they provide solutions with sufficient accuracy to guide the design of combustors and engines. These simulations are cost effective and quick, and thus they are attractive for use in industries.

The LES is in between these two extremes as it explicitly computes large scales in the flow and it is well suited for certain class of flows. Still models are required for quantities related to small scales. The general background of LES is discussed in Chapter ?? of this book. It should be noted that combustion occurs in scales much smaller than those usually captured in LES and thus they need to be modelled. Most LES models are based on RANS type modelling and this chapter presents combustion modelling for the RANS framework. The details of turbulent combustion modelling depend on the combustion regime, which is determined by the relativity of characteristic scales of turbulence and flame chemistry.

1.2 Regimes in turbulent premixed combustion

The characteristic flame scales are defined using the unstrained planar laminar flame speed, s_L^0, and Zeldovich flame thickness given by $\delta = \mathcal{D}/s_L^0$, where $\mathcal{D}$ is a molecular diffusion coefficient. Using these velocity and length scales, one can define the flame time scale as $t_F = \delta/s_L^0$. The turbulence scales are defined using turbulent kinetic energy, $\widetilde{k} = 0.5\overline{\rho\, u_i'' u_i''}/\overline{\rho}$, and its dissipation rate, $\widetilde{\varepsilon} = 2\overline{\rho\, v s_{ij} s_{ij}}/\overline{\rho}$ (strictly $2\overline{\rho\, v(s_{ij}s_{ij} - s_{ii}s_{ii}/3)}/\overline{\rho}$ for combusting flows), where s_{ij} is the symmetric part of the turbulent strain tensor and v is the kinematic viscosity of the fluid. The characteristic turbulence length and time scales are respectively $\Lambda = \widetilde{k}^{1.5}/\widetilde{\varepsilon}$ and $t_T = \widetilde{k}/\widetilde{\varepsilon}$. The RMS of turbulent velocity fluctuation is $u' = \sqrt{2\widetilde{k}/3}$. The viscous dissipation scales are the Kolmogorov scales given by $l_\eta = (v^3/\widetilde{\varepsilon})^{1/4}$ and $t_\eta = (v/\widetilde{\varepsilon})^{1/2}$. For the sake of simplicity, let us take equal molecular diffusivities for all reactive species and the Schmidt number to be unity. One can combine these characteristic scales to form three non-dimensional parameters, turbulence Reynolds, Damköhler and Karlovitz numbers, given respectively by

$$\text{Re} = \frac{u'\Lambda}{s_L^0\delta}, \quad \text{Da} = \frac{t_T}{t_F} = \frac{(\Lambda/\delta)}{(u'/s_L^0)} \quad \text{and} \quad \text{Ka} = \frac{t_F}{t_\eta} = \left(\frac{\delta}{\Lambda}\right)^{1/2}\left(\frac{u'}{s_L^0}\right)^{3/2}. \quad (3)$$

Figure 1 illustrates the relationship between these three parameters in (u'/s_L^0)-(Λ/δ) space. This diagram is commonly known as combustion regime diagram (Peters, 2000). The left

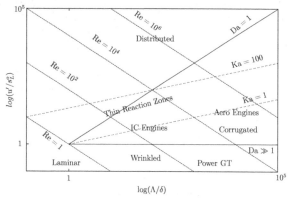

Fig. 1. Turbulent Premixed Combustion Regime Diagram

bottom corner, below $\text{Re} = 1$ line, represents laminar flames. For $u'/s_L^0 < 1$, turbulent fluctuation due to large eddy cannot compete with the flame advancement by the laminar flame propagation mechanism, thus laminar flame propagation dominates over flame front wrinkling by turbulence. For $u'/s_L^0 > 1$ and $\text{Ka} < 1$, the flame scales are smaller than all the relevant turbulent scales. As a result, turbulent eddies cannot disturb the inner reactive-diffusive structure but only wrinkle it and the flame also remains quasi-steady. The Ka $= 1$ line represents that the flame thickness is equal to the Kolmogorov length scale, which is known as the Klimov-Williams line. Although the Kolmogorov scales are smaller than the flame thickness, the inner reaction zone is expected to be undisturbed by the small scale turbulence when Ka < 100. However, the preheat zones are disturbed by the small-scale turbulence. This regime is known as thin reaction zones regime. When Ka > 100, the Kolmogorov eddies can penetrate into the inner reaction zones causing local extinction leading to distributed reaction zones. However, evidence for these reaction zones are sparse (Driscoll, 2008). Also shown in the diagram are the combustion regimes for practical engines (Swaminathan & Bray, 2011). The spark ignited internal combustion engines operate in the border between corrugated flamelets and thin reaction zones regimes, whereas the power gas turbines operate in the border between corrugated and wrinkled flamelets. Aero engines do not operate in premixed mode for safety reasons and if one presumes a premixed mode for them then they are mostly in thin reaction zones regime.

2. Modelling framework for turbulent flames

In the RANS modelling framework, the conservation equations for mass, momentum, energy and the key scalar values are averaged appropriately. These equations are solved along with models and averaged form of state equation. As noted earlier, density weighted average is used for turbulent combustion and the Favre averaged mass and momentum equations are given by

$$\frac{\partial \bar{\rho}}{\partial t} + \frac{\partial \bar{\rho} \widetilde{u}_i}{\partial x_i} = 0, \tag{4}$$

$$\frac{\partial \bar{\rho} \widetilde{u}_i}{\partial t} + \frac{\partial \bar{\rho} \widetilde{u}_i \widetilde{u}_j}{\partial x_j} = -\frac{\partial \bar{p}}{\partial x_i} - \frac{\partial (\overline{\rho u_i'' u_j''})}{\partial x_j} + \frac{\partial \overline{\tau_{ij}}}{\partial x_j}, \tag{5}$$

in the usual nomenclature. In adiabatic and low speed (negligible compressibility effects) combustion problems, the Favre averaged total enthalpy equation given by

$$\frac{\partial \bar{\rho} \widetilde{h}}{\partial t} + \frac{\partial \bar{\rho} \widetilde{h} \widetilde{u}_j}{\partial x_j} = -\frac{\partial (\overline{\rho h'' u_j''})}{\partial x_j} + \frac{\partial}{\partial x_j} \left(\overline{\rho D \frac{\partial h}{\partial x_j}} \right), \tag{6}$$

for the reacting mixture is solved. The total enthalpy $\widetilde{h}$ is defined as $\widetilde{h} = \sum_i \widetilde{Y}_i \widetilde{h}_i$, where $\widetilde{Y}_i$ is the Favre averaged mass fraction of species i. The specific enthalpy of species i is

$$\widetilde{h}_i = h_i^0 + \int_{T_0}^{\widetilde{T}} c_{p,i} \ d\widetilde{T}, \tag{7}$$

where the standard specific enthalpy of formation for species i and its specific heat capacity at constant pressure are respectively h_i^0 and $c_{p,i}$. The equation (7) is used to calculate the temperature, $\widetilde{T}$, from the computed values of $\widetilde{h}$. The state equation is given by $\bar{p} = \bar{\rho} R \widetilde{T}$, where R is the gas constant. One must also know the mean mass fraction fields, which are obtained using combustion modelling.

The two key scalars used in turbulent combustion are the mixture fraction, Z, and reaction progress variable, c. These two scalars are defined later and the transport equation for their

Favre averaged values are respectively given by

$$\frac{\partial \bar{\rho}\tilde{Z}}{\partial t} + \frac{\partial \bar{\rho}\tilde{Z}\tilde{u}_j}{\partial x_j} = -\frac{\partial(\overline{\rho\, Z''u_j''})}{\partial x_j} + \frac{\partial}{\partial x_j}\left(\overline{\rho\mathcal{D}_z \frac{\partial Z}{\partial x_j}}\right). \tag{8}$$

and $$\frac{\partial \bar{\rho}\tilde{c}}{\partial t} + \frac{\partial \bar{\rho}\tilde{c}\,\tilde{u}_j}{\partial x_j} = \bar{\omega}_c - \frac{\partial(\overline{\rho\, c''u_j''})}{\partial x_j} + \frac{\partial}{\partial x_j}\left(\overline{\rho\mathcal{D}_c \frac{\partial c}{\partial x_j}}\right). \tag{9}$$

The Reynolds stress, $\overline{\rho\, u_i''u_j''}$, and fluxes in Eqs. (5) to (9) and the mean reaction rate, $\bar{\omega}_c$, of c need closure models. In principle, all of the above equations are applicable to both RANS and LES, and the correlations must be interpreted appropriately. The Reynolds fluxes are typically closed using gradient hypothesis, which gives, for example, $\overline{u_j''Z''} = -\mathcal{D}_T \partial\tilde{Z}/\partial x_j$, where $\mathcal{D}_T$ is the turbulent diffusivity. This quantity and the Reynolds stress are obtained using turbulence modelling.

2.1 Turbulence modelling

The modelling of turbulence is an essential part of turbulent combustion calculation using CFD. A variety of turbulence models are available in the literature (Libby & Williams, 1994; Swaminathan & Bray, 2011, see for example) and an appropriate choice should be guided by a physical understanding of the flow. A standard two-equation model like k-ε model can be used to model simple free shear flows. In this model, the Reynolds stress is related to the eddy viscosity, μ_T, by

$$\overline{\rho u_i''u_j''} = -\mu_T\left(\frac{\partial \tilde{u}_i}{\partial x_j} + \frac{\partial \tilde{u}_j}{\partial x_i} - \frac{2}{3}\delta_{ij}\frac{\partial \tilde{u}_m}{\partial x_m}\right) + \frac{2}{3}\bar{\rho}\tilde{k}\delta_{ij}. \tag{10}$$

However, complex geometries might require improved models, such as transported Reynolds stress, to capture the relevant flow features such as flow recirculation, the onset of flow separation and its reattachment, etc. These advanced models are known to cause difficulties such as numerical instability during simulations compared to the two-equation models. Furthermore, the two-equation models are commonly used because, (i) they are easy to implement and use, (ii) numerically stable and (iii) provide sufficiently accurate solution to guide the analysis of turbulent flames, provided the mean reaction rate and turbulence-flame interactions are modelled correctly. Two variants of two-equation model commonly used in turbulent premixed flame calculations are discussed briefly next.

2.1.1 k-ε Model

Despite the advancement in understanding and modelling of the turbulence, the standard k-ε model is still one of the most widely used models for engineering calculations. The attractiveness of this model is rooted in its simplicity and favourable numerical characteristics, more importantly, in its surprisingly good predictive capabilities over a fair range of flow conditions. It represents a reasonable compromise between accuracy and cost while dealing with various flows.

Transport equations for the standard k-ε model are (Jones, 1994)

$$\frac{\partial \bar{\rho}\tilde{k}}{\partial t} + \frac{\partial \bar{\rho}\tilde{u}_j\tilde{k}}{\partial x_j} = \frac{\partial}{\partial x_j}\left[\left(\mu + \frac{\mu_T}{Sc_k}\right)\frac{\partial \tilde{k}}{\partial x_j}\right] - \overline{\rho\, u_i''u_j''}\frac{\partial \tilde{u}_i}{\partial x_j} - \overline{u_j''}\frac{\partial \bar{p}}{\partial x_j} + \overline{p'\frac{\partial u_i''}{\partial x_i}} - \bar{\rho}\tilde{\varepsilon}, \tag{11}$$

$$\frac{\partial \overline{\rho}\widetilde{\varepsilon}}{\partial t} + \frac{\partial \overline{\rho}\widetilde{u}_j\widetilde{\varepsilon}}{\partial x_j} = \frac{\partial}{\partial x_j}\left[\left(\mu + \frac{\mu_t}{Sc_\varepsilon}\right)\frac{\partial \widetilde{\varepsilon}}{\partial x_j}\right] - C_{\varepsilon 2}\overline{\rho}\frac{\widetilde{\varepsilon}^2}{\widetilde{k}}$$

$$-C_{\varepsilon 1}\overline{\rho u_i'' u_j''}\frac{\widetilde{\varepsilon}}{\widetilde{k}}\frac{\partial \widetilde{u}_i}{\partial x_j} - C_{\varepsilon 1}\overline{u_j''}\frac{\widetilde{\varepsilon}}{\widetilde{k}}\frac{\partial \overline{p}}{\partial x_j} + C_{\varepsilon 1}\frac{\widetilde{\varepsilon}}{\widetilde{k}}\overline{p'\frac{\partial u_i''}{\partial x_i}}, \tag{12}$$

where μ_t is the turbulent eddy viscosity calculated as $\mu_t = \overline{\rho}C_\mu \widetilde{k}^2/\widetilde{\varepsilon}$ and Sc_m is the turbulent Schmidt number for quantity m. The standard values of these model parameters are $C_\mu = 0.09$, $C_{\varepsilon 1}=1.44$ and $C_{\varepsilon 2}=1.92$. The pressure-dilation term can become important in turbulent premixed flames and it is modelled as (Zhang & Rutland, 1995)

$$\overline{p'\frac{\partial u_i''}{\partial x_i}} = 0.5\widetilde{c}\left(\tau s_L^0\right)^2 \overline{\dot{\omega}}_c, \tag{13}$$

where τ is the heat release parameter defined as the ratio of temperature rise across the flame front normalised by the unburnt mixture temperature. The average value of the Favre fluctuation is obtained as $\overline{u_j''} = \tau \widetilde{u_j''c''}/(1+\tau\widetilde{c})$ (Jones, 1994).

2.1.2 Shear stress transport k-ω model

The Reynolds stress is closed using Eq. (10) in the approach also, but the eddy viscosity is obtained in a different manner as described below. The shear stress transport k-ω model proposed by Menter (1994) aims to combine the advantages in predictive capability of both the standard k-ε model in the free shear flow and the k-ω model, originally proposed by Wilcox (1988), in the near wall region. There are two important ingredients in this model. Firstly, a blending function, F_1, is used to appropriately activate the k-ε model in free shear flow part and the k-ω model in near wall region of the flow. Secondly, the definition of eddy viscosity is modified to include the effects of the principal turbulent stress transport. In this method, the turbulent kinetic energy equation is very similar to Eq. (11) and $\widetilde{\varepsilon}$ equation is replaced by a transport equation for $\widetilde{\omega} = \widetilde{\varepsilon}/\widetilde{k}$. This equation is obtained using Eqs. (11) and (12) and is written as

$$\frac{\partial \overline{\rho}\widetilde{\omega}}{\partial t} + \frac{\partial \overline{\rho}\widetilde{u}_j\widetilde{\omega}}{\partial x_j} = \frac{\partial}{\partial x_j}\left[\left(\mu + \frac{\mu_t}{Sc_\omega}\right)\frac{\partial \widetilde{\omega}}{\partial x_j}\right] - \beta\overline{\rho}\widetilde{\omega}^2 + 2\overline{\rho}(1-F_1)\sigma_{\omega 2}\frac{1}{\widetilde{\omega}}\frac{\partial \widetilde{k}}{\partial x_j}\frac{\partial \widetilde{\omega}}{\partial x_j}$$

$$-\frac{\hat{\alpha}}{\nu_t}\left(\overline{\rho u_i'' u_j''}\frac{\partial \widetilde{u}_i}{\partial x_j} + \overline{u_i''}\frac{\partial \overline{p}}{\partial x_i} - \overline{p'\frac{\partial u_i''}{\partial x_i}}\right), \tag{14}$$

where $\hat{\alpha}$, β and $\sigma_{\omega 2}$ are model constants given by Menter (1994). The turbulent eddy viscosity is obtained using

$$\nu_t = \frac{\mu_t}{\overline{\rho}} = \frac{a_1\widetilde{k}}{\max(a_1\widetilde{\omega};\ \Omega F_2)}, \tag{15}$$

where F_2 is a function taking a value of one in boundary layer and zero in free shear region of wall bounded flows, and $\Omega = \text{abs}(\nabla \times \widetilde{u})$. A number of other turbulence modelling approaches are discussed by Pope (2000) and interested readers can find the detail in there. The combustion sub-modelling is considered next.

3. Combustion sub-modelling

As noted earlier, the governing equations for thermo-chemical state of the reacting mixture can be reduced to one or two key scalars. The reaction progress variable is usually the key scalar for lean premixed flames. It is usually defined using either normalised fuel mass fraction or normalised temperature. Alternative definition using the sensible enthalpy have also been proposed in the literature (Bilger et al., 1991). Here, the progress variable is defined using temperature as $c = (T - T_u)/(T_b - T_u)$ where, T_b and T_u, are the adiabatic and unburnt temperatures respectively. A transport equation for the instantaneous progress variable can be written as (Poinsot & Veynante, 2000)

$$\frac{\partial \rho c}{\partial t} + \frac{\partial \rho c\, u_j}{\partial x_j} = \frac{\partial}{\partial x_j}\left(\rho \mathcal{D}_c \frac{\partial c}{\partial x_j}\right) + \dot{\omega}_c. \tag{16}$$

By averaging the above equation, one gets the transport equation for $\tilde{c}$ given as Eq. (9), which needs closure for the turbulent scalar flux, $\widetilde{u_j'' c''}$, and the mean reaction rate, $\overline{\dot{\omega}}_c$. Many past studies (Echekki & Mastorakos, 2011; Libby & Williams, 1994; Swaminathan & Bray, 2011, see for example) have shown that the scalar flux can exhibit counter-gradient behaviour in turbulent premixed flames. The counter-gradient flux would yield a negative turbulent diffusivity and mainly arises when the local pressure forces accelerate the burnt and unburnt mixtures differentially due to their density difference. This phenomenon is predominant in low turbulence level, when u'/s_L^0 is smaller than about 4, where thermo-chemical effects overwhelms the turbulence. Bray (2011) has reviewed the past studies on the scalar flux and has noted that the transition between gradient and non-gradient transport in complex flows deserves further investigation. However, it is quite common to model this scalar flux as a gradient transport in situations with large turbulence Reynolds number. The closure for $\overline{\dot{\omega}}_c$ in Eq. (9) is central in turbulent premixed flame modelling. Many closure models have been proposed in the past and they are briefly reviewed first before elaborating on the strained flamelet approach.

3.1 Eddy break-up model

This model proposed by Spalding (1976) is based on phenomenological analysis of scalar energy cascade in turbulent flames with $Re \gg 1$ and $Da \gg 1$. The mean reaction rate is given by $\overline{\dot{\omega}} = C_{EBU}\overline{\rho}\,\tilde{\varepsilon}\,\widetilde{c''^2}/\tilde{k}$, where C_{EBU} is the model parameter (Veynante & Vervisch, 2002). The large values of Re and Da implies that the combustion is in the flamelets regime where the flame front is thinner than the small scales of turbulence. In this regime, the variance can be written as $\widetilde{c''^2} = \tilde{c}(1 - \tilde{c})$ and the reaction is assumed to be fast. This fast chemistry assumption does not hold in many practical situations and thus this model tends to over predict the mean reaction rate. Furthermore, this model does not consider the multi-step nature of combustion chemistry. A variant of this approach, known as eddy dissipation concept, is developed with provisions to include complex chemical kinetics (Ertesvag & Magnussen, 2000; Gran & Magnussen, 1996).

3.2 Bray-Moss-Libby model

This model (Bray, 1980; Bray & Libby, 1976; Bray & Moss, 1977) uses a reaction progress variable and its statistics for thermo-chemical closure. An elaborate discussion of this modelling can be found in many books (Bray, 2011; Libby & Williams, 1994, for example). The basic assumption in this approach is that the turbulent flame front is thin and the turbulence scales do not disturb its structure. This allows one to partition the marginal probability density function (PDF) of c into three distinct portions; fresh gases $(0 \leq c \leq c_1)$, burnt mixtures $((1 - c_1) \leq c \leq 1)$ and reacting mixtures $(c_1 \leq c \leq (1 - c_1))$ with probabilities $\alpha(\mathbf{x}, t)$, $\beta(\mathbf{x}, t)$ and $\gamma(\mathbf{x}, t)$ respectively, obeying $\alpha + \beta + \gamma = 1$. In the limit of $c_1 \to 0$, the Reynolds PDF of c can be written as

$$\overline{P}(c; \mathbf{x}, t) = \alpha \delta(c) + \beta \delta(1 - c) + \gamma f(c; \mathbf{x}, t), \tag{17}$$

where the fresh gases and fully burnt mixtures are represented by the Dirac delta functions $\delta(c)$ and $\delta(1 - c)$ respectively. The progress variable is defined as $c = (T - T_u)/(T_b - T_u)$. The interior portion, $f(c; \mathbf{x}, t)$, of the PDF represents the reacting mixture and it must satisfy $\int_0^1 f(c; \mathbf{x}, t) \, dc = 1$. If the flame front is taken to be thin then $\gamma \ll 1$ when $Da \gg 1$. Under this condition, it is straightforward to obtain $\alpha = (1 - \widetilde{c})/(1 + \tau \widetilde{c})$ and $\beta = (1 + \tau) \widetilde{c}/(1 + \tau \widetilde{c})$ using $\overline{\rho} = \int \rho \overline{P} \, dc$ with $\rho/\overline{\rho} = (1 + \tau \widetilde{c})/(1 + \tau c)$. Now, the mean value of any thermo-chemical variable can be obtained simply as $\widetilde{\varphi}(\mathbf{x}) = \int \varphi(\mathbf{x}) \widetilde{P}(c; \mathbf{x}) \, dc = (1 - \widetilde{c}) \varphi_u + \widetilde{c} \varphi_b$, where the Favre PDF is given by $\widetilde{P} = \rho \overline{P}/\overline{\rho}$ and the mean density is $\overline{\rho} = \rho_u/(1 + \tau \widetilde{c})$.

Since the reaction rate is zero everywhere outside the reaction zones, its mean value,

$$\overline{\dot{\omega}}_c = \gamma \int_{c_1}^{1 - c_1} \dot{\omega}_c(c) f(c; \mathbf{x}) \, dc, \tag{18}$$

is proportional to γ. Since γ has been neglected in the above analysis, alternative means are to be devised to estimate the mean reaction rate. One method is to treat the progress variable signal as a telegraphic signal (Bray et al., 1984). This analysis yields that the mean reaction rate is directly proportional to the frequency of undisturbed laminar flame front crossing a given location in the turbulent reacting flow. More detail of this analysis and its experimental verification is reviewed by Bray & Peters (1994). In another approach, the turbulent flame front is presumed to have the structure of unstrained planar laminar flame and this approach (Bray et al., 2006) gives a model for the mean reaction rate as

$$\overline{\dot{\omega}}_c = \frac{\rho_u s_L^0}{\delta_L^*} \frac{\varepsilon \widetilde{c}(1 - \widetilde{c})}{1 + \tau \widetilde{c}}. \tag{19}$$

The symbol δ_L^* is a laminar flame thickness defined as $\delta_L^* = \int c(1 - c)/(1 + \tau c) \, dn$, where n is the distance along the flame normal. The small parameter ε, defined as $\varepsilon = 1 - \overline{c''^2}/[\widetilde{c}(1 - \widetilde{c})]$, is related to γ. This implies that one must also solve a transport equation for the Favre variance, which is given in section 4 as Eq.(30). The Favre variance of c is equal to $\widetilde{c}(1 - \widetilde{c})$ when γ is neglected and the bimodal PDF is used. This simply means that $\varepsilon = 0$ for Eq. (19) which also concurs our earlier observation on the mean reaction rate closure. The variance transport equation, Eq. (30), reduces to a second equation for $\widetilde{c}$ when γ is neglected and a reconciliation of these two transport equations led (Bray, 1979) to the conclusion that the mean reaction rate is directly proportional to the mean scalar dissipation rate, $\widetilde{\varepsilon}_c = \overline{\rho \mathcal{D}_c (\nabla c'' \cdot \nabla c'')}/\overline{\rho}$, and

$$\overline{\dot{\omega}}_c = \frac{2}{2C_m - 1} \overline{\rho} \widetilde{\varepsilon}_c, \tag{20}$$

where C_m typically varies between 0.7 and 0.8 for hydrocarbon-air flames. The mean scalar dissipation rate is an unclosed quantity and if one uses a classical model, $\widetilde{\varepsilon}_c = C_d \overline{c''^2}(\widetilde{\varepsilon}/\widetilde{k})$, for it then one gets

$$\bar{\omega}_c = \frac{2\,C_d}{2C_m - 1} \left(\frac{\tilde{\varepsilon}}{\tilde{k}}\right) \bar{\rho}\, \widetilde{c''^2}, \tag{21}$$

which is very similar to the eddy break-up model discussed in section 3.1.

3.3 Flame surface density model

In this approach (Marble & Broadwell, 1977), the mean reaction rate is expressed as the product of reaction rate per unit flame surface area, $\rho_u\, s_L$, and the flame surface area per unit volume, Σ, which is known as the flame surface density (FSD). The straining, bending, wrinkling and contortion, collectively called as stretching, of the flame surface by turbulent eddies can influence the flame front propagation speed and thus it is quite useful and usual to write $s_L = s_L^0\, I_0$, where I_0 is known as the stretch factor, which is typically of order one. The FSD approach has been studied extensively and these studies are reviewed and summarised by Veynante & Vervisch (2002) and in the books edited by Libby & Williams (1994), Echekki & Mastorakos (2011) and Swaminathan & Bray (2011). Two approaches are normally used to model Σ; in one method an algebraic expression (Bray & Peters, 1994; Bray & Swaminathan, 2006) is used and in another method a modelled differential equation is solved. A simple algebraic model proposed by Bray & Swaminathan (2006) is given as

$$\Sigma_g \delta_L^0 = \delta_L^0 \int_0^1 \Sigma(c;\, \mathbf{x})\, \mathrm{d}c = \frac{2C_{D_c}}{(2C_m - 1)}\, \frac{\bar{\rho}}{\rho_u} \left(1 + \frac{2}{3}\, \frac{C_{\varepsilon_c} s_L^0}{\sqrt{\tilde{k}}}\right) \left(1 + \frac{C_D\, \tilde{\varepsilon}\, \delta_L^0}{C_{D_c}\, \tilde{k}\, s_L^0}\right) \widetilde{c''^2}, \tag{22}$$

with the three model parameters which are of order unity. When $\mathrm{Da} \gg 1$, $C_D\, \tilde{\varepsilon}\, \delta_L^0 / (C_{D_c}\, \tilde{k}\, s_L^0) \ll 1$, suggesting that the generalised FSD scales with the laminar flame thickness rather than the turbulence integral length scale, Λ. Many earlier algebraic models discussed by Bray & Peters (1994) suggest a scaling with Λ. An algebraic FSD model has also been deduced using fractal theories by Gouldin et al. (1989).

The unclosed transport equation for FSD was derived rigorously by Candel & Poinsot (1990) and Pope (1988) and this equation is written, when there is no flame-flame interaction, as

$$\frac{\partial \Sigma}{\partial t} + \frac{\partial \tilde{u}_j \Sigma}{\partial x_j} + \frac{\partial \langle u_j'' \rangle_s \Sigma}{\partial x_j} + \frac{\partial \langle s_d n_j \rangle_s \Sigma}{\partial x_j} = \langle \Phi \rangle_s \Sigma, \tag{23}$$

where $\langle \cdots \rangle_s$ denotes the surface average. The three flux terms on the left hand side are due to the mean flow advection, turbulent diffusion and flame displacement at speed s_d. The last three terms in the above equation require modelling and usually the propagation term is neglected in the modelling, which may not hold at all situations. The turbulent diffusion is usually modelled using gradient flux hypothesis. The term on the right hand side is the source or sink term for Σ due to the effects of turbulence on the flame surface. The quantity Φ is usually called as flame stretch which is a measure of the change in the flame surface area, A, and is given by (Candel & Poinsot, 1990)

$$\Phi \equiv \frac{1}{\Delta A}\, \frac{\mathrm{d}(\Delta A)}{\mathrm{d}t} = (\delta_{ij} - n_i n_j) e_{ij} + s_d \frac{\partial n_i}{\partial x_i} = a_T + 2 s_d k_m, \tag{24}$$

where k_m is the mean curvature of the flame surface, e_{ij} is the symmetric strain tensor and n_i is the component of the flame normal in direction i. Turbulence, generally, has the tendency to increase the surface area implying that the average stretch rate is positive. However, the curvature term is negative because s_d is negatively correlated to k_m, while the tangential strain rate, a_T, is positive (Chakraborty & Cant, 2005; Trouvé & Poinsot, 1994). Thus, to have $\langle \Phi \rangle_s > 0$ the magnitude of the surface averaged tangential strain rate must be larger than or equal to the magnitude of $\langle 2s_d k_m \rangle_s$. The mean curvature of the flame surface was observed to be around zero in a number of studies (Baum et al., 1994; Chakraborty & Cant, 2004; Echekki & Chen, 1996; Gashi et al., 2005). However, fluctuation in the flame surface curvature contributes to the dissipation of the surface area. The modelling of a_T is also typically done by splitting the contributions into mean and fluctuating fields and obtaining accurate models for the various contributions to the flame surface density has been the subject of many earlier studies (Chakraborty & Cant, 2005; Peters, 2000; Peters et al., 1998). The FSD method has also been developed for LES of turbulent premixed flames and these works are summarised by Vervisch et al. (2011) and Cant (2011).

3.4 G-equation, level set approach

A smooth function G, such that $G < 0$ in unburnt mixture, $G > 0$ in burnt mixture, and $G = G_o = 0$ at the flame, is introduced for premixed combustion occurring in reaction-sheet, wrinkled flamelets, regime. A Huygens-type evolution equation can be written for the instantaneous flame element as (Kerstein et al., 1988; Markstein, 1964; Williams, 1985)

$$\frac{\partial G}{\partial t} + u_j \frac{\partial G}{\partial x_j} = s_G \left(\frac{\partial G}{\partial x_i} \frac{\partial G}{\partial x_i} \right)^{1/2}, \tag{25}$$

where s_G is the propagation speed of the flame element relative to the unburnt mixture. This propagation speed may be expressed in terms of s_L^0 corrected for stretch effects using Markstein number. This number is a measure of the sensitivity of the laminar flame speed to the flame stretch (Markstein, 1964). Peters (1992; 1999) has developed this approach for corrugated flamelets and thin reaction zones regimes of turbulent premixed combustion and also proposed (Peters, 2000) s_G expressions suited to these regimes. Using the above instantaneous equation, Peters (1999) deduced transport equations for $\tilde{G}$ and $\widetilde{G''^2}$, which can be used in RANS simulations (Herrmann, 2006, see for example). The development and use of this method for LES has been reviewed by Pitsch (2006). A close relationship between the G field and the FSD is shown by Bray & Peters (1994) as

$$\Sigma(\mathbf{x}) = \left\langle \left(\frac{\partial G}{\partial x_i} \frac{\partial G}{\partial x_i} \right)^{1/2} \middle| G = G_o \right\rangle \overline{P}(G_o; \mathbf{x}), \tag{26}$$

where $\overline{P}(G; \mathbf{x})$ is the PDF of G and an approximate expression has been proposed for $\overline{P}(G_o; \mathbf{x})$ by Bray & Peters (1994).

3.5 Transported PDF approach

In this method, a transport equation for the joint PDF of scalar concentrations is solved along with equations for turbulence quantities. The transport equation for the joint PDF has been presented and discussed by Pope (1985). The attractive aspect of this approach is that the non-linear reaction rate is closed and does not require a model. However, the molecular flux in the sample space, known as micro-mixing, needs a closure model and many models are available in the literature. These models are discussed by Haworth & Pope (2011), Lindstedt (2011) and Dopazo (1994). The micro-mixing is directly related to the conditional dissipation. This dissipation rate, for example for the progress variable, is defined as $\langle N_c | \zeta \rangle = \langle \mathcal{D} \nabla c \cdot \nabla c | \zeta \rangle$, where ζ is the sample space variable for c. The predictive ability of this method depends largely on the quality of the models used for the unclosed terms.

The joint PDF equation is of $(N + 4)$ dimensions for unsteady reacting flows in three spatial dimensions involving $(N - 1)$ reactive scalars and temperature. This high dimensionality of the PDF transport equation poses difficulties for numerical solution. Also the molecular flux term will have a negative sign which precludes the use of standard numerical approaches such as the finite difference or the finite volume methods, and the Monte-Carlo methods (Pope, 1985) are generally used to solve the PDF transport equation. In this method, the computational memory requirement depends linearly on the dimensionality (number of particles used) of problem but one needs a sufficiently large number of particles to get a good accuracy.

3.6 Presumed PDF method

The marginal PDF of the key scalars are presumed to have a known shape, which is determined usually using computed values of the first two moments, mean and variance. A Beta function is normally used and it is given by

$$\widetilde{P}(\zeta; \mathbf{x}) = \frac{\zeta^{(a-1)}(1 - \zeta)^{(b-1)}}{\hat{\beta}(a, b)}, \tag{27}$$

where a and b are related to the first and second moments of the key scalar φ, with sample space variable ζ, by

$$a = \frac{\widetilde{\varphi}^2(1 - \widetilde{\varphi})}{\widetilde{\varphi''^2}} - \widetilde{\varphi} \quad \text{and} \quad b = \frac{a(1 - \widetilde{\varphi})}{\widetilde{\varphi}}. \tag{28}$$

The normalising factor $\hat{\beta}$ is the Beta function (Davis, 1970), which is related to the Gamma function given by

$$\hat{\beta}(a, b) = \int_0^1 \zeta^{(a-1)}(1 - \zeta)^{(b-1)} \, d\zeta = \frac{\Gamma(a)\Gamma(b)}{\Gamma(a + b)}. \tag{29}$$

This presumed form provides an appropriate range of shapes: if a and b approach zero in the limit of large variance then the PDF resembles a bimodal shape of the BML PDF in section 3.2, which requires only the mean value, $\widetilde{\varphi}$. In the limit of small variance, Eq. (27) develops a mono-modal form with an internal peak and it has been shown by Girimaji (1991) that this PDF behaves likes the Gaussian when the variance is very small.

The variance equation is written as

$$\frac{\partial \overline{\rho} \widetilde{c''^2}}{\partial t} + \frac{\partial \overline{\rho} \widetilde{u}_j \widetilde{c''^2}}{\partial x_j} = \frac{\partial}{\partial x_j} \left[\left(D_c + \frac{\mu_t}{Sc_c} \right) \frac{\partial \widetilde{c''^2}}{\partial x_j} \right] - 2\overline{\rho} \widetilde{u_i'' c''} \frac{\partial \widetilde{c}}{\partial x_i} - 2\overline{\rho} \, \widetilde{\varepsilon}_c + 2\overline{\dot{\omega}_c'' c''}. \tag{30}$$

using the standard notations. The contributions from the scalar flux, dissipation rate and the chemical reactions denoted respectively by the last three terms in the above equation, need to be modelled. The modelling of scalar dissipation rate is addressed in many recent studies, which are discussed by Chakraborty et al. (2011). This quantity is typically modelled using turbulence time scale as has been noted in section 3.2 but, this model is known to be inadequate for premixed and partially premixed flames. A recent proposition by Kolla et al. (2009) includes the turbulence as well as laminar flame time scales and this model is given by

$$\tilde{\varepsilon}_c = \frac{1}{\beta'} \left[(2K_c^* - \tau C_4) \frac{s_L^0}{\delta_L^0} + C_3 \frac{\tilde{\varepsilon}}{k} \right] \widetilde{c''^2}, \tag{31}$$

where, β' is 6.7 and K_c^*/τ is 0.85 for methane-air combustion. The parameters C_3 and C_4 are given by

$$C_3 = \frac{1.5\sqrt{\text{Ka}}}{1 + \sqrt{\text{Ka}}} \quad \text{and} \quad C_4 = 1.1(1 + \text{Ka})^{-0.4}. \tag{32}$$

The Karlovitz number, Ka, is defined as

$$\text{Ka} \equiv \frac{t_c}{t_\eta} \simeq \frac{\delta/s_L^0}{\sqrt{\nu/\tilde{\varepsilon}}}, \tag{33}$$

where t_c is the chemical or flame time scale defined earlier, t_η is the Kolmogorov time scale and ν is the kinematic viscosity. The Zeldovich thickness is related to the thermal thickness by $\delta_L^0/\delta \approx 2(1 + \tau)^{0.7}$.

In the presumed PDF approach, the reaction rate is closed as

$$\overline{\dot{\omega}}_c(\mathbf{x}) = \int \dot{\omega}(c) \, \overline{P}(\tilde{c}, \widetilde{c''^2}; \mathbf{x}) \, dc \tag{34}$$

for premixed flames and the function $\dot{\omega}(c)$ is obtained using freely propagating laminar flame having the same thermo-chemical attributes as that of the turbulent flame. In this approach, it is implicit that the flame structure is undisturbed by the turbulent eddies. For partially premixed flames, one can easily extend the above model by including the dependence of the reaction rate on the mixture fraction as $\dot{\omega}(c, Z)$ and thus one must integrate over c and Z space to get the mean reaction rate after replacing the marginal PDF by the joint PDF, $\overline{P}(c, Z)$. In this modelling practice, these two variables are usually taken to be statistically independent. More work is required to address the statistical dependence of Z and c, and its modelling. A closure model for the effects of chemical reaction on the variance transport, see Eq. (30), can be written as

$$\overline{\dot{\omega}''c''} = \int_0^1 \dot{\omega}(c) \, c \, \overline{P}(c, \mathbf{x}) \, dc - \tilde{c} \int_0^1 \dot{\omega}(c) \, \overline{P}(c, \mathbf{x}) \, dc. \tag{35}$$

As noted earlier, the closure in Eq. (34) assumes that the laminar flame structure is undisturbed by the turbulent eddies. The influence of fluid dynamic stretch can also be included in this approach as suggested by Bradley (1992) using

$$\overline{\dot{\omega}}_c = \int \int \hat{\dot{\omega}}_c(\zeta, \kappa) \, d\zeta \, d\kappa = \int \int \dot{\omega}_c(\zeta) \, f(\kappa) \, d\zeta \, d\kappa = P_b \int \dot{\omega}_c(\zeta) \, d\zeta, \tag{36}$$

where P_b is the burning rate factor, which can be expressed in terms of Markstein number (Bradley et al., 2005). Recently, Kolla & Swaminathan (2010a) proposed to use the scalar dissipation rate to characterise the stretch effects on flamelets for the following reasons, viz., (i) the chemical reactions produce the scalar gradient and thus the scalar dissipation rate in premixed and stratified flames and (ii) this quantity signifies the mixing rate between hot and cold mixtures, which are required to sustain combustion in premixed and partially premixed flames. This method is elaborated by Kolla & Swaminathan (2010a) and briefly reviewed in the next section.

4. Strained flamelet model

The closure for the mean reaction rate is given as (Kolla & Swaminathan, 2010a)

$$\overline{\dot{\omega}} = \int_{N_{c1}}^{N_{c2}} \int_0^1 \dot{\omega}_c(\zeta, \psi)\, \overline{P}(\zeta, \psi)\, d\zeta\, d\psi = \int_0^1 \langle \dot{\omega}|\zeta\rangle\, \overline{P}(\zeta)\, d\zeta, \tag{37}$$

where ψ is the sample space variable for the instantaneous scalar dissipation rate, $N_c = \mathcal{D}(\nabla c \cdot \nabla c)$, of the progress variable. The marginal PDF, $\overline{P}(\zeta)$ is obtained from Eq. (27) using the computed values of $\tilde{c}$ and $\widetilde{c''^2}$ from Eqs. (9) and (30). The conditionally averaged reaction rate is given by

$$\langle \dot{\omega}|\zeta\rangle = \int_{N_{c1}}^{N_{c2}} \dot{\omega}(\zeta, \psi)\, \overline{P}(\psi|\zeta)\, d\psi. \tag{38}$$

The conditional PDF $P(\psi|\zeta)$ is presumed to be log-normal and $\dot{\omega}(\zeta, \psi)$ is obtained from calculations of strained laminar flames established in opposed flows of cold reactant and hot products. This flamelet configuration seems more appropriate to represent the local scenario in turbulent premixed flames (Hawkes & Chen, 2006; Libby & Williams, 1982). The log-normal PDF is given by

$$\overline{P}(\psi|\zeta) = \frac{1}{(\psi|\zeta)\sigma\sqrt{2\pi}} \exp\left\{ \frac{-[\ln(\psi|\zeta) - \hat{\mu}]^2}{2\sigma^2} \right\}. \tag{39}$$

The mean, $\hat{\mu}$, and variance, σ^2, of $\ln(\psi|\zeta)$ are related to conditional mean $\langle N|\zeta\rangle$ and variance of the scalar dissipation rate G_N^2 via $\langle N|\zeta\rangle = \exp(\hat{\mu} + 0.5\sigma^2)$ and $G_N^2 = \langle N|\zeta\rangle^2 [\exp(\sigma^2) - 1]$. The conditional mean of scalar dissipation rate is related to the unconditional mean through

$$\langle N|\zeta\rangle \approx \frac{\tilde{\epsilon}_c f(\zeta)}{\int_0^1 f(\zeta)\tilde{P}(\zeta)d\zeta}, \tag{40}$$

where $f(\zeta)$ is the variation of N_c normalised by its value at the location of peak heat release rate in unstrained planar laminar flame. A typical variation of $f(\zeta)$ is shown in Fig. 2. It has been shown by Kolla & Swaminathan (2010a) that $f(\zeta)$ is weakly sensitive to the stretch rate for ζ values representing intense chemical reactions and $f(\zeta)$ has got some sensitivity to the stretch rate in thermal region of the flamelet. Despite this, the variation shown in Fig. 2 is sufficiently accurate for turbulent premixed flame calculation and it must be also be noted that $f(\zeta)$ will strongly depend on the thermo-chemical conditions of the flamelet. The unconditional mean dissipation rate, $\tilde{\epsilon}_c$, is modelled using Eq. (31) and it is to be noted that the model parameters and their numerical values are introduced to represent the correct physical behaviour of $\tilde{\epsilon}_c$ in various limits of turbulent combustion and thus they are not arbitrary. The robustness of this model for $\tilde{\epsilon}_c$ has been shown in earlier studies (Darbyshire et al., 2010; Kolla et al., 2009; 2010; Kolla & Swaminathan, 2010a;b). The influence of Lewis number on this modelling is also addressed in a previous study (Chakraborty & Swaminathan, 2010).

The mean reaction rate can now be obtained using Eq. (37) for given values of $\tilde{c}$, $\widetilde{c''^2}$ and $\tilde{\epsilon}_c$ and thus a three dimensional look up table can be constructed for use during turbulent flame calculations. However, care must be exercised to cover a range of fully burning flamelets to a nearly extinguished one. Such a turbulent flame calculation has been reported recently (Kolla & Swaminathan, 2010b) and this study aims to implement this approach in a complex, commercial type, CFD code and validate it by comparing the simulation results to the previously published results.

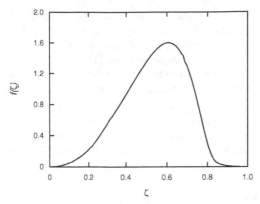

Fig. 2. Typical variation of $f(\zeta)$ with ζ from unstrained planar laminar flame. The curve is shown for stoichiometric methane-air combustion at atmospheric conditions.

4.1 Partially premixed flame modelling

Partially premixed flame occurs if the fuel and oxidiser were mixed unevenly. As a result, the mixture fraction Z is required to describe the local mixture composition and it is defined following Bilger (1988) as

$$Z = \frac{2Z_C/W_C + 0.5Z_H/W_H + (Z_O^{ox} - Z_O)/W_O}{2Z_C^f/W_C + 0.5Z_H^f/W_H + Z_O^{ox}/W_O}, \qquad (41)$$

where Z_i is the mass fraction of element i with atomic mass W_i. The superscripts f and ox refer to reference states of the fuel and oxidiser respectively. The subscripts C, H and O refer to carbon, hydrogen and oxygen.

The transport equations for the Favre mean mixture fraction, $\widetilde{Z}$, given as Eq. (8), and its variance $\widetilde{Z''^2}$, given by

$$\frac{\partial \overline{\rho}\widetilde{Z''^2}}{\partial t} + \frac{\partial \overline{\rho}\widetilde{u}_j\widetilde{Z''^2}}{\partial x_j} = \frac{\partial}{\partial x_j}\left[\left(D_Z + \frac{\mu_t}{Sc_Z}\right)\frac{\partial \widetilde{Z''^2}}{\partial x_j}\right] - 2\overline{\rho u_i''Z''}\frac{\partial \widetilde{Z}}{\partial x_i} - 2\overline{\rho}\widetilde{\epsilon}_Z, \qquad (42)$$

are usually solved in the presumed PDF approach to obtain the local mixing related information. The turbulent flux of the variance, $\widetilde{u_i''Z''^2}$, is expressed using gradient hypothesis in the above equation. The scalar dissipation rate is modelled by assuming a constant ratio of turbulence to scalar time scales and this model is given as

$$\widetilde{\epsilon}_Z = C_\xi \left(\frac{\widetilde{\epsilon}}{\widetilde{k}}\right) \widetilde{Z''^2}, \qquad (43)$$

where C_ξ is a model constant and it is typically unity. The strained flamelet modelling discussed in the previous section can be extended to turbulent partially premixed flame by considering this flame as an ensemble of strained premixed flamelets with mixture fraction ranging from the lean to rich flammability limits. Then, the mean reaction rate can be written as

$$\overline{\dot{\omega}} = \int \int \int \dot{\omega}(\zeta, \psi, \xi)\, \overline{P}(\zeta, \psi, \xi)\, d\zeta\, d\psi\, d\xi, \qquad (44)$$

where ζ is the sample space variable for Z. Modelling the joint PDF $\overline{P}(\zeta, \psi, \xi)$ is a challenging task and as a first approximation, it is common to assume that ζ and ξ are statistically independent leading to $\overline{P}(\zeta, \psi, \xi) = \overline{P}(\zeta, \psi)\overline{P}(\xi)$. One can then follow the approach suggested by Eq. (37) to model $\overline{P}(\zeta, \psi)$. The marginal PDF, $\overline{P}(\xi)$, can be obtained using a Beta PDF. Now, the flamelet library will have five controlling parameters instead of three for the purely premixed case. The assumption of statistical independence of Z and c may not be valid and can be easily removed by including the covariance, $\widetilde{Z''c''}$, by solving its transport equation. However, no reliable modelling is available yet to close this equation.

One can also extend the unstrained flamelet model, given in Eq. (34), to partially premixed flames by including the mixture fraction in this equation. This is same as Eq. (44) after removing ψ and the associated integral from it.

4.1.1 Assessment using DNS data

A priori assessment of the unstrained and strained flamelet modelling for partially premixed flames is discussed in this subsection. The DNS data for a hydrogen turbulent jet lifted flame (Mizobuchi et al., 2002) is used for this analysis. The time averaged reaction rate $\overline{\omega}_c$ of the progress variable c, which is defined using the equilibrium value of H_2O mass fraction as $c = Y_{H_2O}/Y_{H_2O}^{eq}$ has been extracted from the DNS data. Figure 3 presents the radial variation of $\overline{\omega}_c$ at two axial positions. The radial distance, r, is normalised using the fuel jet nozzle diameter d. The values of $\overline{\omega}_c$ computed using the unstrained and strained flamelet modellings are also shown in this figure. The means and variances required to construct the PDFs are obtained from the DNS data for this analysis. In figure 3(a), it is clear that unstrained flamelet model agrees well with the DNS results in the region close to the centreline, but it generally over predicts the mean reaction rate for $r/d > 1$. The strained flamelet model under predicts $\overline{\omega}_c$ for $r/d < 2.5$ while giving a good agreement for $r/d > 2.5$. At a downstream location, figure 3(b) shows a similar trend where unstrained flamelet model over predicts $\overline{\omega}_c$ while strained flamelet model gives a reasonable agreement. It is likely that the strain effects are important and need to be included to give correct mean reaction rate depending on axial and radial positions and unstrained flamelet model is insufficient. A note of caution is that the strained flamelet model used in this assessment only includes the strain effects for rich mixture and it is constructed with only 12 rich flamelets up to the fuel rich flammability limit. Further work needs to be done to include the strain effects for lean flamelets and to examine the effects of using more than 12 fuel rich flamelets. Exploring a way to combine the unstrained and strained flamelets for partially premixed flames in a unified modelling framework need to be addressed. Whether this approach would be sufficient or a completely different approach would be required, is an open question. Also, the cross dissipation, $\widetilde{\epsilon}_{cZ} = \overline{\rho \mathcal{D}(\nabla Z'' \cdot \nabla c'')}/\overline{\rho}$, can play important role in these kind of closure modelling for partially premixed flames (Bray et al., 2005). It is clear that more works need to be done for partially premixed flames.

4.2 Model implementation

The implementation of the strained flamelet model into a commercial CFD software (for example, FLUENT), which can handle complex geometries that are common in industrial scenarios are discussed in this section. The flow and turbulence models available in the software are utilised to provide the required information for combustion calculation. Additional transport equations for $\widetilde{Z}$, $\widetilde{Z''^2}$, $\widetilde{c}$ and $\widetilde{c''^2}$ are included as user defined scalars (UDS). A transport equation for total enthalpy, $\widetilde{h}$, is also included to obtain spatial temperature distribution using Eq. (7). Various sources and sinks appearing in these transport equations,

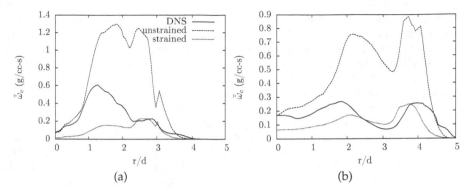

Fig. 3. Comparison of radial variation for mean reaction rate for DNS (solid), unstrained flamelet model (dash) and strained flamelet model (dotted) at two axial positions.

Eqs. (8), (9), (30) and (42), are calculated using user defined functions (UDFs) and the turbulent transports are modelled using gradient hypothesis. The mean density is obtained using the equation of state. Some discrepancies in the modelling of the Reynolds stress in the Fluent for the mean momentum and the k-ε equations were noted and corrected for the results reported in this study. Instructions to incorporate the UDS transport equations and UDFs are provided in the theory and user guides of FLUENT.

It is to be noted that the choice of the progress variable is guided broadly by the flame configuration. Progress variable definitions based on temperature or species mass fraction are popular choices for most premixed combustion calculations as noted earlier in this chapter. These choices are equally applicable for open as well as enclosed flames. However, if there are heat losses to the boundary then the fuel or product mass fractions can be used to define the progress variable. A prudent decision on the choice of the progress variable can go a long way in obtaining accurate CFD predictions of flame related quantities.

Once an appropriate progress variable is chosen, the flamelets reaction rate, $\dot{\omega}_c(\zeta, \psi)$, is calculated using unstrained and strained laminar flames. An arbitrarily complex chemical kinetics mechanism can be used for these calculations and GRI-3.0 is used for the flames computed and discussed in this chapter. The PREMIX and OPPDIF suites of Chemkin software is used for the flamelet computations. As noted earlier, reactant-to-product configuration is used for the OPPDIF calculations. These flamelets reaction rates are then used in Eq. (37) to obtain the mean reaction rate, $\bar{\omega}$, as explained in section 4. This mean reaction rate, $\overline{\omega''c''}$ (see Eq. 35), $\overline{c_p}$, Δh_f^0 for the mixture, and $\tilde{Y}_i$ are tabulated for $0 \leq \tilde{c} \leq 1$, $0 \leq g \leq 1$ and $\tilde{\varepsilon}_{c,min} \leq \tilde{\varepsilon}_c \leq \tilde{\varepsilon}_{c,max}$, where $g = \widetilde{c''2}/[\tilde{c}(1-\tilde{c})]$. These tabulated values are read during a CFD calculation for the computed values of $\tilde{c}$, $\widetilde{c''2}$ and $\tilde{\varepsilon}_c$ in each computational cell. The converged fields of these three quantities can then be used to obtain the species mass fractions, $\tilde{Y}_i$, from the tables as a post-processing step. For a purely premixed flames, there is no need to solve for $\tilde{Z}$ and $\widetilde{Z''2}$ and for partially premixed flames one must solve for these two quantities.

5. Sample results

Pilot stabilised Bunsen flames (Chen et al., 1996) of stoichiometric methane-air mixture with three Reynolds number, based on bulk mean jet velocity and nozzle diameter, of 52000, 40000, and 24000 designated respectively as F1, F2 and F3 have been computed using the strained flamelet model. These flames are in the thin reaction zones regime of turbulent combustion. As noted earlier, the implementation of the strained flamelet model in a commercial CFD software is validated here by comparing the results with those published in an earlier study (Kolla & Swaminathan, 2010b). It is to be noted that the earlier study used a research type CFD code with HLPA (hybrid linear parabolic approximation) discretisation schemes (Zhu, 1991) and TDMA solver. A pressure correction based technique was used in that study while the current study uses a density based method with Roe scheme (Roe, 1981) and a second order accurate upwind discretisation scheme available in Fluent. The model constant $C_{\varepsilon 1}$ in the k-ε model is changed from its standard value of 1.44 to 1.52 to correct for the round jet anomaly (Pope, 1978). The turbulent Schmidt numbers for the scalar transport equations are taken to be unity and the turbulent Prandtl number for the enthalpy equation is 0.7 for this study. Mean axial velocity profiles at the nozzle exit measured and reported by Chen et al. (1996) are used as the boundary condition for the inlet velocity. The profiles of RMS of turbulent velocity fluctuations along with longitudinal length scale reported in the experimental study are used to specify the boundary conditions for $\bar{k}$ and $\tilde{\varepsilon}$. The numerical values for the various model parameters for turbulence and combustion models, turbulent Schmidt and Prandtl numbers, and the boundary conditions used in this study are consistent with those used by Kolla & Swaminathan (2010b).

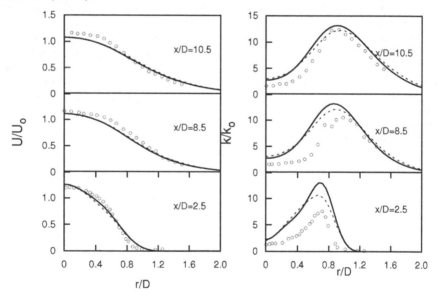

Fig. 4. The normalised mean axial velocity and turbulent kinetic energy in flame F1. Fluent (—) results are compared with experimental data (o) of Chen et al. (1996) and previously published results (- -) of Kolla & Swaminathan (2010b).

The computational results for the F1 and F3 flames are compared with previously published results along with experimental measurements here. Figure 4 shows the normalised mean axial velocity and turbulent kinetic energy with radial distance for three axial locations for the flame F1. The distances are normalised by the nozzle diameter D.

The mean axial velocity profiles computed in this study compare well with the experimental measurements and previously published values. The centreline values of $\tilde{k}/k_0$ from the Fluent simulation are slightly over predicted compared to the experimental values, but they are almost the same as the values computed by Kolla & Swaminathan (2010b). However, the Fluent calculation over predicts the shear generation of turbulent kinetic energy, which is evident in the higher peak values seen in Fig. 4. The results shown here are grid independent and the differences between the Fluent and previous results are acceptable, given the difference in the numerical schemes, solutions methods and solvers used.

Radial variation of the normalised Reynolds mean temperature, $\bar{c} = (\overline{T} - T_u)/(T_b - T_u)$ and the fuel mass fraction are plotted for three axial locations in Fig. 5. The Fluent results are compared to the calculations of Kolla & Swaminathan (2010b) and the experimental measurements of Chen et al. (1996). The mean methane mass fraction shows good comparison with experimental data for flame F1 and the centreline values computed using Fluent agree with previously published values indicating that the flame length is predicted accurately. The computed values of peak mean temperatures at $x/D = 2.5$ is consistently higher than the experimental measurements, However, the peak mean temperature agrees well for $x/D = 8.5$ and is under predicted for $x/D = 10.5$. These trends are consistent with those reported by Kolla & Swaminathan (2010b). Note that the Fluent solution predicts a higher rate of turbulent diffusion of mean progress variable $\bar{c}$, which results in lower mean temperature for $r/D > 1.0$. This could be explained by the higher peak values of turbulence quantities computed by Fluent. Higher values of turbulence would result in increase in the turbulent diffusivity, thus increasing the rate of turbulent diffusion of passive scalars. Also, the Fluent code seem to over predict the rate at which the ambient air is entrained into the reacting jet compared to the solution of Kolla & Swaminathan (2010b).

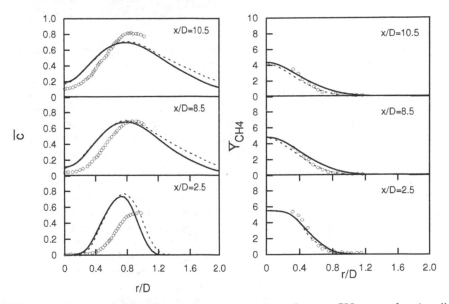

Fig. 5. The variation of normalised mean temperature, $\bar{c}$, and mean CH_4 mass fraction (in %) with r/D in flame F1; — Fluent results, o experimental data (Chen et al., 1996), - - published results of (Kolla & Swaminathan, 2010b).

Figure 6 shows the typical values of $\bar{c}$ and $\overline{Y}_{CH_4}$ computed using Fluent for flame F3 at $x/D = 8.5$ as a function of normalised radial distance, r/D. The experimental measurements and the results computed earlier are also shown for comparison. This flame has lower Reynolds and Karlovitz numbers compared to F1 and hence thermo-chemical effects are dominant compared to turbulence effects. The experimental data clearly shows that the peak value of mean temperature in F3 is larger compared to F1 (cf. Figs. 5 and 6). The relative role of the turbulence and thermo-chemistry is supposed to be naturally captured by the scalar dissipation rate based modelling of turbulent premixed flames, which is reflected well in the results shown in Fig. 6. There is some under prediction of the mean temperature in the calculations using Fluent compared to the previous results, which is due to, as noted earlier, over prediction of the entrainment rate. However, the agreement is good for $r/D \leq 1.0$ and the trend is captured correctly for $r/D > 1$ for the flame F3.

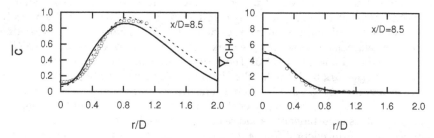

Fig. 6. The radial variation of $\bar{c}$ and $\overline{Y}_{CH_4}$ (in %) in flame F3; Fluent results (—), experimental measurements (o) and previously published results (- -).

6. Summary and future scope

In this chapter, a brief overview of various combustion modelling approaches to simulate lean premixed and partially premixed flames is given. The focus is limited to RANS framework because of its high usage in industry currently. The strained flamelet formulation developed recently is discussed in some detail and important details in implementing this model into a commercial CFD code are discussed. The results obtained for pilot stabilised turbulent Bunsen flames using Fluent with strained flamelet model are compared to experimental measurements and earlier CFD results. These published CFD results (Kolla & Swaminathan, 2010b) are obtained using another CFD code employing different numerical schemes and solver methodologies. A good comparison among the Fluent and previous CFD results and the experimental measurements is observed. These comparisons, gives good confidence on the implementation of the strained flamelets model and the associated source and sink terms in the commercial CFD code, Fluent. This initial work serves as a foundation for further studies of lean premixed, partially premixed combustion in industry relevant combustor geometries and, turbulence and thermo-chemical conditions using this modelling framework. Also, this implementation provides opportunities to study self induced combustion oscillations, interaction of flame and sound, interaction of flame generated sound waves with combustor geometries, etc., since a compressible formulation is used in the implementation. The influence of non-unity Lewis number on this combustion modelling framework is yet to be addressed.

7. Acknowledgements

The financial supports from Mitsubishi Heavy Industries, Japan, UET, Lahore, Pakistan and CCT, Cambridge, UK are acknowledged. Dr. Yasuhiro of JAXA is acknowledged for making the DNS data available through Cambridge-JAXA collaborative research programme and a part of this work was conducted under this programme.

8. References

Anastasi, C., Hudson, R. & Simpson, V. J. (1990). Effects of future fossil fuel use on CO_2 levels in the atmosphere, *Energy Policy* pp. 936–944.

Baum, M., Poinsot, T. J., Haworth, D. C. & Darabiha, N. (1994). Direct numerical simulation of $H_2/O_2/N_2$ flames with complex chemistry in two-dimensional turbulent flows, *J. Fluid Mech.* 281: 1–32.

Bilger, R. W. (1988). The structure of turbulent nonpremixed flames, *Proc. Combust. Inst.* 22: 475–488.

Bilger, R. W., Esler, M. B. & Starner, S. H. (1991). Reduced kinetic mechanisms and asymptotic approximations for methane-air flames, *in* M. D. Smooke (ed.), *On reduced mechanisms and for methane-air combustion*, Springer-Verlag, Berlin, Germany.

Bradley, D. (1992). How fast can we burn?, *Proc. Combust. Inst* 24: 247–262.

Bradley, D., Gaskell, P. H., Gu, X. J. & Sedaghat, A. (2005). Premixed flamelet modelling: Factors influencing the turbulent heat release rate source term and the turbulent burning velocity, *Combust. Flame* 143: 227–245.

Bray, K. N. C. (1979). The interaction between turbulence and combustion, *Seventeenth Symposium (International) on Combustion* pp. 223–333.

Bray, K. N. C. (1980). Turbulent flows with premixed reactants, *in* P. A. Libby & F. A. Williams (eds), *Turbulent Reacting flows*, Springer-Verlag, Berlin, pp. 115–183.

Bray, K. N. C. (2011). Laminar flamelets and the Bray, Moss, and Libby model, *in* N. Swaminathan & K. N. C. Bray (eds), *Turbulent premixed flames*, Cambridge University Press, Cambridge, UK, pp. 41–60.

Bray, K. N. C., Champion, M., Libby, P. A. & Swaminathan, N. (2006). Finite rate chemistry and presumed PDF models for premixed turbulent combustion, *Combust. Flame* 146: 665–673.

Bray, K. N. C., Domingo, P. & Vervisch, L. (2005). Role of the progress variable in models for partially premixed tubulent combustion, *Combust. Flame* 141: 431–437.

Bray, K. N. C. & Libby, P. A. (1976). Interaction effects in turbulent premixed flames, *Phys. Fluids* 19: 1687–1701.

Bray, K. N. C., Libby, P. A. & Moss, J. B. (1984). Flamelet crossing frequencies and mean reaction rates in premixed turbulent combustion, *Combust. Sci. Tech.* 41: 143–172.

Bray, K. N. C. & Moss, J. B. (1977). A unified statistical model of the premixed turbulent flame, *Acta Astronautica* 4: 291–319.

Bray, K. N. C. & Peters, N. (1994). Laminar flamelets in turbulent flames, *in* P. A. Libby & F. A. Williams (eds), *Turbulent Reacting flows*, Academic Press, London, pp. 63–113.

Bray, K. N. C. & Swaminathan, N. (2006). Scalar dissipation and flame surface density in premixed turbulent combustion, *C. R. Mecanique* 334: 466–473.

Candel, S. M. & Poinsot, T. J. (1990). Flame stretch and the balance equation for the flame area, *Combust. Sci. Tech.* 70: 1–15.

Cant, R. S. (2011). RANS and LES modelling of premixed turbulent combustion, *in* T. Echekki & E. Mastorakos (eds), *Turbulent combustion modelling*, Springer-Verlag, Heidelberg, Germany, pp. 63–90.

Carbon Dioxide Information Analysis Center (2007). Carbon Dioxde Emission Analysis, *http://cdiac.ornl.gov/*.

Chakraborty, N. & Cant, R. S. (2004). Unsteady effects of strain rate and curvature on turbulent premixed flames in an inflow-outflow configuration, *Combust. Flame* 137: 129–147.

Chakraborty, N. & Cant, R. S. (2005). Effects of strain rate and curvature on surface density function transport in turbulent premixed flames in the thin reaction zones regime, *Phys. Fluids* 17(6): 065108.

Chakraborty, N., Champion, M., Mura, A. & Swaminathan, N. (2011). Scalar dissipation rate approach, *in* N. Swaminathan & K. N. C. Bray (eds), *Turbulent premixed flames*, Cambridge University Press, Cambridge, UK, pp. 74–102.

Chakraborty, N. & Swaminathan, N. (2010). Efects of Lewis number on scalar dissipation transport and its modelling in turbulent premixed combustion, *Combus. Sci. Tech.* 182: 1201–1240.

Chen, Y. C., Peters, N., Schneeman, G. A., Wruck, N., Renz, U. & Mansour, M. S. (1996). The detailed flame structure of highly stretched turbulent-premixed methane-air flames, *Combust. Flame* 107: 223–244.

Darbyshire, O. R., Swaminathan, N. & Hochgreb, S. (2010). The effects of small-scale mixing models on the prediction of turbulent premixed and stratified combustion, *Combus. Sci. Tech.* 182: 1141–1170.

Davis, P. J. (1970). Gamma functions and related functions, *in* M. Abramowitz & I. A. Stegun (eds), *Handbook of mathematical functions*, Dover Publications Inc., New York.

Dopazo, C. (1994). Recent developments in pdf methods, *in* P. A. Libby & F. A. Williams (eds), *Turbulent Reacting flows*, Academic Press, London, pp. 375–474.

Driscoll, J. F. (2008). Turbulent premixed combustion: Flamelet structure and its effect on turbulent burning velocities, *Prog. Energy Combust. Sci.* 34: 91–134.

Echekki, T. & Chen, J. H. (1996). Unsteady strain rate and curvature effects in turbulent premixed methane-air flames, *Combust. Flame* 106: 184–202.

Echekki, T. & Mastorakos, N. (2011). *Turbulent combustion modelling*, Springer-Verlag, Heidelberg, Germany.

Ertesvag, I. S. & Magnussen, B. F. (2000). The eddy dissipation turbulence energy cascade model, *Combust. Sci. Tech.* 159: 213–235.

Gashi, S., Hult, J., Jenkins, K. W., Chakraborty, N., Cant, R. S. & Kaminski, C. F. (2005). Curvature and wrinkling of premixed flame kernels - comparisons of OH PLIF and DNS data, *Proc. Combust. Inst.* 30: 809–817.

Girimaji, S. S. (1991). Assumed β-pdf model for turbulent mixing: Validation and extension to multiple scalar mixing, *Combust. Sci. Technol.* 78: 177–196.

Gouldin, F., Bray, K. N. C. & Chen, J. Y. (1989). Chemical closure model for fractal flamelets, *Combust. Flame* 77: 241–259.

Gran, I. R. & Magnussen, B. F. (1996). A numerical study of a bluff-body stabilized diffusion flame. part 2. influence of combustion modeling and finite-rate chemistry, *Combust. Sci. Tech.* 119: 191–217.

Hawkes, E. R. & Chen, J. H. (2006). ??, *Combust. Flame* 144: 112–125.

Haworth, D. C. & Pope, S. B. (2011). Transported probability density function methods for Reynolds-Averaged and Large-Eddy simulations, *in* T. Echekki &

E. Mastorakos (eds), *Turbulent combustion modelling*, Springer-Verlag, Heidelberg, Germany, pp. 119–142.

Herrmann, M. (2006). Numerical simulation of turbulent Bunsen flames with a level set flamelet model, *Combust. Flame* 145: 357–375.

Heywood, J. B. (1976). Pollutant formation and control in spark ignition engines, *Prog. Energy Combust. Sci.* 1: 135–164.

Jones, W. P. (1994). Turbulence modelling and numerical solution methods for variable density and combusting flows, *in* P. A. Libby & F. A. Williams (eds), *Turbulent Reacting flows*, Academic Press, London, pp. 309–374.

Kerstein, A. R., Ashurst, W. T. & Williams, F. A. (1988). Field equation for interface propagation in an unsteady homogeneous flow field, *Phys. Rev. A* 37: 2728–2731.

Kolla, H., Rogerson, J. W., Chakraborty, N. & Swaminathan, N. (2009). Scalar dissipation rate modeling and its validation, *Combus. Sci. Tech.* 181: 518–535.

Kolla, H., Rogerson, J. W. & Swaminathan, N. (2010). Validation of a turbulent flame speed model across combustion regimes, *Combus. Sci. Tech.* 182: 284–308.

Kolla, H. & Swaminathan, N. (2010a). Strained flamelets for tubulent premixed flames,I: Formulation and planar flame results, *Combust. Flame* 157: 943–954.

Kolla, H. & Swaminathan, N. (2010b). Strained flamelets for tubulent premixed flames,II: Laboratory flame results, *Combust. Flame* 157: 1274–1289.

Libby, P. A. & Williams, F. A. (1982). ??, *Combust. Flame* 44: 287–303.

Libby, P. A. & Williams, F. A. (1994). Fundamentals aspects and a review, *in* P. A. Libby & F. A. Williams (eds), *Turbulent Reacting flows*, Academic Press, London, pp. 1–61.

Lindstedt, P. (2011). Transported probability density function methods, *in* N. Swaminathan & K. N. C. Bray (eds), *Turbulent premixed flames*, Cambridge University Press, Cambridge, UK, p. ??

Marble, F. E. & Broadwell, J. E. (1977). The coherent flame model for turbulent chemical reactions, *Technical report*, TRW-9-PU.

Markstein, G. H. (1964). *Non-steady flame propagation*, Pergamon.

Menter, F. R. (1994). Two-equation eddy-viscosity turbulence models for engineering applications, *AIAA J.* 32(8): 1598–1605.

Mizobuchi, Y., Tachibana, S., Shinio, J., Ogawa, S. & Takeno, T. (2002). A numerical analysis of the structure of a turbulent hydrogen jet lifted flame, *Proc. Combust. Inst.* 29: 2009–2015.

Peters, N. (1992). A spectral closure for premixed turbulent combustion in the flamelet regime, *J. Fluid Mech.* 242: 611–629.

Peters, N. (1999). The turbulent burning velocity for large-scale and small-scale turbulence, *J. Fluid Mech.* 384: 107–132.

Peters, N. (2000). *Turbulent Combustion*, Cambridge University Press, Cambridge, UK.

Peters, N., Terhoeven, P., Chen, J. H. & Echekki, T. (1998). Statistics of flame displacement speeds from computations of 2-D unsteady methane-air flames, *Proc. Combust. Inst.* 27: 833–839.

Pitsch, H. (2006). Large-eddy simulation of turbulent combustion, *Annu. Rev. Fluid Mech.* 38: 453–482.

Poinsot, T. & Veynante, D. (2000). *Theoretical and Numerical Combustion*, Edwards, Philadelphia.

Pope, S. B. (1978). An explanation of the turbulent round-jet/plane-jet anomaly, *AIAA J.* 16(3): 279–281.

Pope, S. B. (1985). PDF methods for turbulent reactive flows, *Prog. Energy Combust. Sci.* 11: 119–192.

Pope, S. B. (1988). The evolution of surfaces in turbulence, *Int. J. Eng. Sci.* 26: 445.

Pope, S. B. (2000). *Turbulent flows*, Cambridge University Press, Cambridge, UK.

Roe, P. L. (1981). Approximate riemann solvers, parameter vectors and difference schemes, *J. Comp. Phys.* 43: 357–372.

Sawyer, R. F. (2009). Science based policy for addressing energy and environmental problems, *Proc. combust. Inst.* 32: 45–56.

Spalding, D. B. (1976). Development of Eddy-Break-Up model for turbulent combustion, *16th Symposium (International) on Combustion*, The Combustion Institute, Pittsburgh.

Swaminathan, N. & Bray, K. N. C. (2011). *Turbulent Premixed Flames*, Cambridge University Press, Cambridge, UK.

Trouvé, A. & Poinsot, T. (1994). The evolution equation for the flame surface density in turbulent premixed combustion, *J. Fluid Mech.* 278: 1–31.

US Department of Commerce (2011). NOAA, Earth System Research Laboratory, Global Monitoring Division, *http://www.esrl.noaa.gov/ gmd/ccgg/trends/global.html* .

US Energy Information Administraion (2010). International energy outlook 2010(IEO2010).

Vervisch, L., Moureau, V., Domingo, P. & Veynante, D. (2011). Flame surface density and the G equation, *in* N. Swaminathan & K. N. C. Bray (eds), *Turbulent premixed flames*, Cambridge University Press, Cambridge, UK, pp. 60–74.

Veynante, D. & Vervisch, L. (2002). Turbulent combustion modelling, *Prog. Energy Combust. Sci.* 28: 193–266.

Wilcox, D. C. (1988). Re-assessment of the scale-determining equation for advanced turbulence models, *AIAA J.* 26(11): 1299–1310.

Williams, F. A. (1985). Turbulent combustion, *in* J. D. Buckmaster (ed.), *The mathematics of combustion*, SIAM, Philadelphia, pp. 97–131.

Zhang, S. & Rutland, C. J. (1995). Premixed flame effects on turbulence and pressure-related terms, *Combus. Flame* 102: 447–461.

Zhu, J. (1991). A low-diffusive and oscillation-free convection scheme, *Commun. App. Num. Method* 7: 225–232.

8

Multiscale Window Interaction and Localized Nonlinear Hydrodynamic Stability Analysis

X. San Liang*

¹Harvard University, School of Engineering and Applied Sciences, Cambridge, MA
²Central University of Finance and Economics, Beijing
³Stanford University, Center for Turbulence Research, Stanford, CA
⁴Nanjing Institute of Meteorology, Nanjing
¹,³USA
²,⁴China

1. Introduction

Hydrodynamic stability has been a subject extensively investigated in fluid dynamics (see Lin 1966, Drazin & Reid 1982; Godreche & Manneville 1998; Schmid & Henningson 2001, and references therein). We still lack, however, a stability analysis capable of operationally handling results from experiments and numerical simulations, or data taken directly from nature. These "real problems", as we will hereafter refer to, are in general highly nonlinear and localized in space and time. In other words, the signal tends to be temporally intermittent, the regions of interest may be finitely and irregularly defined, and the definition domain could be on the move. Specific examples include atmospheric cyclogenesis, ocean eddy shedding, vortex shedding behind bluff bodies, emergence of turbulent spots, among many others. In this study, we present a new approach to address this old issue, and show subsequently how this approach can be conveniently used for the investigation of a variety of fluid flow problems which would otherwise be very difficult, if not impossible, to investigate.

Localization and admissibility of finite amplitude perturbation are the two basic requirements for the approach. Classically, stability in terms of normal modes (e.g., Drazin & Reid 1982) organizes the whole domain together to make one dynamical system; stability defined in the sense of Lyapunov is measured by a norm (energy) of the perturbation over the whole spatial domain (cf. section 2). These definitions do not retain local features. On the other hand, many analyses have been formulated aiming at localized features, among which the geometrical optics stability method (Lifschitz 1994) and the Green function approach for convective/absolute instability study (Briggs 1964; Huerre & Monkewitz 1990; Pierrehumbert & Swanson 1995) now become standard. These approaches, though localized, usually rely on small perturbation approximation to make linearization possible.

We integrate the philosophies of the above two schools to build our own methodology, which retains full physics, and admits arbitrary perturbation, particularly perturbation of local dynamics, finite amplitude and variable spatial scales. The basic idea is that: The Lyapunov type norm (energy) could be "localized" to make a spatio-temporal field-like metric. In doing

---

*URL:http://people.seas.harvard.edu/~san

so the stability of a system gains a "structure" (which is a more reasonable representation of nature), and the system is organized into structures of distinct processes. The approach is Eulerian. It does not rely on trajectory calculation for local dynamics.

The problem now becomes how to achieve the Lyapunov norm localization. In the next section, we show how a hydrodynamic stability is defined in the Lyapunov sense, and what the definition really implies. We then show that it could be connected to our previous work on mean-eddy interaction within classical framework (section 3), provided that a suitable multiscale decomposition is used. The decomposition is fulfilled with a new mathematical machinery called *multiscale window transform*, (section 4), on the basis of which our instability analysis is formulated. Sections 5-8 are devoted to the establishment of the formalism. Toward the end of this study, we present two real problem applications. The first is a dynamical interpretation of a rather complicated oceanic circulation which has been a continuing challenge since 70 years ago (section 9); we will see in there that the problem becomes straightforward within our framework. The other one is about the vortex shedding, turbulence production, laminarization, and structure sustaining in a turbulent wake behind a circular cylinder (section 10). This study is summarized in section 11.

2. Lyapunov stability

A fundamental definition of hydrodynamic stability was introduced by Lyapunov (cf. Godreche and Manneville, 1998). Given a flow, let $\xi = \xi(\mathbf{x}, t)$ stand for a snapshot of its state, and $\Xi = \Xi(\mathbf{x}, t)$ for a basic solution, which could be a fixed point or any time-varying equilibrium (limiting cycle, for example). In the sense of Lyapunov, the stability of Ξ can be defined as follows: For any $\varepsilon > 0$, if there is a $\delta = \delta(\varepsilon) > 0$, such that

$$\|\xi(\mathbf{x}, t_1) - \Xi(\mathbf{x}, t_1)\| < \varepsilon, \tag{1}$$

$$\text{as} \quad \|\xi(\mathbf{x}, t_0) - \Xi(\mathbf{x}, t_0)\| < \delta, \tag{2}$$

for all $t_1 > t_0$, then Ξ is stable; otherwise it is unstable. Here the norm is defined to be such that $\|\xi\| = \left[\int_\Omega \xi^T \xi d\mathbf{x}\right]^{1/2}$ for ξ over the whole domain Ω, where the superscript T stands for a transpose when ξ is a vector.

Observe that, given a time interval Δ, the basic solution Ξ may be approximately understood as a reconstruction of ξ in time on some slow manifold. In this sense, the terms in the norms of (1) and (2) are actually the perturbations from the slow reconstruction at time instants t_1 and t_0, respectively. Note the norm in the definition can be essentially replaced by its square, so the left hand sides are related to the perturbation energy growth on this interval, recalling how the Parseval relation connects $\int_\Delta (\xi - \Xi)^T (\xi - \Xi)\, dt$ to the perturbation energy. (Note $(\xi - \Xi)^T (\xi - \Xi)$ itself is not the perturbation energy.)

Another observation is about the spatial integration of the norm, which is taken over the whole domain Ω for a closed system. It is this very integration that eliminates the local features and accordingly makes the Lyapunov stability analysis inappropriate for localized events. We need to see what it really means.

It has been observed in real fluid problems, particularly in atmosphere-ocean problems (e.g., Gill 1982), that a fluid system, though complex, often displays a combination of two independent components: a transport and a transfer. The transport component reveals itself in a form like advection and propagation, while the transfer results in the local growth of disturbances. One may intuitively argue that these two processes are both a kind of energy redistribution process. The former redistributes energy in physical space, while the latter redistributes energy between the basic state and the perturbed state. This implies that a transport should integrate to zero over a closed domain; in other words, it has a divergence form in mathematical expression. The integration in the Lyapunov definition serves to eliminate any transport process that may exist, resulting in a pure transfer component over the domain. Therefore, underlying the Lyapunov definition is essentially about the transfer from the basic state to the perturbation. This makes sense, as physically an instability is actually a process of energy transfer between the two states.

Therefore, the problem of hydrodynamic stability is fundamentally a problem about the interaction between the basic state and the perturbation, which is measured by the energy transfer between them. The above physical clarification implies that we may localize the Lyapunov norm if we can separate the transfer from the transport. In that case, there is no need to take the integration, and hence the transfer thus obtained is a spatio-temporal field. All the problem is now reduced to how to achieve the separation.

3. Mean-eddy interaction within the Reynolds decomposition framework

Liang (2007) achieved the separation within the framework of Reynolds decomposition. Originally his formalism was developed in the statistical context, i.e., a Reynolds average is understood as an ensemble mean or mathematical expectation with probability measure. The idea can be best elucidated with the evolution of a passive scalar advected by an incompressible flow $\mathbf{v}$:

$$\frac{\partial T}{\partial t} + \nabla \cdot (\mathbf{v}T) = 0. \tag{3}$$

Decompose T into a mean part and an eddy part:: $T = \bar{T} + T'$. It is easy to obtain the equations governing the mean energy and eddy energy (variance) (e.g., Lesieur, 1990)

$$\frac{\partial \bar{T}^2/2}{\partial t} + \nabla \cdot (\bar{\mathbf{v}}\bar{T}^2/2) = -\bar{T}\nabla \cdot (\overline{\mathbf{v}'T'}) \tag{4}$$

$$\frac{\partial \overline{T'^2}/2}{\partial t} + \nabla \cdot (\overline{\mathbf{v}T'^2}/2) = -\overline{\mathbf{v}'T'} \cdot \nabla \bar{T}. \tag{5}$$

It has been a tradition to classify the energetics so that terms in a divergence form stand out, just like what is done here. The divergence terms are conventionally understood as the transports of the mean and eddy energies, while the other terms singled out from the nonlinear processes are the "transfers". However, as pointed out by Liang (2007), the right hand sides of (4) and (5) do not cancel out—in fact, they sum to $\nabla \cdot (\bar{T}\overline{\mathbf{v}'T'})$, which is in general not zero.[1] As a result, they cannot represent the transfer process, which is by physics a redistribution of energy between the decomposed subspaces.

In his study, Liang (2007) established a rigorous formalism for the transfer. By his result, corresponding to (4) and (5) are the following two equations:

$$\frac{\partial \bar{T}^2/2}{\partial t} + \nabla \cdot \mathbf{Q}_0 = -\Gamma, \tag{6}$$

---

[1] A term $\nabla \cdot \overline{\mathbf{v}'T'}\bar{T}$ may be added to both sides of Eq. 4 to ensure that cancellation (see, for example, Pope, 2003). But physically it is not clear where this extra term comes from, and why it should be there.

$$\frac{\partial \overline{T'^2/2}}{\partial t} + \nabla \cdot \mathbf{Q}_1 = \Gamma, \tag{7}$$

where

$$\mathbf{Q}_0 = \frac{1}{2}\left[\bar{\mathbf{v}}\bar{T}^2 + \bar{T}\overline{\mathbf{v}'T'}\right], \qquad \mathbf{Q}_1 = \frac{1}{2}\left[\mathbf{v}\overline{T'^2} + \bar{T}\overline{\mathbf{v}'T'}\right], \tag{8}$$

$$\Gamma = \frac{1}{2}\left[\bar{T}\nabla \cdot \overline{(\mathbf{v}'T')} - \overline{(\mathbf{v}'T')} \cdot \nabla\bar{T}\right]. \tag{9}$$

This way the energy processes are separated into a transport, which is in a divergence form, and a transfer Γ, which sums to zero over the decomposed subspaces. The separation is unique. Because of the "zero-sum" property, and its similarity in form to the Poisson bracket in Hamiltonian mechanics, Γ has been called *canonical transfer*,[2] in distinction from other transfers one may have encountered in the literature.

A hydrodynamic stability analysis is thus fundamentally a problem of finding the canonical transfer. But the above Γ still cannot make the metric of hydrodynamic stability. As we argued before, a hydrodynamic stability in the Lyapunov sense is defined with respect to a decomposition in time. Within the Reynolds decomposition framework, this was fulfilled with a time averaging, as Liang (2007) discussed. But the time averaging applies only when a system is stationary, while unstable processes are in nature not stationary at all. Besides, if a decomposition is achieved through a time averaging, the eddy energy [e.g., the $\frac{1}{2}\overline{T'^2}$ in (7)] does not have time dependence. As a result, there would be no eddy energy growth in the Lyapunov definition. We need to generalize the traditional mean-eddy decomposition to resolve these issues.

4. Scale window and multiscale window transform

The *multiscale window transform* developed by Liang and Anderson (2007) is such a generalization. It extends the traditional mean-eddy decomposition to retain local physics, and to allow for interactions beyond the mean and eddy processes, e.g., the mean-eddy-turbulence interaction. This section gives it a brief introduction.

4.1 Scale window and multiscale window transform
An MWT organizes a signal into several distinct time scale ranges, while retaining its track in physical space. These time scale ranges form mutually exclusive *scale windows* which we hereafter define. The definition could be over a univariate interval, or a multi-dimensional domain; in the context of this study, it is univariate as we only deal with time. Consider a Hilbert space $V_{\varrho,j} \subset L_2[0,1]$. It is generated by $\{\phi_n^j(t)\}_{n=0,1,\dots,2^j\varrho-1}$, where

$$\phi_n^j(t) = \sum_{\ell=-\infty}^{+\infty} 2^{j/2}\phi[2^j(t + \varrho\ell) - n + 1/2]$$

(n runs through $\{0, 1, \dots, 2^{j\varrho-1}\}$), $\phi(t)$ is a scaling function constructed via orthonormalization from cubic splines (Fig. 1, see Strang and Nguyen 1997), and $\varrho = 1$ and $\varrho = 2$ corresponding to the periodic and symmetric extension schemes, respectively. The periodic extension and symmetric extension are two commonly used schemes (see Liang and Anderson, 2007). In the spanning basis the scheme dependence is suppressed for notational simplicity, but one should be aware of the fact that different extension schemes may give different results. For

---

[2] Originally it was termed "perfect transfer" in Liang (2007).

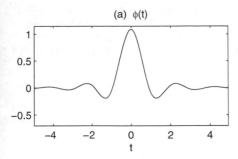
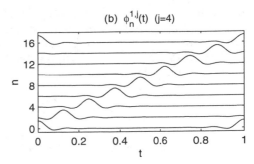

Fig. 1. (a) Scaling function ϕ constructed via cubic spline orthonormalization (Liang and Anderson, 2007). (b) Basis periodized from ϕ.

any integers $j_0 < j_1 < j_2$, it has been shown that the inclusion $V_{\varrho,j_0} \subset V_{\varrho,j_1} \subset V_{\varrho,j_2}$ holds (see Liang and Anderson, 2007). Thus a decomposition can be made such that

$$
\begin{aligned}
V_{\varrho,j_2} &= V_{\varrho,j_1} \oplus W_{\varrho,j_1-j_2} \\
&= V_{\varrho,j_0} \oplus W_{\varrho,j_0-j_1} \oplus W_{\varrho,j_1-j_2}
\end{aligned}
\tag{10}
$$

where W_{ϱ,j_1-j_2} is the orthogonal complement of V_{ϱ,j_1} in V_{ϱ,j_2}, and W_{ϱ,j_0-j_1} the orthogonal complement of V_{ϱ,j_0} in V_{ϱ,j_1}. It has been shown that V_{ϱ,j_0} contains scales larger than 2^{-j_0} only, while in W_{ϱ,j_0-j_1} and W_{ϱ,j_1-j_2} live the scale ranges between 2^{-j_0} to 2^{-j_1} and 2^{-j_1} to 2^{-j_2}, respectively (Liang and Anderson, 2007). These three subspaces of V_{ϱ,j_2} are referred to as, respectively, large-scale window (or mean window), meso-scale window (or eddy window), and sub-mesoscale window (or turbulence window). More windows can be likewise defined, but in this context, three are enough; in fact only two are concerned in most cases.

Suppose $p(t)$ is a realization of some function in $L_2[0,1]$, Liang and Anderson (2007) justified that p always lies in V_{ϱ,j_2} for some j_2 large enough. With the above basis, there is a scaling transform:

$$
\widehat{p}_n^j = \int_0^\varrho p(t)\phi_n^j(t)\, dt,
\tag{11}
$$

for any scale level j. Given window bounds $j_0 < j_1 < j_2$, p then can be reconstructed on the three windows formed above:

$$
p^{\sim 0}(t) = \sum_{n=0}^{2^{j_0}\varrho - 1} \widehat{p}_n^{j_0} \phi_n^{j_0}(t),
\tag{12}
$$

$$
p^{\sim 1}(t) = \sum_{n=0}^{2^{j_1}\varrho - 1} \widehat{p}_n^{j_1} \phi_n^{j_1}(t) - p^{\sim 0}(t),
\tag{13}
$$

$$
p^{\sim 2}(t) = p(t) - p^{\sim 0}(t) - p^{\sim 1}(t),
\tag{14}
$$

with the notations ~ 0, ~ 1, and ~ 2 signifying respectively the corresponding large-scale, meso-scale, and sub-mesoscale windows. As V_{ϱ,j_0}, W_{ϱ,j_0-j_1}, W_{ϱ,j_1-j_2} are all subspaces of V_{ϱ,j_2}, functions $p^{\sim 0}$ and $p^{\sim 1}$ can be transformed using the spanning basis of V_{ϱ,j_2}:

$$
\widehat{p}_n^{\sim \omega} = \int_0^\varrho p^{\sim \omega}(t)\, \phi_n^{j_2}(t)\, dt,
\tag{15}
$$

for windows $\omega = 0, 1, 2$, and $n = 0, 1, ..., 2^{j_2\varrho} - 1$, while keeping information only for their corresponding windows. In doing so, the transform coefficients $\widehat{p}_n^{\sim\omega}$, though discretely defined with n, has the finest resolution permissible in the sampling space on $[0,1]$. We call (15) a *multiscale window transform*, or MWT for short. With this, (12), (13), and (14) can be written in a unified way:

$$p^{\sim\omega}(t) = \sum_{n=0}^{2^{j_2\varrho}-1} \widehat{p}_n^{\sim\omega}\phi_n^{j_2}(t), \qquad \omega = 0, 1, 2. \tag{16}$$

Eqs. (15) and (16) form the transform-reconstruction pair for the MWT.

4.2 Marginalization and multiscale energy

As proved in Liang and Anderson (2007), an important Parseval relation-like property of the MWT is

$$\sum_n \widehat{p}_n^{\sim\omega}\widehat{q}_n^{\sim\omega} = \overline{p^{\sim\omega}(t)\, q^{\sim\omega}(t)}, \tag{17}$$

for $p, q \in V_{\varrho, j_2}$, where the overline indicates an averaging over time, and $\sum_n$ is a summation over the sampling set $\{0, 1, 2, ..., 2^{j_2} - 1\}$. $\sum_n$ is also called "marginalization" as in a localized analysis, properties with locality dependence summed out are usually referred to as marginal properties (see Huang et al., 1999). Eq. (17) states that, a product of two multiscale window transforms followed by a marginalization is equal to the product of their corresponding reconstructions averaged over the duration. We will henceforth refer to it as **property of marginalization**.

The property of marginalization ensures an efficient representation of energy in terms of the MWT transform coefficients. In (17), let $p = q$, the right hand side is then the energy of p (up to some constant factor) averaged over $[0, 1]$. It is equal to a summation of $N = 2^{j_2}$ individual objects $(\widehat{p}_n^{\sim\omega})^2$ centered at time $t_n = 2^{-j_2}n + \frac{1}{2}$, with a characteristic influence interval $\Delta t = t_{n+1} - t_n = 2^{-j_2}$. The energy represented at time t_n then should be the mean over the interval: $\frac{(\widehat{p}_n^{\sim\omega})^2}{\Delta t} = 2^{j_2}(\widehat{p}_n^{\sim\omega})^2$. Note the constant factor 2^{j_2} is essential for physical interpretation. But for notational succinctness, we will neglect it in the following derivations.

5. Canonical transfer with respect to the multiscale window transform

We now carry the canonical transfer (9) to a more generic framework, the MWT framework. Consider again the evolution of a passive scalar T with a governing equation (3). Take an MWT on both sides, and multiply by $\widehat{T}_n^{\sim\omega}$, the resulting left hand side can be shown to be $\dot{E}_n^{\omega}$, time rate of change of the (generalized) multiscale energy (variance) $\frac{1}{2}\left(\widehat{T}_n^{\sim\omega}\right)^2$. Here ω and n represent the window and the time location, respectively. We use a dot to indicate the time rate of change because in performing the transform time t has been translated to the discrete time location n, and hence the time rate of change is actually in an approximate sense. The resulting energy equation is

$$\dot{E}_n^{\omega} = -\widehat{T}_n^{\sim\omega}\widehat{\nabla \cdot (\mathbf{v}T)}_n^{\sim\omega}, \tag{18}$$

which should be equal to $-\nabla \cdot \mathbf{Q}_n^{\omega} + \Gamma_n^{\omega}$ for some flux $\mathbf{Q}_n^{\omega}$ and transfer Γ_n^{ω} on window ω at time step n, based on what we argued before.

Following the same strategy adopted in Liang (2007), we first find $\mathbf{Q}_n^\varpi$, the multiscale flux of $\frac{1}{2}T^2$ by flow $\mathbf{v}$ on window ϖ and time step n. In the MWT framework, it can be obtained through reconstructing the decomposed objects. Notice that

$$\frac{1}{2}T^2 = \sum_{n_1,\varpi_1}\sum_{n_2,\varpi_2} \frac{1}{2}[\widehat{T}_{n_1}^{\sim\varpi_1}\phi_{n_1}^{j_2}(t)\,\widehat{T}_{n_2}^{\sim\varpi_2}\phi_{n_2}^{j_2}(t)], \tag{19}$$

where the summations are over all the possible time steps and windows. We look at the flux of the "atom" $\frac{1}{2}[\widehat{T}_{n_1}^{\sim\varpi_1}\phi_{n_1}^{j_2}(t)\,\widehat{T}_{n_2}^{\sim\varpi_2}\phi_{n_2}^{j_2}(t)]$ by a flow $\mathbf{v}(t)$ (spatial dependence suppressed for clarity) over $t \in [0,1]$ on window ϖ at step n. It is

$$\int_0^1 \mathbf{v}(t) \cdot \frac{1}{2}[\widehat{T}_{n_1}^{\sim\varpi_1}\phi_{n_1}^{j_2}(t)\,\widehat{T}_{n_2}^{\sim\varpi_2}\phi_{n_2}^{j_2}(t)]$$
$$\cdot\,\delta(n-n_2)\delta(\varpi-\varpi_2)\,dt,$$

where δ is the Kronecker function. [Note the arguments of the δ's can be replaced with $(n-n_1)$ and $(\varpi-\varpi_1)$ without affecting the final result, as all the time steps and window indices are to be summed out.] A flux of $\frac{1}{2}T^2$ by $\mathbf{v}$ on ϖ at n is then the sum of it over all the possible n_1, n_2 and ϖ_1, ϖ_2. By the definition of MWT, this is

$$\mathbf{Q}_n^\varpi = \sum_{n_1,\varpi_1}\sum_{n_2,\varpi_2}\int_0^1 \frac{1}{2}\mathbf{v}(t)\cdot\widehat{T}_{n_1}^{\sim\varpi_1}\phi_{n_1}^{j_2}(t)\cdot\widehat{T}_{n_2}^{\sim\varpi_2}\phi_{n_2}^{j_2}(t)$$
$$\cdot\,\delta(n-n_2)\delta(\varpi-\varpi_2)\,dt$$
$$= \frac{1}{2}\int_0^1 \mathbf{v}(t)T(t)\cdot\widehat{T}_n^{\sim\varpi}\phi_n^{j_2}(t)\,dt.$$

But $\widehat{T}_n^{\sim\varpi}\phi_n^{j_2}(t)$ lies in window ϖ, and all the windows are mutually orthogonal, so the above equation is equal to

$$\frac{1}{2}\int_0^1 (\mathbf{v}T)^{\sim\varpi}\cdot\widehat{T}_n^{\sim\varpi}\phi_n^{j_2}(t)\,dt.$$

Notice $\widehat{T}_n^{\sim\varpi}$ is a constant with respect to t. Factoring it out and we get, by the definition of MWT,

$$\mathbf{Q}_n^\varpi = \frac{1}{2}\widehat{T}_n^{\sim\varpi}\widehat{(\mathbf{v}T)}_n^{\sim\varpi}. \tag{20}$$

The transfer Γ is obtained by subtracting $-\nabla\cdot\mathbf{Q}_n^\varpi$ from the right hand side of (18):

$$\Gamma_n^\varpi = -\widehat{T}_n^{\sim\varpi}\nabla\cdot\widehat{(\mathbf{v}T)}_n^{\sim\varpi} + \nabla\cdot\left[\frac{1}{2}\widehat{T}_n^{\sim\varpi}\widehat{(\mathbf{v}T)}_n^{\sim\varpi}\right]$$
$$= \frac{1}{2}\left[\widehat{(\mathbf{v}T)}_n^{\sim\varpi}\cdot\nabla\widehat{T}_n^{\sim\varpi} - \widehat{T}_n^{\sim\varpi}\nabla\cdot\widehat{(\mathbf{v}T)}_n^{\sim\varpi}\right]. \tag{21}$$

There is an important property with the transfer thus obtained. Mathematically, it is expressed as

$$\boxed{\sum_n\sum_\varpi \Gamma_n^\varpi = 0.} \tag{22}$$

In fact, by the property of marginalization, (21) gives

$$\sum_n \Gamma_n^{\omega} = \frac{1}{2} \int_0^1 \left[(\mathbf{v}T)^{\sim\omega} \cdot \nabla T^{\sim\omega} - T^{\sim\omega} \nabla \cdot (\mathbf{v}T)^{\sim\omega} \right] dt,$$

and because of the orthogonality between different windows, this followed by a summation over ω results in

$$\frac{1}{2} \int_0^1 \left[(\mathbf{v}T) \cdot \nabla T - T\nabla \cdot (\mathbf{v}T) \right] dt = 0.$$

In the above derivation, the incompressibility assumption of the flow has been used. Property (22) states that the transfer (21) vanishes upon summation over all the windows and marginalization over the time sampling space. Because of its similarity in form to the Poisson bracket in Hamiltonian mechanics, we will refer it to as *canonical* in the future to distinguish it from other transfers one might have met in the literature. Canonical transfers only re-distribute energy among scale windows, without generating or destroying energy as a whole.

The canonical transfer (21) may be further simplified in expression when $\widehat{T}_n^{\sim\omega}$ is nonzero:

$$\Gamma_n^{\omega} = -E_n^w \nabla \cdot \left(\frac{\widehat{(\mathbf{v}T)}_n^{\sim\omega}}{\widehat{T}_n^{\sim\omega}} \right), \text{ if } \widehat{T}_n^{\sim\omega} \neq 0, \tag{23}$$

where $E_n^{\omega} = \frac{1}{2} \left(\widehat{T}_n^{\sim\omega} \right)^2$ is the energy on window ω at step n, and is hence always positive. Note that (23) defines a field variable which has the dimension of velocity in physical space:

$$\mathbf{v}_T^{\omega} = \frac{\widehat{(T\mathbf{v})}_n^{\sim\omega}}{\widehat{T}_n^{\sim\omega}}. \tag{24}$$

It may be loosely understood as a weighted average in time, with the weights derived from the MWT of the scalar field T. For convenience, we will refer to $\mathbf{v}_T^{\omega}$ as *T-coupled velocity*. The growth rate of energy on window ω is now totally determined by $-\nabla \cdot \mathbf{v}_T^{\omega}$, the convergence of $\mathbf{v}_T^{\omega}$, and

$$\Gamma_n^{\omega} = -E_n^{\omega} \nabla \cdot \mathbf{v}_T^{\omega}. \tag{25}$$

Note Γ_n^{ω} makes sense even though $\widehat{T}_n^{\sim\omega} = 0$ and hence $\mathbf{v}_T^{\omega}$ does not exist. In this case, (25) should be understood as (21). We may keep using (23) and (25) for notational simplicity and physical clarity.

6. Connection to the formalism with respect to Reynolds decomposition

It is of interest to connect the transfer of (21) to that obtained by Liang (2007) within the framework of Reynolds decomposition, i.e., that of (9), with an ensemble mean understood as a time average. As the Reynolds formalism does not allow for time dependence in Γ, we perform a marginalization on the transforms. A basic property of the MWT is that, when $j_0 = 0$ and a periodic extension is used for the time sequence, the field in a two-window decomposition is reconstructed to a mean (over time) and the deviation from the mean (see Liang and Anderson, 2007). In that case,

$$\sum_n \Gamma_n^1 = \frac{1}{2} \left[\overline{(\mathbf{v}T)' \cdot \nabla T'} - \overline{T' \nabla \cdot (\mathbf{v}T)'} \right]$$
$$= \frac{1}{2} \left[\bar{T} \nabla \cdot (\overline{\mathbf{v}'T'}) - (\overline{\mathbf{v}'T'}) \cdot \nabla \bar{T} \right], \tag{26}$$

where in the derivation, the decomposed version of continuity equation $\nabla \cdot \bar{\mathbf{v}} = 0$ and $\nabla \cdot \mathbf{v}' = 0$ has been used. This is the very Γ in (7). Likewise, $\sum_n \Gamma_n^0$ is the $-\Gamma$ in (6). So the canonical transfer of Liang (2007) is a particular case of the present formalism (21).

7. Interaction analysis

In contrast to the canonical transfer (9) resulting from the Reynolds decomposition, the localized Γ_n^ϖ of (21) involves not only inter-window energy transfers, but also transfers from within the same window. As an instability/stability is by definition a process between different windows, we need to eliminate the information other than the inter-window transfer from the Γ_n^ϖ obtained above. This is achieved through a technique called interaction analysis, which has been used by Liang and Robinson (2005) in the MWT framework to single out the desired processes from quadratic energetic terms.

It is observed in (21) that Γ_n^ϖ is made of terms in the form $\Gamma_{\mathcal{R}pq,n}^\varpi = \widehat{\mathcal{R}_n^{\sim\varpi}} \widehat{pq}_n^{\sim\varpi}$, for $\mathcal{R}, p, q \in V_{\varrho, j_2}$. Using the transform-analysis pair (15) and (16),

$$\Gamma_{\mathcal{R}pq,n}^\varpi = \sum_{\varpi_1, \varpi_2} \sum_{n_1, n_2} Tr(n, \varpi \mid n_1, \varpi_1; n_2, \varpi_2), \tag{27}$$

with the *basic transfer function* of $\Gamma_{\mathcal{R}pq,n}^\varpi$

$$Tr(n, \varpi \mid n_1, \varpi_1; n_2, \varpi_2) = \widehat{\mathcal{R}_n^{\sim\varpi}} \frac{\widehat{p}_{n_1}^{\sim\varpi_1} \widehat{q}_{n_2}^{\sim\varpi_2} + \widehat{p}_{n_2}^{\sim\varpi_2} \widehat{q}_{n_1}^{\sim\varpi_1}}{2} \widehat{(\phi_{n_1}^{\varrho, j_2} \phi_{n_2}^{\varrho, j_2})_n^{\sim\varpi}}. \tag{28}$$

Following the terminology of Iima and Toh (1995), $Tr(n, \varpi \mid n_1, \varpi_1; n_2, \varpi_2)$ represents an interaction between the *receiving mode* (n, ϖ), and *giving modes* (n_1, ϖ_1), and (n_2, ϖ_2). It gives the energy made to the receiving mode from the two giving modes. Here both the index pairs (n_1, ϖ_1) and (n_2, ϖ_2) are dummy in (27). We write in (28) $\frac{1}{2} \left[\widehat{p}_{n_1}^{\sim\varpi_1} \widehat{q}_{n_2}^{\sim\varpi_2} + \widehat{p}_{n_2}^{\sim\varpi_2} \widehat{q}_{n_1}^{\sim\varpi_1} \right]$ instead of $(\widehat{p}_{n_1}^{\sim\varpi_1} \widehat{q}_{n_2}^{\sim\varpi_2})$ to ensure symmetry. Using this definition it is easy to show that the basic transfer function of our canonical transfer Γ_n^ϖ defined in (21) satisfies a detailed balance relation which is found in interaction analyses in a variety of contexts (cf. Lesieur 1990, Pope 2003). See Appendix 13 for details.

Every transfer can now be viewed as an installment of the basic transfer functions. The purpose of interaction analysis is to extract the cross-window transfer from these functions. For instability analysis, particularly, we need to find the transfer from the mean window (or window 0) to the eddy window (or window 1) in a two-window decomposition. For example, if we are dealing with $\Gamma_{\mathcal{R}pq,n}^1 = \widehat{\mathcal{R}_n^{\sim 1}} \widehat{(pq)}_n^{\sim 1}$, the summation in (27) over $\varpi_1, \varpi_2, n_1, n_2$ organizes the product pq into four parts (see Liang and Robinson, 2005):

$$p^{\sim 0} q^{\sim 0}, \quad p^{\sim 0} q^{\sim 1}, \quad p^{\sim 1} q^{\sim 0}, \quad p^{\sim 1} q^{\sim 1}.$$

The last part $p^{\sim 1}q^{\sim 1}$, while combined with $\widehat{\mathcal{R}}_n^{\sim 1}$, gives the energy transferred within the eddy window. So it must be removed if only stability/instability is concerned. Using superscript $0 \rightarrow 1$ to signify an operator that selects out the transfer from window 0 to window 1, we have

$$\left(\Gamma^1_{\mathcal{R}pq,n}\right)^{0\rightarrow 1} = \widehat{\mathcal{R}}_n^{\sim 1}\left(\widetilde{(p^{\sim 0}q^{\sim 0})}_n^{\sim 1} + \widetilde{(p^{\sim 0}q^{\sim 1})}_n^{\sim 1} + \widetilde{(p^{\sim 1}q^{\sim 0})}_n^{\sim 1}\right). \tag{29}$$

This operator can be easily applied to the canonical transfer. For example, an application to (21) with $\varpi = 1$ gives

$$(\Gamma^1_n)^{0\rightarrow 1} = \frac{1}{2}\left\{\nabla\widehat{T}_n^{\sim 1}\cdot\left[\widetilde{(\mathbf{v}^{\sim 0}T^{\sim 0})}_n^{\sim 1} + \widetilde{(\mathbf{v}^{\sim 0}T^{\sim 1})}_n^{\sim 1} + \widetilde{(\mathbf{v}^{\sim 1}T^{\sim 0})}_n^{\sim 1}\right]\right.$$
$$\left. - \widehat{T}_n^{\sim 1}\nabla\cdot\left[\widetilde{(\mathbf{v}^{\sim 0}T^{\sim 0})}_n^{\sim 1} + \widetilde{(\mathbf{v}^{\sim 0}T^{\sim 1})}_n^{\sim 1} + \widetilde{(\mathbf{v}^{\sim 1}T^{\sim 0})}_n^{\sim 1}\right]\right\}. \tag{30}$$

8. Hydrodynamic stability analysis

Hydrodynamic stability usually concerns with the transfer process between two windows, the large-scale window and the eddy window. But sometimes the transfer to a smaller scale window, e.g., a turbulence window, might also be of interest. In this section, we first give the instability identification criterion with a two-window decomposition, then in section 8.4 briefly touch the formalism with three windows.

8.1 Idealized flow
We proceed to find the criterion for instability. For an ideal (frictionless) incompressible fluid flow, the governing equations are

$$\frac{\partial\mathbf{v}}{\partial t} = -\nabla\cdot(\mathbf{vv}) - \frac{\nabla P}{\rho}, \tag{31}$$

$$\nabla\cdot\mathbf{v} = 0. \tag{32}$$

Here we do not consider the acceleration of gravity; thus only kinetic energy is involved. This is useful in practice unless buoyancy is perturbed. Following the procedure in section 5, it is easy to know that the pressure term makes no contribution to the energy transfer across scale windows in an incompressible flow. Actually, an MWT of (31) followed by a dot product with $\widehat{\mathbf{v}}_n^{\sim 1}$ results in a pressure working rate proportional to $\widehat{\mathbf{v}}_n^{\sim 1}\cdot\nabla\widehat{P}_n^{\sim 1}$, which is equal to $\nabla\cdot\left(\widehat{P}_n^{\sim 1}\widehat{\mathbf{v}}_n^{\sim 1}\right)$ by the MWTed continuity equation. Only the nonlinear terms require some thought for the canonical transfer.

Let the scalar T in (25) be u, v, w. We have a u-coupled velocity, a v-coupled velocity, and a w-coupled velocity, and hence a canonical transfer on the eddy window:

$$\Gamma^1_n = -\frac{1}{2}\left[(\widehat{u}_n^{\sim 1})^2\nabla\cdot\mathbf{v}_u^1 + (\widehat{v}_n^{\sim 1})^2\nabla\cdot\mathbf{v}_v^1 + (\widehat{w}_n^{\sim 1})^2\nabla\cdot\mathbf{v}_w^1\right].$$

This is well-defined even when the coupled velocities vanish. In that case, one only needs to do an expansion with the above equation. Applying the interaction analysis to select out the process from window 0 to window 1, the expanded equation results in a metric

$$\mathcal{P}_{x,n} = \left(\Gamma^1_n\right)^{0\rightarrow 1} = -\frac{1}{2}\left[\widetilde{(\mathbf{vv})}_n^{\sim 1} : \nabla\widehat{\mathbf{v}}_n^{\sim 1} - \nabla\cdot\widetilde{(\mathbf{vv})}_n^{\sim 1}\cdot\widehat{\mathbf{v}}_n^{\sim 1}\right]^{0\rightarrow 1} \tag{33}$$

Based on the argument in the foregoing sections, a localized stability criterion consistent with the Lyapunov definition is naturally obtained:

The system under consideration is

(i) stable at $(\mathbf{x}, n)$ if $\mathcal{P}_{\mathbf{x},n} < 0$;

(ii) unstable otherwise, and

(iii) the (algebraic) growth rate is $\mathcal{P}$.

This stability criterion is stated with $\mathcal{P}$ in the form of a spatio-temporal field, albeit the time dependence is discrete (n), in contrast to the bulk form in the original Lyapunov definition.

8.2 Real fluids

For a real fluid flow, one generally needs to take into account more physical effects in the governing equations. Buoyancy, dissipation, compressibility, for example, may not be negligible in many cases. As different problems usually have different governing equations, it is not our intention in this study to give a universal expression of the localized stability criterion. We just present the general strategy to obtain the metric.

The commonality of these factors is that they usually do not involve nonlinear interaction in the multiscale energetics, so it is generally easy to incorporate their inputs into our formulation. The key is to have the transport singled out, which is straightforward should there be only linear processes. If buoyancy is perturbed, then potential energy must be counted in. In that case, the system stability is dependent on the growth of perturbation mechanic energy (kinetic energy plus potential energy), instead of just kinetic energy alone.

As an example, we show how this works when dissipation is included in (31). We choose this example not just because of its ubiquity, but also because it will be needed in an application later on.

Dissipation appears in Eq. (31) as an extra term $\nabla \cdot (\nu \nabla \mathbf{v})$. In forming the energetics for window 1, this term results in

$$\hat{\mathbf{v}}_n^{\sim 1} \cdot \nabla \cdot (\nu \nabla \hat{\mathbf{v}}_n^{\sim 1}) = \nabla \cdot (\nu \nabla \hat{\mathbf{v}}_n^{\sim 1} \cdot \hat{\mathbf{v}}_n^{\sim 1}) - \nu \nabla \hat{\mathbf{v}}_n^{\sim 1} : \nabla \hat{\mathbf{v}}_n^{\sim 1}.$$

So the instability criterion metric now should be Eq. (33) plus the non-transport part on the right hand side of the above equation:

$$\begin{aligned}
\mathcal{P}_{\mathbf{x},n} &= -\frac{1}{2} \left[\widehat{(\mathbf{v}\mathbf{v})}_n^{\sim 1} : \nabla \hat{\mathbf{v}}_n^{\sim 1} - \nabla \cdot \widehat{(\mathbf{v}\mathbf{v})}_n^{\sim 1} \cdot \hat{\mathbf{v}}_n^{\sim 1} \right]^{0 \to 1} - \nu \nabla \hat{\mathbf{v}}_n^{\sim 1} : \nabla \hat{\mathbf{v}}_n^{\sim 1} \\
&= -\frac{1}{2} \left[\left(\widehat{(\mathbf{v}\mathbf{v})}_n^{\sim 1} + 2\nu \nabla \hat{\mathbf{v}}_n^{\sim 1} \right) : \nabla \hat{\mathbf{v}}_n^{\sim 1} - \nabla \cdot \widehat{(\mathbf{v}\mathbf{v})}_n^{\sim 1} \cdot \hat{\mathbf{v}}_n^{\sim 1} \right]^{0 \to 1}
\end{aligned} \tag{34}$$

The dissipative input in $\mathcal{P}_{\mathbf{x},n}$ always appears negative; it thus functions to reduce the local energy transfer, as is expected. On the other hand, because of its localized feature, it might modify the stability metric distribution, and hence lead to the emergence of some new stability pattern.

8.3 Instability identification from a large-scale point of view

The above criterion is established based on the eddy window or window 1. Instability can also be described from a large-scale point of view. This makes sense as a canonical transfer is a protocol between two scale windows: If the eddy window works, we may as well equally view the problem from the large-scale window. The advantage of a large-scale formulation is to have the eddy scale features filtered, for the transfer thus derived is on the balance of the large-scale energetics. For an idealized fluid, we may define:

$$\mathcal{P}_{x,n} = +\frac{1}{2}\left[\widehat{(\mathbf{vv})}_n^{\sim 0} : \nabla\widehat{\mathbf{v}}_n^{\sim 0} - \nabla\cdot\widehat{(\mathbf{vv})}_n^{\sim 0}\cdot\widehat{\mathbf{v}}_n^{\sim 0}\right]^{1\to 0}. \tag{35}$$

Notice the positive sign here, compared to the negative sign in (33), as in this case, a positive Γ_n^0 is toward the large-scale window. Also notice the different interaction analysis operator $1 \to 0$. Eqs. (35) and (33) are equivalent on the large-scale window because the transfer is canonical. This can be easily proved with the definition of canonical transfer (22).

For real fluids, the problem is not as simple as that in the preceding subsection: The canonical transfer and other terms to be combined are not defined on the same scale window. To circumvent the difficulty, we first take averaging for the extra eddy scale energetic terms over an interval in the sampling space commensurate with the large-scale window transform, before adding these terms to the canonical transfer. As an example, consider the dissipative case. The metric in (35) now becomes

$$\mathcal{P}_{x,n} = +\frac{1}{2}\left[\widehat{(\mathbf{vv})}_n^{\sim 0} : \nabla\widehat{\mathbf{v}}_n^{\sim 0} - \nabla\cdot\widehat{(\mathbf{vv})}_n^{\sim 0}\cdot\widehat{\mathbf{v}}_n^{\sim 0}\right]^{1\to 0} - \nu\overline{\nabla\widehat{\mathbf{v}}_\ell^{\sim 1} : \nabla\widehat{\mathbf{v}}_\ell^{\sim 1}}, \tag{36}$$

where the averaging over ℓ is taken in the sampling space over interval $[n - 2^{j_2-j_0-1}, n + 2^{j_2-j_0-1}]$. The instability identification criterion with $\mathcal{P}$ is the same as that stated in the previous subsections. When other effects are taken into account, $\mathcal{P}$ can be defined likewise as it is defined in (36).

8.4 Mean-eddy-turbulence interaction

Sometimes just two scale windows are not enough to characterize the fluid processes. For instance, one often encounters problems involving interactions between mean, eddy, and turbulent windows. In that case, a system could not only lose stability to fuel the growth of the eddy events, but also transfer energy from the eddy window to the turbulence window through another instability.

We remark that this situation is in fact already taken into account in the formalism of canonical transfer. By (21) it is easy to obtain

$$\Gamma_n^{\varpi} = -\frac{1}{2}\left[\widehat{(\mathbf{vv})}_n^{\sim\varpi} : \nabla\widehat{\mathbf{v}}_n^{\sim\varpi} - \nabla\cdot\widehat{(\mathbf{vv})}_n^{\sim\varpi}\cdot\widehat{\mathbf{v}}_n^{\sim\varpi}\right] \tag{37}$$

for the governing equations (31) and (32). Different instability metric can be formed by taking the interaction analysis. For example, $\left(\Gamma_n^2\right)^{1\to 2}$ and $\left(\Gamma_n^2\right)^{0\to 2}$ represent, respectively, the transfer from the eddy window to the turbulence window, and the transfer from the mean window directly to the turbulence window.

9. Application to oceanographic studies

The above theory and methodology have been applied to problems in different fields in fluid dynamics, such as oceanography, turbulence, fluid control, to name a few. In this section we briefly summarize the results of Liang and Robinson (2009) on a successful application to the dynamical interpretation of a rather complicated ocean phenomenon, the Monterey Bay circulation.

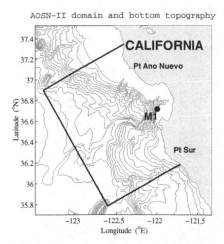

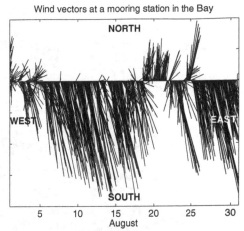

Fig. 2. Left: The research domain and bottom topography of Monterey Bay and its adjacent regions for the August 2003 AOSN-II Experiment. Also shown is the location of a mooring station (M_1) within the Bay. Right: The wind vectors in August 2003 at Station M_1.

Monterey Bay is a large embayment located to the south of San Francisco, California (Fig. 2). Distinguished by its high productivity and marine life diversity, it has been arousing enormous interest in the oceanographic community (see Liang and Robinson, 2009, and references therein). Among the existing issues, how the Monterey Bay circulation is excited and sustained has been of particular interest ever since the 1930s, which had been a continuing challenge until a breakthrough was made recently by Liang and Robinson (2009), with the aid of the afore-mentioned new machinery namely the multiscale window transform (MWT), and an earlier version of the MWT-based stability analysis tailored for atmosphere-ocean problems (Liang and Robinson, 2007). In their study, Liang and Robinson (2009) examined an unprecedented dataset acquired in August 2003 during the Second Autonomous Ocean Sampling Network (AOSN-II) Experiment, a program with the involvement of more than 10 major institutions nationwide in the United States. Out of the observational data the 4D flow field is reconstructed in an optimal way for the duration in question using the Harvard Ocean Prediction System (Haley et al., 2009). Shown in Fig. 3 are several snapshots of the surface flow. As expected, it looks very complex; there seems to be no way to dynamically interpret it with the geophysical fluid dynamics (GFD) theories available then.

Cherishing the hope that underlying the seemingly chaotic phenomena the dynamical processes could be tractable, let us look at the canonical transfers and their evolutions. To do this, first we need to determine the scale windows where the respective processes occur. This is fulfilled by analyzing the wavelet spectra of some typical time series. Shown in Fig. 4 is such a series and its spectrum. Note here we have to use orthonormal wavelets to allow for a definition of energy in the physical sense (see Liang and Anderson, 2007). By observation there is a clear gap between scale levels 2 and 3; accordingly j_0 is chosen to be 2, which corresponds to a time scale of $2^{-j_0} \times$ duration $= 2^{-2} \times 32 = 8$ days. Likewise, a reasonable choice for j_1 is 5, corresponding to a time scale of 1 day. (This essentially keeps the mesoscale window free of tides.) With these parameters, the canonical transfers and hence the stability criterion can be computed in a straightforward way. In Liang and Robinson (2007), we have established that here the criterion actually corresponds to that for the barotropic

Surface Flow

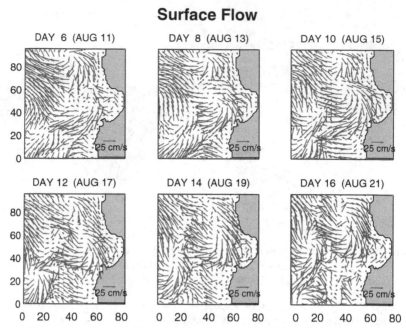

Fig. 3. Snapshots of the surface flow reconstructed from the 2003 AOSN-II dataset. The domain in Fig. 2 has been rotated by $30°$ clockwise.

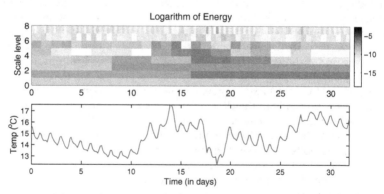

Fig. 4. A time series of the surface temperature at a point off Pt. Ano Nuevo (lower panel) and its orthonormal wavelet spectrum (upper panel). In performing the spectral analysis, the mean has been removed for clarity.

instability in GFD. For convenience in this context, we write it as BT. Meanwhile, In (23), replacing T by density anomaly ρ we get the canonical potential energy transfers. We are particularly interested in the transfer from window 0 to window 1 which, if the interaction analysis operator $0 \rightarrow 1$ is applied, has been proved to correspond to the baroclinic instability identification criterion in GFD (Liang and Robinson, 2007). We shorthand it as BC henceforth. Surprisingly, though the flow is very chaotic, both BT and BC follow a similar and quite regular evolutionary pattern. Contoured in Fig. 5 are the 10-meter BCs at several time instants.

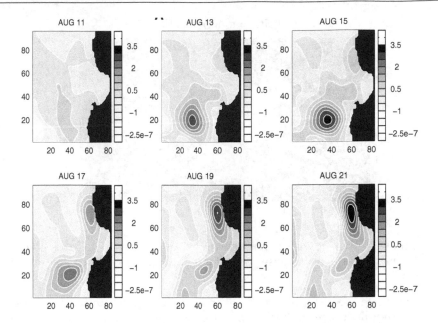

Fig. 5. A time sequence of the 10-meter BC (units: m^2/s^3). Positive values indicate baroclinic instability.

Clearly there are two centers of instability, the northern one lying off Pt. Ano Nuevo, another off Pt. Sur. Generally, during the experimental period, the former gets strengthened as time moves on, while the latter varies the opposite way. Comparing to the wind vectors (Fig. 2), when the northwesterly (southeastward), i.e., the upwelling-favorable wind, prevails, an instability (baroclinic and barotropic) occurs off Pt. Sur; when the wind is relaxed, the Pt. Sur instability disappears, but the relaxation triggers another instability off Pt. Ano Nuevo. This instability is also baroclinic and barotropic, i.e., mixed, in nature. Liang and Robinson (2009) showed that this bimodal structure supplies two sources of mesoscale processes. The resulting mesoscale eddies propagate northward in the form of coastal-trapped waves, with a celerity of about 0.09 m/s. It is the mesoscale activities that make the flow pattern complex.

The above discovery is remarkable; it shows that, during the AOSN-II experimental period, large-scale winds actually do not directly excite the Monterey Bay region circulation, in contrast to predictions with classical theories. Rather, it stores energy in the large-scale window, and then release to the mesoscale window through baroclinic and barotropic instabilities. The mesoscale disturbances are organized in the form of eddies, which, once formed, propagate northward as coastal-trapped waves. Liang and Robinson (2009) also found that a significant upwelling event in this region is driven through nonlinear instabilities, distinctly different from the classical coastal upwelling paradigm.

10. Application to turbulence structure studies

The above theory and methodology have also been applied to turbulence research. Here we briefly summarize the study by Liang and Wang (2004) with a benchmark simulation of a saturated turbulent wake behind a circular cylinder (Reynolds number: Re=3900). The

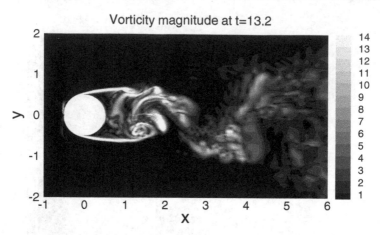

Fig. 6. A saturated turbulent wake at Re=3900 (from Liang and Wang, 2004). Shown here is the vorticity magnitude on a spanwise plane at $t = 13.2$ (arbitrary starting time after a statistical equilibrium).

configuration is referred to Fig. 6, with x, y, and z indicating the streamwise, cross-stream, and spanwise coordinates, respectively ((x, y) originated at the cylinder center). The variables are nondimensionalized with the cylinder diameter d as the length scale, free-stream velocity U_∞ as the velocity scale, and d/U_∞ as the time scale. The dataset for analysis is generated using an energy-conserving, hybrid finite-difference/spectral model, as described in Mittal and Moin (1997). Incompressible Navier-Stokes equations are solved on a C-type mesh using a numerical scheme with second-order central differences in the streamwise and cross-stream directions, and Fourier collocation in the spanwise direction. The subgrid processes are parameterized with the dynamical model described in Germano et al. (1991) and Lilly (1992). The time advancement is of the fractional step type in combination with the Crank-Nicholson method for viscous terms and third order Runge-Kutta scheme for the convective terms. The resulting Poisson equation for pressure is solved using a multigrid iterative procedure at each Runge-Kutta substep. More details about the numerics are referred to Liang and Wang (2004). Shown in Fig. 6 is a snapshot of the computed instantaneous vorticity magnitude.

To perform stability analysis for the simulation, we first determine the scale window bounds. This is fulfilled through orthonormal wavelet spectral analysis. The needed time series are chosen from the velocity components at several typical points within the wake. Following the same procedure as for the above oceanographic problem, it is justified that $j_0 = 1$ and $j_1 = 2$, together with a symmetric extension, serve our purpose well. Shown in Fig. 7 is the multiscale decomposition of a typical time series–the time series of of u at point $(2, -0.5, 0)$. On the right hand side of the equality are the large-scale, mesoscale, and sub-mesoscale reconstructions. Generally, turbulence problems involve complicated scale window structures. Here the mesoscale window is characterized by amplifications of vortex shedding processes, and the process within the sub-mesoscale window appears turbulent.

We now investigate the window interactions with the above theory on localized hydrodynamic stability. As before, we continue to analyze from a longer-time span point of view. That is to say, we will use (36) rather than (34) for the analysis. The results are presented in a spanwise plane only, albeit the analysis is three-dimensional. This is justified by the fact that in a statistical sense the flow field is equivalent spanwise.

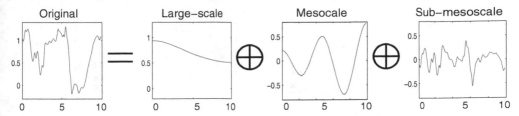

Fig. 7. Multiscale window decomposition of the time series of u at point $(2, -0.5, 0)$. Window bounds used are $j_0 = 1$, and $j_1 = 2$. The abscissa and ordinate are time (scaled by d/U_∞) and u (scaled by U_∞), respectively.

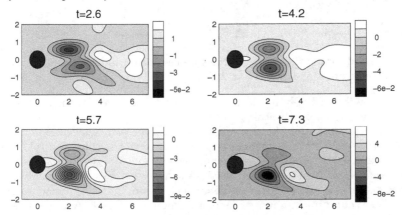

Fig. 8. The canonical kinetic energy transfer from the mesoscale window to the large-scale window for the saturated turbulent wake (adapted from Liang and Wang, 2004). Negative values (in black) indicate instability. The abscissa and ordinate are x and y, respectively.

Figure 8 is the canonical kinetic energy transfer from the meso-scale window to the large-scale window originally computed by Liang and Wang (2004). It differs from (36) by a minus sign; that means here negative values indicate instability. From it obviously there are two centers of instability, each located at one side of the x-axis. These centers are permanent, though their magnitudes do oscillate with time. In the episode as shown, the oscillations are out of phase. The transfer center on the top flank weakens as time goes on, while on the bottom the transfer strengthens. Another observation is that, some inverse transfers (white regions in the figure) are found in the near and far wake, or sandwiched between the instability lobes.

It is these two centers of hydrodynamic instability that cause the shedding of the vortices. This, however, does not correspond to what one may observe about the perturbation growth in vortex shedding. Indeed, by computation both the maximal disturbance and the largest disturbance growth are in the near wake along the x-axis (e.g., Liang, 2007). The discrepancy in location between instability and disturbance growth reveals a fundamental fact in fluid dynamics which has been mostly overlooked: Rapid amplification of perturbation does not necessarily correspond to instability; the eddy energy, when generated, may be transported away instantaneously and lead the perturbation to grow elsewhere. From our results, this observation is in fact a rule rather than an exception. In other words, the causality of

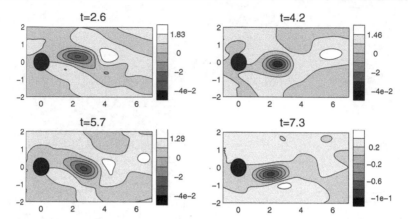

Fig. 9. As Fig. 8, but for the transfer from the sub-mesoscale window to the mesoscale window (adapted from Liang and Wang, 2004).

perturbation growth is usually not local; what we have observed by visual inspection may not reflect the underlying dynamical processes.

The above is about the canonical transfer between the large-scale window and the mesoscale window. We go on to investigate that between the mesoscale and sub-mesoscale windows. In the decomposition in Fig. 7, we have seen that this accounts for how turbulence is produced. As above, the analysis is performed from the stance on the mesoscale window, instead of the sub-mesoscale window, to ensure a better understanding of the dynamics on a longer time span. The result on a spanwise plane is contoured in Fig. 9. Again, this is what was originally obtained by Liang and Wang (2004), and differ from our previous instability metric by a negative sign. That is to say, in the figure negative values (in black) imply instability. Notice that here the canonical transfer is mainly within a monopole lying near the x-axis, in contrast to the dipole structure in Fig. 8. This is interesting, for it shows that the energy gained by the mesoscale process from the basic flow does not go directly to the sub-mesoscale window where the turbulent structures reside. Instead, it is first transported from the two side lobes to the center and then released to fuel the turbulence. As in Fig. 8, inverse transfers are found in the far wake, indicating that some re-laminarization is occurring there.

The instability structures in Figs. 8 and 9 indicate that, in a saturated turbulent wake, there are two primary instabilities, either on one side of the x-axis, followed by a secondary instability lying in between on the axis. The primary instabilities form the mechanism for the vortex shedding; the secondary instability derives the energy from the mesoscale window and funnels energy to sustain the turbulence. These processes are summarized and schematized in Fig. 10.

11. Discussion and conclusions

A localized hydrodynamic stability analysis was developed to relate stability theory to experimental data and direct observations of nature, which are in general highly nonlinear and intermittent in space and time. The theory was applied to the study of the wake behind a circular cylinder, and the suppression of the vortex street formation.

We established that the Lyapunov definition of hydrodynamic stability is essentially about the energy transfer between the mean and eddy states. A transfer is a nonlinear process in

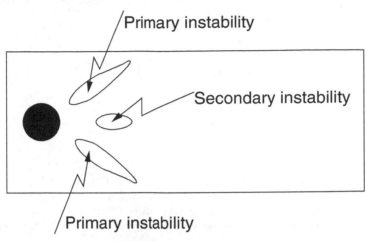

Fig. 10. A schematic of the instability structure in a turbulent wake. The basic flow loses energy to the mesoscale window to shed vortices via the two primary instabilities, while the turbulent processes derive their energy from the vortices through a secondary instability lying in between.

a flow which does not generate nor destroy energy as a whole. In the text it is referred to as "canonical transfer", in distinction from other transfers one may have encountered in the literature. A hydrodynamic instability analysis is thus fundamentally a problem of finding the canonical transfer.

The canonical transfer has been rigorously formulated within the framework of a function analysis apparatus newly developed by Liang and Anderson (2007), i.e., the multiscale window transform, or MWT for short. The MWT is a generalization of the classical mean-eddy decomposition, representing a signal on subspaces or scale windows with distinct time scale ranges. We particularly introduced a large-scale window, a mesoscale window, plus a sub-mesoscale window when needed. Symbolically they are denoted as $\varpi = 0, 1, 2$ for notational convenience. In special cases, these windows may also be referred to as mean window, eddy window, and turbulence window, respectively. The basic idea of the formulation is that the nonlinear process in a flow can be uniquely decomposed into a transport (expressed in a divergence form) and a canonical transfer; if the former is found, the latter follows accordingly. The resulting canonical transfer has a concise form in expression in terms of MWT. In the case of a passive scalar T advected by an incompressible flow, the energy transferred to window ϖ can be written as $\Gamma_n^\varpi = -E_n^\varpi \nabla \cdot \mathbf{v}_T$, where E_n^ϖ is the energy on window ϖ, $\mathbf{v}_T = \dfrac{\widehat{(\mathbf{v}T)}_n^{\sim\varpi}}{\widehat{T}_n^{\sim\varpi}}$, and $\widehat{(\cdot)}_n^{\sim\varpi}$ indicates an MWT on window ϖ. The complicated multiscale window interaction (mean-eddy-turbulence interaction in particular) in fluid flows is therefore nicely characterized by this concise formula, as schematized in Fig. 11.

Hydrodynamic instability is a particular case of the above multiscale window interaction. Take a two-window decomposition, and consider the eddy window $\varpi = 1$. It is shown that the energy transferred from the background to fuel the variance of T is $\Gamma_n^1 = -E_n^1 \nabla \cdot \mathbf{v}_T$. The instability is then totally determined by the convergence of $\mathbf{v}_T$, a velocity weighted by T in the MWT framework. The same derivation applies to the momentum equations, and a metric $\mathcal{P}$ for hydrodynamic instability can be easily obtained. Note that the so-obtained $\mathcal{P}$ is

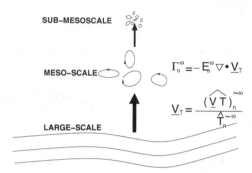

Fig. 11. A schematic of the multiscale window interaction in fluid flows. The intricate multiscale window interaction, and mean-eddy-turbulence interaction in particular, is characterized by a quantity namely the canonical transfer Γ_n^ϖ (for scale window ϖ and time step n) which, in terms of multiscale window transform, can be succinctly written in a form as Eq. (23) or Eq. (21).

a spatio-temporal field, and hence localized events are naturally represented. For a flow, one can conveniently identify the instability structure, and the corresponding growth rate, simply by visual inspection.

With this convenience we examined an oceanographic problem. For nearly 70 years, how the complex circulation in the Monterey Bay, California, is excited has been a continuing challenge in physical oceanography. Numerous efforts have been invested to investigate its dynamical origins without much luck. This problem, however, becomes straightforward in our framework. With an unprecedented dataset collected in 2003, it is found that, though the circulation seems to be chaotic, the underlying dynamical processes turn out to be tractable. More specifically, the complexity is mainly due to two mixed instabilities which are located outside Pt. Sur and Pt. Ano Nuevo, respectively. The resulting mesoscale eddies propagate northward in the form of coastally trapped waves, orchestrating into a flow complex in pattern. In this study, we see how winds may instill energy into the ocean, first stored in the large-scale window, then releasing to fuel the mesoscale eddies through barotropic and baroclinic instabilities. We have also seen that intense upwelling events may not be directly driven by winds, nor by topography variation, but may have their origin intrinsically embedded in the nonlinearity of the system. These dynamical scenarios are distinctly different from their corresponding classical paradigms.

We have also examined a turbulent wake behind a circular cylinder. It is found the processes are organized into three distinct scale windows: On the two extremes are the basic flow and the turbulence; lying in between are the shedding vortices. The vortex shedding is sustained through two instabilities either on a side of the x-axis, and the mesoscale shedding further releases energy into the sub-mesoscale window to produce turbulence. The former two are what we call primary instabilities, while the latter is a secondary instability. The locations of the instabilities are relatively steady. The primary instabilities are within two lobes, one at a side of the axis; the secondary instability is in the form of a monopole, located mainly along the axis, as shown in Fig. 10. This structure implies that the canonical energy transfer may not be local. In this case, for example, the energy acquired during the primary instabilities must be first transported from the two side lobes to the middle before being utilized for turbulence production. This nonlocal transfer may also be reflected in the discrepancy between perturbation growth and its corresponding canonical transfer. That is to say, what one sees about the growth at one location may have its origin elsewhere. From the mathematical expressions shown in this study this phenomenon is actually a rule rather than an exception.

The nonlocality of energy transfer raises an issue about causality and accordingly poses a challenge to turbulence simulation in the parameterization of subgrid processes. Since the rapid amplification of perturbation need not be instability, schemes based solely on local perturbation or perturbation growth may not serve the modeling purpose well. We hope the notion of canonical transfer may come to help in this regard.

On both the primary and secondary instability maps for the turbulent wake, a remarkable feature is the inverse transfer spots/centers sandwiched between the instability structures. This phenomenon tells that, even though the flow is turbulent, there exist processes which introduce orders rather than chaos to the system. This phenomenon has profound implications; we look at how it may be utilized for flow control, particularly turbulence control. Turbulence control is an important applied field which has been widely investigated. Several reviews can be found in Huerre and Monkewitz (1990), Oertel (1990), Williamson (1996), Pastoor et al. (2008), to name a few. Generally speaking, to control turbulence is to inhibit perturbation energy from being generated, and accordingly, the traditional control is designed based on suppression of turbulence growth. This goes back to what we have observed above: *Looking solely at the turbulence growth could be misleading, as energy increase does not necessarily occur in accordance with transfer.* Particularly, in a region with perturbation growing there could be actually an ongoing inverse transfer or laminarization lying beneath. If one puts a control in this region to inhibit the perturbation growth, he probably also defeats the laminarization which is actually helpful to control. To illustrate, consider a two-point system as shown in Fig. 12. We have points 1 and 2, with eddy energies K_1^{eddy} and K_2^{eddy}, respectively. Both K_1^{eddy} and K_2^{eddy} grow, but their sources of growth are different. The former is from the *in situ* large-scale window through instability ($\Gamma_1 > 0$), while the latter

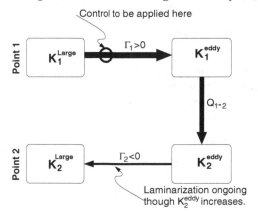

Fig. 12. Schematic of the energetics in a two-point system. We use K, Γ, and Q to signify energy, transfer, and transport, respectively. In this case, both K_1^{eddy} and K_2^{eddy} grow, but their mechanisms are different. The former is due to an instability, while the latter comes from the positive offset $|Q_{1\to2}| - |\Gamma_2|$, although the transfer $\Gamma_2 < 0$ is in the inverse direction. To take advantage of the inverse transfer or laminarization, the control should be placed at point 1 only. (Adapted from Liang and Wang, 2004.)

is transported from the former. The transfer Γ_2 at point 2 is toward the large-scale window, i.e., a laminarization is taking place there, but since $Q_{1\to2} - |\Gamma_2| > 0$, K_2^{eddy} still grows. Of course, control of the perturbation energy growth at both points 1 and 2 does help to suppress the onset of turbulence, but it is not optimal in terms of energy saving. Suppression of K_2^{eddy} also suppresses the intrinsic trend of laminarization at point 2, and therefore reduces the control performance. An optimal control strategy should take advantage of this trend, implying that the control should be applied at point 1 only. Besides, the optimal objective functional should be chosen to be Γ_1, rather than $K_1^{eddy} + K_2^{eddy}$. With this, we have proposed a strategy to harness vortex shedding behind a circular cylinder, and obtained satisfactory, albeit preliminary, results (e.g., Liang, 2007). We shall explore more about this in future studies.

12. Acknowledgments

The author has benefited from several discussions with Howard A. Stone, Donald G.M. Anderson, and Pierre F.J. Lermusiaux. Fritz Busse read through an early version of the manuscript, and his remarks are appreciated. The concept of localized hydrodynamic stability analysis was originally inspired when the author worked at Harvard University with Allan R. Robinson on MS-EVA namely the Multi-Scale Energy and Vorticity Analysis, and part of the work was conducted during a visit at the Center for Turbulence Research at Stanford University and NASA Ames Research Center, thanks to the kind invitation by Parviz Moin and Nagi Monsour. The author is especially indebted to Meng Wang, who hosted the visit and provided the wake simulations (shown in Fig. 6 is an example), for many important scientific discussions with him in analyzing the resulting canonical transfers. He also read through several versions of this manuscript, and his insightful comments have been incorporated into the text. This work was partially supported by the Office of Naval Research under Contract N00014-02-1-0989 to Harvard University, and by the Ministry of Finance of China through the Basic Research Funding to China Institute for Advanced Study.

13. Appendix: Detailed balance relation

It has been a common practice to check if the interaction analysis satisfies a Jacobian identity-like *detailed balance relation* (e.g., Lesieur 1990; Iima and Toh 1995). We show this is true with our formulation with the incompressibility assumption. For a scalar field T, and flow $\mathbf{v}$, the transfer at n on window ϖ, Γ_n^{ϖ}, is given by (21). The basic transfer function of Γ_n^{ϖ} is, by our definition in section 7,

$$
\begin{aligned}
\mathit{Tr}\,(n_1,\varpi_1 \mid n_2,\varpi_2; n_3,\varpi_3) = \\
\frac{1}{2}\left[-\widehat{T}_{n_1}^{\sim\varpi_1}\nabla\cdot\left(\widehat{\mathbf{v}}_{n_2}^{\sim\varpi_2}\widehat{T}_{n_3}^{\sim\varpi_3}\right)+\frac{1}{2}\nabla\cdot\left(\widehat{T}_{n_1}^{\sim\varpi_1}\widehat{\mathbf{v}}_{n_2}^{\sim\varpi_2}\widehat{T}_{n_3}^{\sim\varpi_3}\right)\right]\widetilde{(\phi_{n_2}^{j_2}\phi_{n_3}^{j_3})}_{n_1}^{\sim\varpi_1} \\
+\frac{1}{2}\left[-\widehat{T}_{n_1}^{\sim\varpi_1}\nabla\cdot\left(\widehat{\mathbf{v}}_{n_3}^{\sim\varpi_3}\widehat{T}_{n_2}^{\sim\varpi_2}\right)+\frac{1}{2}\nabla\cdot\left(\widehat{T}_{n_1}^{\sim\varpi_1}\widehat{\mathbf{v}}_{n_3}^{\sim\varpi_3}\widehat{T}_{n_2}^{\sim\varpi_2}\right)\right]\widetilde{(\phi_{n_2}^{j_2}\phi_{n_3}^{j_3})}_{n_1}^{\sim\varpi_1}, \quad (38)
\end{aligned}
$$

for windows ϖ_1, ϖ_2, ϖ_3, and locations n_1, n_2, n_3 in the time sampling space. When the flow is incompressible (hence $\nabla\cdot\widehat{\mathbf{v}}_n^{\sim\varpi} = 0$ for any ϖ and n), it is straightforward check that

$$
\begin{aligned}
\mathit{Tr}\,(n_1,\varpi_1 \mid n_2,\varpi_2; n_3,\varpi_3) + \mathit{Tr}\,(n_2,\varpi_2 \mid n_3,\varpi_3; n_1,\varpi_1) \\
+ \mathit{Tr}\,(n_3,\varpi_3 \mid n_1,\varpi_1; n_2,\varpi_2) = 0. \quad (39)
\end{aligned}
$$

This is the detailed balance relation.

14. References

[1] Briggs, R.J. (1964). Criteria for identifying amplifying waves and absolute instabilities. *Electro-Stream Interaction with Plasamas*. The MIT Press, 8-46.

[2] Carnevale, G.F. & Frederiksen, J.S. (1987). Nonlinear stability and statistical mechanics of flow over topography. *J. Fluid. Mech.*, Vol. 175:157-181.

[3] Drazin, P.G. & Reid, W. H. (1982). *Hydrodynamic stability*, Cambridge University Press.

[4] Germano, M., Piomelli, U., Moin, P., & Cabot, W.H. (1991). A dynamic subgrid-scale eddy viscosity model. *Phys. Fluids A*, Vol. 3:1760-1765.

[5] Gill, A.E. (1982). *Atmosphere-Ocean Dynamics*, Academic Press.

[6] Godreche, C. & Manneville, P. (1998). *Hydrodynamics and Nonlinear Instabilities*, Cambridge University Press.

[7] Haley, P.J., Lermusiaux, P.F.J., Robinson, A.R., Leslie, W.G., Logoutov, O., Cossarini, G., Liang, X.S., Moreno, P., Ramp, S.R., Doyle, J.D., Bellingham, J., Chavez, F., & Johnston, S. (2009). Forecasting and reanalysis in the Monterey Bay/California Current region for the Autonomous Ocean Sampling Network-II experiment, *Deep-Sea Res.*, Part II, Vol. 56 (3-5):127-148.

[8] Huang, N. E., Shen, Z. & Long, S.R. (1999). A new view of nonlinear water waves: The Hilbert spectrum, *Annu. Rev. Fluid Mech.*, Vol. 31:417-457.

[9] Huerre, P. & Monkewitz, P. A. (1990). Local and global instabilities in spatially developing flows, *Annu. Rev. Fluid Mech.*, Vol. 22:473-537.

[10] Iima, M. & Toh, S. (1995). Wavelet analysis of the energy transfer caused by convective terms: Application to the Burgers shock, *Phys. Rev. E*, Vol. 52(6):6189-6201.

[11] Lesieur, M. (1990). *Turbulence in Fluids: Stochastic and Numerical Modeling*, 2nd ed., Klumer Academic Publishers.

[12] Liang, X. San (2007). Perfect transfer and mean-eddy interaction in incompressible fluid flows. arXiv:physics/0702002. http://arxiv.org/abs/physics/0702002.

[13] Liang, X. San & Anderson, Donald G. M. (2007). Multiscale window transform, *SIAM J. Multiscale. Model. Simu.*, Vol 6(2):437-467.

[14] Liang, X. San & Robinson, Allan R. (2005). Localized multiscale energy and vorticity analysis: I. Fundamentals, *Dyn. Atmos. Oceans*, Vol 38:195-230.

[15] Liang, X. San & Robinson, Allan R. (2007). Localized multiscale energy and vorticity analysis: II. Finite-amplitude instability theory and validation, *Dyn. Atmos. Oceans*, Vol 44:51-76.

[16] Liang, X. San & Wang, Meng (2004). A study of turbulent wake dynamics using a novel localized stability analysis. Center for Turbulence Research, *Proceedings of the Summer Program 2004*, 211-222.

[17] Lifschitz, A. (1994). On the instability of certain motions of an ideal incompressible fluid, *Advances in Applied Mathematics*, Vol 15:404-436.

[18] Lilly, D.K. (1992): A proposed modification of the Germano subgrid scale closure method, *Phys. Fluids A*, Vol 4:633-635.

[19] Lin, C.C. (1966): *The theory of hydrodynamic stability*, Cambridge U Press.

[20] Mittal, R. & Moin, P. (1997). Suitability of upwind-biased finite difference schemes for large-eddy simulation of turbulence flows, *AIAA J.*, Vol. 35:1415-1417.

[21] Oertel, H., Jr. (1990). Wakes behind blunt bodies, *Annu. Rev. Fluid Mech.*, Vol 22:539-558.

[22] Pierrehumbert, R. T. & Swanson, K. L. (1995). Baroclinic Instability, *Ann. Rev. Fluid Mech.*, Vol. 27:419-467.

[23] Pope, S. (2003). *Turbulent Flows*, Cambridge University Press.

[24] Schmid, P.J. & Henningson, D. S. (2001). *Stability and transition in shear flows*, Springer, New York.

[25] Strang, G. & Nguyen, T. (1997). *Wavelets and Filter Banks*, Wellesley-Cambridge Press.

[26] Pastoor, M., Henning, L., Noack, B.R., King, R., & Tadmor, G., (2008). Feedback shear layer control for bluff body drag reduction, *J. Fluid Mech.*, Vol. 608:161-196.

[27] Williamson, C. H. K. (1996). Vortex dynamics in the cylinder wake, *Ann. Rev. Fluid Mech.*, Vol. 28:477-526.

9

Stability Investigation of Combustion Chambers with LES

Balázs Pritz and Martin Gabi
Karlsruhe Institute of Technology, Department of Fluid Machinery
Germany

1. Introduction

Our primary energy consumption is supported in 81% by the combustion of fossil energy commodities (IEA, 2010). The demand on energy will grow by about 60% in the near future (Shell, 2008). The efficiency of the combustion processes is crucial for the environment and for the use of the remaining resources. At the Karlsruhe Institute of Technology the long-term project Collaborative Research Centre (CRC) 606: "Non-stationary Combustion: Transport Phenomena, Chemical Reactions, Technical Systems" was founded to investigate the basics of combustion and for the implementations relevant processes coupled to combustion (Bockhorn et al., 2003; SFB 606, 2002).

Modern combustion concepts comprise lean premixed (LP) combustion, which allows for the reduction of the pollutant emissions, in particular oxides of nitrogen (NO_x) (Lefebvre, 1995). Lean premixed combustors are, however, prone to combustion instabilities with both low and high frequencies. These instabilities result in higher emission, acoustical load of the environment and even in structural damage of the system.

A subproject in CRC 606 was dedicated to investigate low frequency instabilities in combustion systems. The main goal of this subproject was to validate an analytical model, which was developed to describe the resonant characteristics of combustion systems consisting of Helmholtz resonator type components (burner plenum, combustion chamber) (Büchner, 2001). The subproject included experimental and numerical investigations as well. The goal of the numerical part was to find a reliable tool in order to predict the damping ratio of the system. The damping ratio is a very important input of the analytical model. The combination of the numerical prediction of the damping ratio and the analytical model enables the stability investigation of a system during the design phase.

In the numerical part Large Eddy Simulation (LES) was used to predict the damping ratio as previous investigations with unsteady Reynolds-averaged Navier-Stokes simulation (URANS) failed to predict the damping ratio satisfactorily (Rommel, 1995). The results of LES showed a very good agreement with the experimentally measured damping ratio. The focus of this chapter is to show results of further numerical investigations, which sheds light on a very important source of self-excited combustion instabilities, and to show how can provide LES the eigenfrequencies of a system.

In this chapter firstly a short description to combustion instabilities is given. After it the experimental and the numerical investigations of the resonant characteristics of the combustion systems will be shown briefly. In these investigations the system was excited with a sinusoidal mass flow rate at the inlet and the system response was captured at the outlet. Contrarily in the ensuing numerical investigations there is no excitation at the inlet and the system is still pulsating. The source of this pulsation and the consequences will be discussed.

It is important to notice that in these investigations the flow is non-reacting. There is no combustion, thus no flame in the combustion chamber. Hence there is no self-excited thermo-acoustic oscillation. In the subproject of CRC 606 the investigations of the low-frequency oscillations in the range of a few Hz up to several $100\ Hz$ were focused on the passive parts of the system: the combustion chamber and the burner plenum. The determination of the flame resonant characteristics is the object of other works (Büchner, 2001; Giauque et al., 2005; Lohrmann et al., 2004; Lohrmann & Büchner, 2004, 2005), and also of an other subproject within the CRC 606.

It is also important to clarify here that in these investigations the ignition stability of the flame will not be concerned. The combustion instabilities mentioned here are driven by thermo-acoustic self-excited oscillations. If there is no pulsation in the combustion chamber the flame is stable. Furthermore pulse combustors designed for oscillations are also not dealt within this chapter (Reynst, 1961; Zinn, 1996).

On the other hand, if the flow in a combustion system without flame is investigated the mostly used terms to express this are "cold flow", "non-reacting flow" or "isothermal condition". The last one neglects any changes in the temperature of the gas beyond the one occurred by the heat release of the flame. This is however misleading for peoples who do not investigate flames and physically incorrect. The LES results showed temperature changes due to the pulsation nearly $100\ K$ in the exhaust gas pipe, which is then in the range of 10% of the temperature changes produced by the flame.

2. Combustion instabilities

It is an indispensable prerequisite for the successful implementation of advanced combustion concepts to avoid periodic combustion instabilities in combustion chambers of turbines and in industrial combustors (Büchner et al., 2000; Külsheimer et al., 1999). For the elimination of the undesirable oscillations it is important to know the mechanisms of feedback of periodic perturbations in the combustion system. If the transfer characteristics of the subsystems (in a simple case burner, flame and chamber) furthermore of the coupled subsystems are known, the oscillation disposition of the combustion system can be evaluated during the design phase for different, realistic operation conditions (desired load range, air ratio, fuel type, fuel quality and temperature).

In order to get a high density of heat release flux i.e. power density and simultaneously low NO_x emission highly turbulent lean premixed or partially premixed flames are mostly used (Lefebvre, 1995). Significant property of these flames is that any disturbances in the equivalence ratio through turbulence or in the air/fuel mixture supply produce a very fast change in the heat release. Compared to axial jet flames the premixed swirl flames can significantly amplify the disturbances (Büchner & Külsheimer, 1997). The combustion process is increasingly sensitive to perturbation in the equivalence ratio under lean operating conditions.

Unsteady heat release involves pressure and velocity pulsation in the combustion chamber. These can result in thrust/torque oscillation, enhanced heat transfer and thermal stresses to combustor walls and other system components, oscillatory mechanical loads that results in low- and high-cycle fatigue of system components (Joos, 2006; Lieuwen & Yang, 2005). The oscillation of flow parameters can increase the amplitude of flame movements. This can cause blowoff of the flame or, in worst case, a flashback of the flame into the burner plenum. There are several mechanisms suspected of leading to combustion instabilities, such as periodic inhomogeneities in the mixture fraction, pressure sensitivity of the flame speed and the formation of large-scale turbulent structures.

The coupling of flame and acoustics can produce self-excited thermo-acoustic pulsation. The pulsation will be amplified then to the "limit cycle". Thermo-acoustic or thermal acoustic oscillations (TAO) were observed at first by Higgins in 1777 during his investigation of a "singing flame" (Higgins, 1802). The computation of self-excited thermo-acoustic oscillations began with the investigation of the Rijke-tube in (Lehmann, 1937). A short overview about the history of simulations of TAO is given in (Hantschk, 2000). It shows that most of the investigators wanted to compute oscillations excited by the flame or the system with flames excited by an external force at least. Because of the complexity of the problem many computations could not predict the limit cycle.

Lord Rayleigh proposed for the first time a criterion, which, regardless of the source of the instabilities, describes the necessary condition for instabilities to occur (Rayleigh, 1878). The criterion expresses that a pressure oscillation is amplified if heat is added at a point of maximum amplitude or extracted at a point of minimum amplitude. If the opposite occurs, a pressure oscillation is damped. The mathematical representation of this criterion was first proposed in (Putnam, 1971) as:

$$\int_0^T \tilde{q}(t) \cdot \tilde{p}(t) dt > 0 \qquad (1)$$

where $\tilde{q}$ and $\tilde{p}$ are the fluctuating parts of the heat release rate and the pressure, respectively, t is the time and T is the period of the pulsation. The condition will be satisfied for a given frequency if the phase difference between the heat release oscillation and the pressure oscillation is less than ±90°. Additionally, the amplitude of the pressure oscillation will be amplified if the losses through the damping effects are less than the energy fed into the oscillation. More appropriate forms of the Rayleigh criterion and similar criterions can be found in (Poinsot & Veynant, 2005).

2.1 Suppression of combustion-driven oscillations

In combustion systems of highly complex shape there can be more various modes: low frequency bulk mode, transversal, tangential, radial and longitudinal modes. In such a combustion system it is almost impossible today to predict all the unstable operating points. There are more strategies in practice to suppress the combustion oscillations in the unstable operating points. These can be grouped into passive and active control methods.

Passive or static control methods tune the resonance characteristics of the combustion system with additional devices as quarter-wave tube, Helmholtz resonators, sound-absorbing batting, orifice, ports and baffles (Putnam, 1971). Resonators can be placed in the fuel system (Richards & Robey, 2008), in the combustor (Gysling et al., 2000) or in other components. Perforates can be used at the premixer inlet (Tran et al., 2009), which is also an additional resonator to tune the resonant characteristics of the system. Instabilities can also be suppressed by means of injection of aluminium (Heidmann & Povinelli, 1967). Passive or static control strategies methods are more robust and need a minimum of maintenance. Their disadvantage is that while an unstable operating point is removed, another may arise. An overview about theory and practice of active control methods is given in (Annaswamy & Ghoniem, 2002). Active control methods can be subdivided into open-loop (Richards et al., 2007) and closed-loop design (Kim et al., 2005). Active control is achieved by a sensor in the combustion chamber, which measures frequency and phase of the combustion oscillation. The measured signal is analyzed and a proper periodic response is determined. The response is either an acoustic perturbation (Sato et al., 2007) or a modulation of the fuel injection (Guyot et al., 2008). Active control is able to suppress combustion instabilities substantially and is already in use for numerous practical applications. However, the apparatus is rather expensive and needs continuous maintenance. A failure of the control system can lead to a break down of the combustion system.

Based on the investigations of combustion instabilities (Culick, 1971; Zinn, 1970) there is also an approach to keep off unstable regimes during altering operation conditions. Online prediction of the onset of the combustion instabilities can help the operator to avoid, that the system becoming unstable (Johnson et al., 2000; Lieuwen, 2005; Yi & Gutmark, 2008). This technique is very useful if the ambient conditions vary in wide range e.g. for aircraft gas turbine. For stationary gas turbines with approximately constant ambient conditions, however, this cannot help to design the system for operation conditions, where combustion instabilities are not present.

2.2 System analysis

In order to analyse the stability of the system control theory can be used. The combustion system can be divided in subsystems as burner plenum, flame and combustion chamber (Baade, 1974; Büchner, 2001; Lenz, 1980; Priesmeier, 1987). The simplified feedback loop of these subsystems is depicted in Fig. 1. A perturbation of the pressure in the combustion chamber influences the mass flow rate at the burner outlet. This changes the heat release rate of the flame, which results in an alteration of the pressure in the combustion chamber. The transfer function of this closed loop and the subsystems can be determined by system identification furthermore the stability can be investigated by e.g. the Nyquist criterion (Deuker, 1994; Sattelmayer & Polifke, 2003a, 2003b).

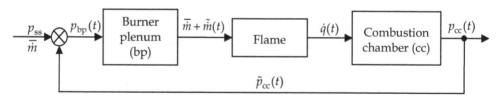

Fig. 1. Feedback loop of a combustion system with mass flow rate, pressure and heat release rate signals

If the system is built from these elements, a thermoacoustic network can be modelled to predict the unstable modes (Bellucci, 2005). Here, however, some information from measurement is needed.

If the phase shift and gain of the components is known the amplification of the pulsation can be predicted by means of the Rayleigh criterion. This shows that the accurate knowledge of the phase and gain relationship between pressure and heat release oscillation is a key issue to design stable combustion systems.

2.3 Helmholtz resonator

Helmholtz resonators are mostly used as passive devices for attenuations of pulsations in combustion systems. Furthermore the resonance behaviour of the combustion system can be described if it bears analogy to this resonator.

If a cavity is coupled to the ambient through a port (Fig. 2), the gas in this system can be forced into resonance if excited with a certain frequency. Such a geometrical configuration is named Helmholtz resonator after Hermann von Helmholtz, who investigated such devices in the 1850s. The port is the resonator neck, the cavity is the resonator.

The mechanical counterpart of the Helmholtz resonator is a mass-spring-damper system (Fig. 2). The gas in the neck acts as the mass, the gas in the cavity acts as the spring. The identification of the damping is more difficult. There are linear and non-linear effects in the flow. Damping is provided by the bulk viscosity during the pressure-volume work, the laminar viscosity in the oscillating boundary layer in the resonator neck, the vortex shedding at the ends of the resonator neck at the inflow and outflow and the dissipation of the kinetic energy through turbulence generation. Which source is dominating in the pulsating flow in the combustion system is discussed in (Pritz, 2010).

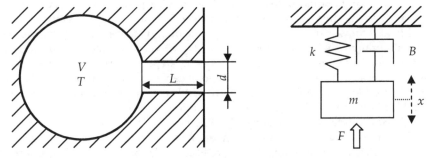

Fig. 2. The Helmholtz resonator and a mass-spring-damper system

The eigenfrequency of the Helmholtz resonator can be predicted as:

$$f_0 = \frac{c}{2 \cdot \pi} \cdot \sqrt{\frac{A}{V \cdot \left(L + \frac{\pi}{4} \cdot d\right)}} \tag{2}$$

where c is the speed of sound and can be calculated from the temperature T, the specific heat ratio γ and the specific gas constant R of an ideal gas as:

$$c = \sqrt{\gamma R T} \ . \tag{3}$$

Furthermore in Eq. (2) d is the diameter, L is the length and A is the cross section area of the neck, V is the volume of the resonator. The second term in the parenthesis in the denominator is a length correction term, which can be different for Helmholtz resonators with different geometries.

In order to describe the resonance behaviour of combustion systems, they can be treated as single or coupled Helmholtz resonators. In combustion systems the combustion chamber, the burner plenum or other components with larger volume act as resonators. The exhaust gas pipe and the components coupling the resonator volumes together are resonator necks. In industrial combustors the identification of the components of the Helmholtz resonators is easier, in gas turbines more difficult. It is very important which components are assumed to be coupled and which are decoupled. Wrong assumptions can lead to predicting modes incorrectly or even it is impossible to predict certain modes.

2.4 The reduced physical model

The suppression of the combustion oscillations is not a universal solution. The main goal is to design the combustion system not to be prone to combustion instabilities.

For the prediction of the stability of combustion systems regarding the development and maintaining of self-sustained combustion instabilities the knowledge of the periodic-non-stationary mixing and reacting behaviour of the applied flame type and a quantitative description of the resonance characteristics of the gas volumes in the combustion chamber is conclusively needed. In order to describe the periodic combustion instabilities many attempt have been made to assign the dominant frequency of oscillation to the geometry of the combustion chamber. For the description of the geometry-dependent resonance frequency of the system the equations were derived under the assumption of undamped oscillation (e.g. $\frac{1}{4}$ wave resonator, Helmholtz resonator). These models predict the resonance frequency quite accurate since the shift due to the moderate damping in the system is negligible. Such a simplified model, however, is not applicable for a quantitative prediction of the stability limit of a real combustion system. On one hand it predicts infinite amplification at the resonance frequency. On the other hand the frequency-dependent phase shift between input and output is described by a step function, hence it cannot be used for the application of a phase criterion (Rayleigh or Nyquist criterion), which is used to predict the occurrence of pressure and heat release oscillations in real combustion systems.

A reduced physical model was developed in (Büchner, 2001), which is able to describe the resonance characteristics of combustion chambers, if their geometry satisfies the geometrical conditions of a Helmholtz resonator (Arnold & Büchner, 2003; Büchner, 2001; Lohrmann et al., 2001; Petsch et al., 2005; Russ & Büchner, 2007). The reduced physical model was derived similar to the resonance behaviour of a mass-spring-damper system, which provides a continuous transfer function of the amplification and the phase shift. First the model was developed to describe a single resonator, later it was extended to a coupled system of two resonators. For this reduced physical model scaling laws were developed based on experimental data. The influence of the amplitude of pulsation, the mean mass flow rate, the temperature of the gas and the geometry were investigated.

In this model the damping in the system is expressed by an integral value. The damping factor cannot be determined by analytical solution. The accurate determination of the damping based on the 2nd Rayleigh-Stokes problem is not possible because of the complexity and non-linearity of the flow motion in the chamber and in the exhaust gas pipe. It was, however, possible to derive a scaling law for the damping in function of the gas temperature. A scaling law for the dependency of the damping on the length of the exhaust gas pipe could be also derived but its prediction is less accurate (Büchner, 2001).

There is a possibility to determine the integral value of the damping ratio by one measurement e.g. at the resonance frequency predicted by the undamped Helmholtz resonator model. This is, however, feasible only if the combustion system already exists. In order to determine the value of the damping factor in the design stage numerical simulation should be carried out.

3. Resonant characteristics of combustion systems

As mentioned in the Introduction the investigations focused on the passive parts of the system: burner plenum and combustion chamber (including exhaust gas pipe). Here two configurations will be discussed. A single combustion chamber as a single resonator, and a coupled system of burner plenum and combustion chamber as coupled resonators.

3.1 Experimental setup

Former experimental investigations showed that the combustion chamber has specific impact on the stability of the overall system. As first approximation, if the components upstream to the combustion chamber are decoupled by the pressure loss of the coupling element (e.g. burner), the only vibratory component is the combustion chamber, and the system can be treated as a single resonator.

In Fig. 3 the sketch of the experimental setup is shown. In the experiments the transfer function of the combustion chamber was calculated from the input signal measured with the hot-wire probe 1 at the inlet of the chamber and from the output signal measured with hot-wire probe 2 at the exit cross section of the exhaust gas pipe (Arnold & Büchner, 2003). An alternative output signal was the pressure measured with a microphone probe at the middle of the side wall in the combustion chamber (Büchner, 2001).

The model of the single Helmholtz resonator describes combustion systems sufficiently precise only in a first approximation, since real combustion systems in general have more vibratory gas volumes in addition to the combustion chamber (mixing device, air/fuel supply, burner plenum and exhaust gas system). The linking of these vibratory subsystems results in a significantly more complex vibration behaviour of the overall system compared to the single combustion chamber. To get closer to real combustion systems the model of the single Helmholtz resonator must be extended to describe more resonators coupled to each other.

For modelling a coupled system the burner plenum was added upstream to the combustion chamber. The reduced physical model was extended for the coupled system of burner and combustion chamber (Russ & Büchner, 2007). In order to prove the prediction of the model for the coupled system different geometric parameters (burner volume, resonator geometry) and operating parameters (mean mass flow rate) were varied in the experimental part. In each case the flow was non-reacting. The transfer function was calculated from the input signal (inlet of the burner plenum) and output signal (exit cross section of the exhaust gas pipe) similar to the case of the single resonator. The sketch of the experimental setup and the analogy of a mass-spring-damper system are shown in Fig. 4.

In order to excite the system at different discrete frequencies a pulsator unit was used. This unit could produce a sinusoidal component of the mass flow rate with prescribed amplitude and frequency (Büchner, 2001). For example in the case of the coupled Helmholtz-resonators in Fig. 4 the mean volume flow rate is partially pulsated by the pulsator unit. The pulsating flow passes through the burner plenum (bp), reaches the combustion chamber (cc) through the resonator neck and leaves the system at the end of the exhaust gas pipe (egp).

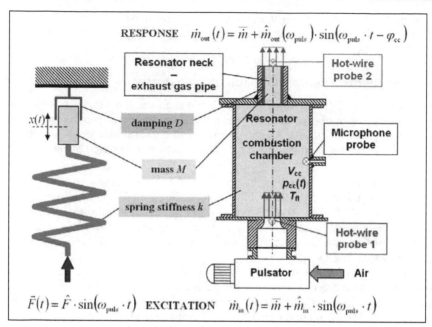

Fig. 3. The sketch of the test rig and the analogy of the mass-spring-damper system and the combustion chamber as Helmholtz resonator

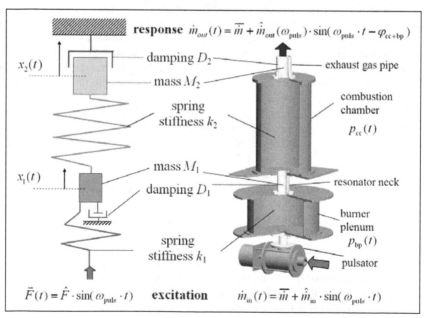

Fig. 4. Coupled Helmholtz-resonators and oscillating masses connected with springs and damping elements

3.2 Numerical setup

In order to compute the resonance characteristics of the system a series of LESs at discrete forcing frequencies had to be completed.

In the case of the single resonator these were taken for a basic configuration corresponding to the experiments, for variation of the geometry of the resonator neck and for variation of the fluid temperature. The compressible flow in the chamber of the basic geometry was simulated for five different frequencies in the vicinity of the resonance frequency. A detailed description of the investigated cases is omitted here as it is not in the focus of this chapter and it can be found in (Magagnato et al., 2005).

In the case of the coupled resonators one configuration was investigated. Ten LESs were calculated at different excitation frequencies, because the domain of interest is a broader frequency range than in the case of the single resonator. A detailed description of this investigation can be found in (Pritz et al., 2009).

3.2.1 Numerical method

The main goal of the numerical investigation was to predict the damping coefficient of the system which is an important input for the reduced physical model. In order to provide an insight into the flow mechanics inside the system LES were carried out. LES is an approach to simulate turbulent flows based on resolving the unsteady large-scale motion of the fluid while the impact of the small-scale turbulence on the large scales is accounted for by a sub-grid scale model. By the prediction of flows in complex geometries, where large, anisotropic vortex structures dominate, the statistical turbulence models often fail. The LES approach is for such flows more reliable and more attractive as it allows more insight into the vortex dynamics. In recent years the rapid increase of computer power has made LES accessible to a broader scientific community. This is reflected in an abundance of papers on the method and its applications.

The solution of the fully compressible Navier-Stokes equations was essential to capture the physical response of the pulsation amplification, which is mainly the compressibility of the gas volume in the chamber. Viscous effects play a crucial role in the oscillating boundary layer in the neck of the Helmholtz resonator and, hereby, in the damping of the pulsation. The pulsation and the high shear in the resonator neck produce highly anisotropic swirled flow. Therefore it is improbable that a URANS can render such flow reliably.

The LESs of this system were carried out with the in-house developed parallel flow solver called SPARC (Structured PArallel Research Code) (Magagnato, 1998). The code is based on three-dimensional block structured finite volume method and parallelized with the message passing interface (MPI).

In the case of the combustors the fully compressible Navier-Stokes equations are solved. The spatial discretization is a second-order accurate central difference formulation. The temporal integration is carried out with a second-order accurate implicit dual-time stepping scheme (Zou & Xu, 2000). For the inner iterations the 5-stage Runge-Kutta scheme was used. The time step was $\Delta t = 2 \cdot 10^{-5}\,s$ and $\Delta t = 2 \cdot 10^{-6}\,s$ for the single resonator and for the coupled resonators, respectively. This was a compromise in order to resolve the turbulent scales and compute the pulsation cycles within the permitted time. The Smagorinsky-Lilly model was chosen as subgrid-scale model (Lilly, 1967). Later investigations with MILES approach and dynamic Smagorinsky model show no significant difference in the results. This proofs that the mesh was sufficiently fine in the regions which are responsible for the damping of the pulsation, thus the modelling of the SGS structures has a minor influence there.

The Full Multigrid (FMG) method is used with four grid levels to achieve faster the statistically stationary state. The FMG method implies grid sequencing and a convergence acceleration technique. The number of cells on a grid level is eight time less then on the next finer grid level.

3.2.2 Computational domain and boundary conditions

If the flow in the combustion chamber and the resonator neck has to be simulated (grey area in Fig. 5) attention should be paid to some difficulties by the definition of the boundary conditions.

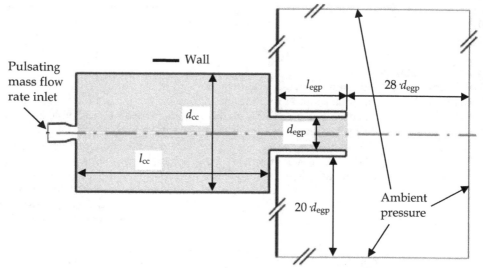

Fig. 5. Sketch of the computational domain and boundary conditions of the single resonator

Even though the geometry of the chamber is axisymmetric no symmetry or periodic condition could be used because the vortices in the flow are three-dimensional and they are mostly on the symmetry axis of the chamber. In the present simulations an O-type grid is used to avoid singularity at the symmetry axis.

At the inflow boundary the fluctuation components should be prescribed for a LES. Furthermore, the boundary must not produce unphysical reflections, if the pressure fluctuations, which move in the chamber back and forth, go through the inlet. A conventional boundary condition can reflect up to 60% of the incident waves back into the flow area. One can avoid these reflections only by the use of a non-reflecting boundary condition. If the inlet would be set at the boundary of the grey area, this problem can be solved hardly. In the experimental investigation a nozzle was used at the inflow into the chamber. The pressure drop of the nozzle ensures that the gas volume in the test rig components upstream of the combustion chamber does not affect the pulsation response of the resonator. It was decided to use this nozzle in the computation also. Although the additional volume of the nozzle increases the number of computing cells, a non-reflecting boundary condition is no more necessary. In addition, the fluctuation components at the inlet can be neglected, since the nozzle decreases strongly the turbulence level downstream.

At the inlet a partially pulsated mass flow rate was prescribed. The rate of pulsation was set to 25%.

The definition of the outflow conditions at the end of the exhaust pipe is particularly difficult. The resolved eddies can produce a local backflow in this cross section occasionally. In particular, by excitation frequencies in the proximity of the resonant frequency there is a temporal backflow through the whole cross section, which has been observed by the experimental investigations as well.

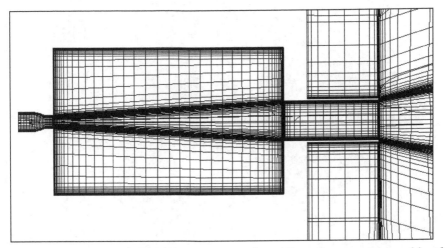

Fig. 6. Third finest mesh extracted to the symmetry plane (distortions were caused by the extraction in Tecplot)

The change of the direction of the flow changes the mathematical character of the set of equations. For compressible subsonic flow four boundary values must be given at the inlet and one must be extrapolated from the flow area. At the outlet one must give one boundary value and extrapolate four others. Since these values are a function of the space and time, their determination from the measurement is impossible. Further the reflection of the waves must be avoided also at the outlet. For these reasons the outflow boundary was set not at the end of the exhaust gas pipe, but in the far field. In order to damp the waves in direction to the outlet boundary mesh stretching was used.

At the solid surfaces the no-slip boundary condition and an adiabatic wall were imposed. For the first grid point $y^+<1$ was obtained, the turbulence effect of the wall was modelled with the van Driest type damping function. The geometry of the computational domain and the boundary conditions are shown in Fig. 5. The entire computational domain contains about $4.3 \cdot 10^6$ grid points in 111 blocks. A coarsened mesh is shown in Fig. 6.

The definition of the computational domain and the boundary conditions in the case of the coupled resonators were very similar. The geometry of the configuration chosen for the numerical investigation of the coupled resonators is illustrated in Fig. 7. The observation windows (for operations with flame) and the inserted baffle plates increased the complexity of the geometry and hence the generation of the mesh significantly. There were baffle plates placed in the burner plenum and in the combustion chamber to avoid the jet of the nozzle and of the resonator neck to flow directly through the system, furthermore to achieve a

homogeneous distribution of the velocity in the cross-section of the measuring point at the
end of the exhaust gas pipe.

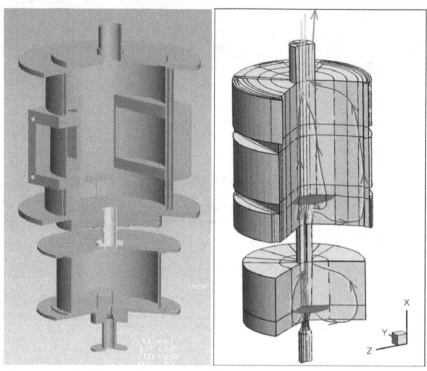

Fig. 7. Geometry of the test rig (left) and the 3D block-structure of the mesh (right)

The outlet boundary had to be modified somewhat compared to the case of the single
resonator. The size of this outflow region is 50 d_{egp} in axial direction and 40·d_{egp} in radial
direction. At the outlet surface at $x=5\ m$ the static pressure outlet condition is used and the
surface is inclined based on the observation explained next (Fig. 8). In order to obtain a
statistically steady solution before applying the excitation at the inlet a long time calculation
on the multigrid level 4 (coarsest mesh) and 3 was carried out. The entropy waves generated
by the transient of the initialization must be advected through the burner plenum and the
combustion chamber and finally out of the system. This needed a relative long time as the
convection velocity behind the baffle plates is quite small. After the acoustic waves
generated also by the transient of the initialization were decayed, it was detected, that
acoustic waves of a discrete frequency were amplified to extreme high amplitudes. The
wave length coincided width the length of the computational domain. After the outlet
surface was slanted these standing waves decayed.

For the distribution of the control volumes a very important aspect was to apply the
findings of the investigations of the single resonator. Thus much more computational cells
were arranged in the regions of the resonator neck and of the exhaust gas pipe, respectively,
and in this case around the baffle plates. The final version of the mesh consists of approx.
27 ·10^6 control volumes distributed among 612 blocks.

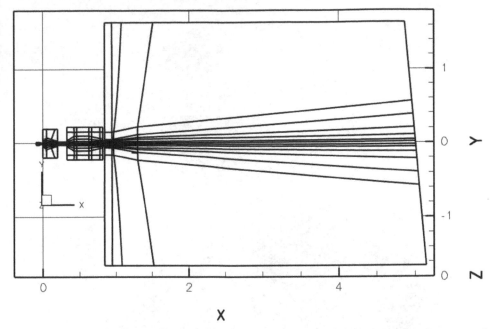

Fig. 8. The computational domain with block structure in the symmetry plane of the coupled resonators

3.3 Comparison of the results

The aim of the investigations of the single resonator was to identify the main damping mechanisms and estimate their effect on the stability of the system. In order to get an impression about the flow in the resonators iso-surfaces of the Q-criterion are plotted in Fig. 9. A detailed investigation of the pulsating flow is shown in (Pritz, 2010).

In this section the resonance characteristics of the combustion chamber obtained from experiments and computations are compared by means of the amplitude and phase transfer functions. The amplitude ratio of the mass flow rates is defined as:

$$A = \frac{\hat{\dot{m}}_{out}}{\hat{\dot{m}}_{in}} \qquad (4)$$

The amplitude ratios and phase shift were identified in the numerical simulations if the cycle limit was reached.

In Fig. 10 experimental data sets with the analytical model and the results of the computation are exhibited. In one case of the experiments the exhaust pipe was manufactured from a turned steel tube, in the other case the tube was polished. The LES data compare more favourable with the experimental data of polished tube, because the wall in the simulation was aerodynamically smooth, just like the polished resonator neck. The computation predicts the damping factor quite well; the deviation is about 7%. If the results of the measurement of the turned steel tube are compared with the simulation, the deviation is about 40%.

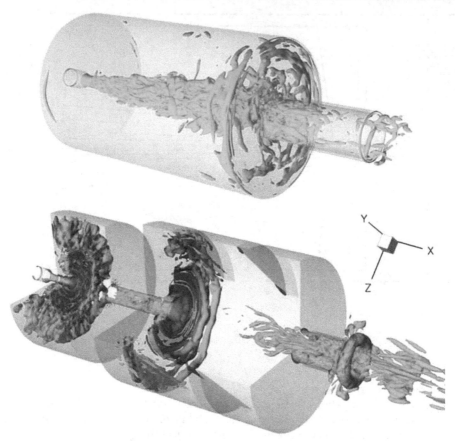

Fig. 9. Flow pattern in the resonators: iso-surfaces of the Q-criterion at $5 \cdot 10^4 \ s^{-2}$ in the single resonator (top) and at $10^4 \ s^{-2}$ in the coupled resonators (bottom)

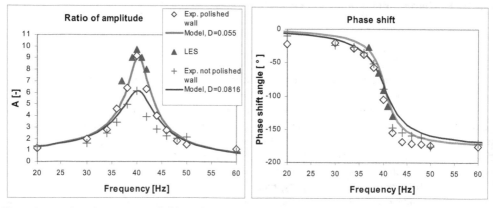

Fig. 10. Amplitude response (left) and phase transfer function (right) of the single resonator

The results of the coupled resonators on different grid levels are plotted in Fig. 11. The difference in the resonance characteristics on the finest and second finest grid is negligible. It was tested only at the lower resonant frequency, at the highest amplitude ratio, because the calculation on the finest mesh was very time consuming. The higher is the amplitude ratio the higher are the demands on the mesh. This result shows that the flow phenomena, which influence the damping, are adequate resolved on the second finest mesh. It is important to take into consideration that the mesh was optimized on the results of the investigation of the single resonator.

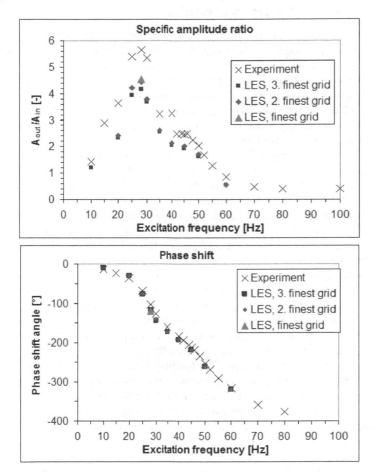

Fig. 11. Amplitude response (top) and phase shift function (bottom) of the coupled resonators

The plotted results in Fig. 11 show generally a very good prediction of the resonance frequencies and of the phase shift, respectively. In the gain, however, there is a discrepancy of approx. 20% in the prediction of the amplitude ratio at the highest peak, at f_{ex}=28 Hz. It was mentioned at the experimental setup that baffle plates were implemented in the burner plenum and in the combustion chamber. On these plates the flow is strongly deflected, there is a significant shearing (see Fig. 9). Unfortunately, in the experiments the plates were perforated. This was necessary to achieve the best velocity distribution at the outlet for the measurement with hot wire. In the simulations the wall condition was used for the plates. The resolution of the holes would yield a tremendous number of grid points. A boundary condition which can model this effect was not available. By the time the geometry data of the configuration were received, it was not possible to replace the plates any more. Probably this difference plays the major role in the underprediction of the amplitude ratio.

4. Investigation without external excitation

In the previous section the resonance characteristics of the system was measured experimentally and predicted numerically. In order to reduce the investigation on a few discrete excitation frequencies the eigenfrequency of the system must be approximated firstly. If the geometry is simple Eq. (2) can be used. In order to determine the amplification and the phase shift of the system well defined excitation had to be prescribed. Therefore a partially pulsated mass flow rate with a prescribed frequency near to the eigenfrequency was used in the numerical simulation at the inlet. In this section the simulations were carried out with a constant mass flow rate at the inlet.

The experiences of the investigations showed that the transient waves generated at the start of the computation should be decayed before the excitation with given amplitude and frequency was started. This was necessary to get the real system response at the outlet. A calculation was initialized with homogeneous distribution of each variable. This produces quite strong transient waves. The mass flow rate signal at the outlet of the exhaust gas pipe was used to monitoring the decaying of these waves. As soon as an almost constant mass flow rate was reached the computation could be continued on the second coarsest grid level. The extrapolation of the solution from the coarser on the next finer grid level produces also transient waves because of the sudden change of the shear stress at the walls. These waves are much smaller than the waves generated at the initialization but they are still considerable on the second coarsest grid level. The mass flow rate signal at the outlet of the exhaust gas pipe showed the decaying of these waves but later a certain amount of pulsation was observed and it decayed not at all. The amplitude of this pulsation was not negligible. As the mass flow rate was computed through integration over the whole cross section of the exhaust gas pipe the turbulent fluctuations were mostly filtered out.

At the description of the computational domain it was mentioned that the outlet boundary in the case of the coupled resonators had to be inclined to eliminate standing waves. These standing waves produced a dominant pulsation with large amplitude in the mass flow rate signal therefore the identification of the frequency was relative simple. In the case of the single resonator the computation was shorter so the standing waves were not amplified to a noticeable value. In the mass flow rate signal the frequency of the dominating wave was approximately at the eigenfrequency of the combustion chamber. In order to analyze this signal better the computation without excitation at the inlet was continued to get enough sample for a Fourier transformation.

The frequency spectrum plotted in Fig. 12 was computed from the mass flow rate signal on the second finest and finest mesh. For this computation the time step was increased to $\Delta t=10^{-4}\,s$, to a relative high value to get a better resolution of the spectra in the low frequency range. The samples were taken in each time step, thus the sampling frequency was 10 kHz furthermore the sampling length was 32768. The peak at 39 Hz agrees very good with the response function of the combustion chamber in Fig. 10. Recent investigations showed that the resonance frequency can be captured already on the coarsest grid level and the signal of the solution on the second coarsest grid can already predict the resonance frequency quite accurately.

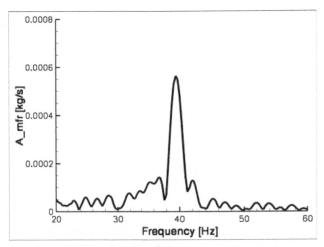

Fig. 12. Frequency spectrum of the outlet mass flow rate of the single resonator

It was shown in (Büchner, 2001) that the mass flow rate signal at the outlet and the pressure signal in the combustion chamber can be used as output signal equivalently i.e. the pulsation of the mass flow rate indicates a pulsation of the pressure in the chamber. The Fourier transform of the pressure signal measured at the middle of the side wall of the chamber gives the same result.

The mass flow rate at the inlet for this calculation was kept on a constant value. There was no external excitation in this computation and no turbulence at the inlet was described. The only possible forcing of the pulsation could arise from the turbulent motions inside the combustion chamber. The inflow into the chamber is a jet with strong shear layer which generates a broad band spectrum of turbulent fluctuations (Fig. 9). The combustion chamber then amplifies the pressure fluctuations generated by the turbulence at its eigenfrequency.

In order to investigate the effect of periodic flow instabilities further calculations with different mass flow rate at the inlet were carried out. It was changed to 200% and to 80% of the original value, respectively. The spectra of the mass flow rate of these calculations gave the same distribution in the low frequency range except the amplitude of the pulsation was changing proportional to the mean mass flow rate.

Based on these results the mass flow rate signal in the case of the coupled resonators was also investigated. In Fig. 13 the frequency spectrum of the mass flow rate signal on the second finest mesh is exhibited. The peaks at 27 Hz and 54 Hz correspond with the eigenfrequencies of the coupled system, which can be read e.g. from Fig. 11 at the phase shift angle 90° and 270°, respectively.

There are some possible mechanisms listed in the literature, which could trigger self-excited instabilities in combustion systems, but they are not sufficiently understood (Büchner, 2001; Joos, 2006; Poinsot & Veynant, 2005; Reynst, 1961). An important achievement of these simulations is that the pressure in the combustion chamber can pulsate already without any external excitation e.g. compressor or other incoming disturbances from ambient or even periodic flow instabilities depending on the design of the burner. Thus the flame is also pulsating. The amplitude of this pulsation will be amplified to the limit cycle if the time lag of the flame changed so that the pressure fluctuation and the heat release fluctuation meet the Rayleigh criterion.

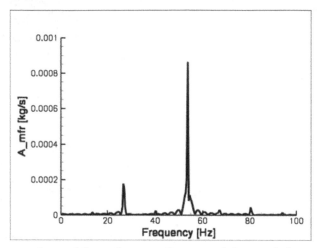

Fig. 13. Frequency spectrum of the outlet mass flow rate of the coupled resonators

The results of the earlier investigations show that the pulsation and the high shear in the resonator neck produce highly anisotropic swirled flow. Therefore it is unlikely that a URANS simulation can render such flow reliably. Furthermore if the turbulence is modelled statistically, it cannot excite the flow in the combustion chamber. The use of LES for the investigation of combustion instabilities is essential.

For the analytical model the eigenfrequency of the system is an important input parameter. If the geometry is rather simple the undamped Helmholtz resonator model can be used. Further important achievement of the present computations is that the eigenfrequency of the system with geometry of high complexity can be predicted without an additional modal analysis. The calculation with constant mass flow rate is a preparation for the investigation with excitation at the inlet. As the amplitude of the excitation is not well defined in the former case only the latter calculation can provide the damping ratio for the analytical model.

5. Conclusion

The lean premixed combustion allows for reducing the production of thermal NO_x, therefore it is largely used in stationary gas turbines and for other industrial combustion. Lean premixed combustors are, however, prone to combustion instabilities with both low and high frequencies. For the prediction of the stability of technical combustion systems the knowledge of the periodic-non-stationary mixing and reacting behaviour of the applied flame type and a quantitative description of the resonance characteristics of the gas volumes in the combustion chamber is conclusively needed.

In this chapter the numerical investigation of the non-reacting flow in a Helmholtz resonator-type model combustion chamber and in a coupled system of burner and combustion chamber is presented briefly. The work was a part of series of investigations to determine the stability limits of combustion systems. The resonance characteristics of the combustion systems were calculated using Large Eddy Simulation. The results are in good agreement with the experimental data and a reduced physical model, which was developed to describe the resonant behaviour of a damped Helmholtz resonator-type combustion chamber (Büchner, 2001).

The solution of the fully compressible Navier-Stokes equations was essential to capture the physical response of the pulsation amplification, which is mainly the compressibility of the gas volume in the chamber. Viscous effects play a crucial role in the oscillating boundary layer in the neck of the Helmholtz resonator and, hereby, in the damping of the pulsation. The pulsation and the high shear in the resonator neck produce highly anisotropic turbulent flow. Therefore it is improbable that an URANS simulation can render such flow reliably.

The investigation of the case without external excitation showed that the frequency spectrum of the mass flow rate signal at the outlet of the exhaust gas pipe provides a peak at the eigenfrequency of the combustion chamber. The only possible forcing of this pulsation was the turbulent fluctuations generated by the jet in the combustion chamber. The broadband excitation of the turbulent flow can be amplified by the flame and can produce a broadband background heat release rate oscillation as detected also in (Yi & Gutmark, 2008). If the eigenfrequency of the combustion chamber or other vibratory component is in the range of the frequencies of the energetic turbulent eddies a dominant pulsation can occur. The amplitude of this pulsation will be amplified to the limit cycle if the time lag of the flame changed so that the pressure fluctuation and the heat release fluctuation meet the Rayleigh criterion. If the turbulence is modelled statistically (URANS), it cannot excite the flow in the combustion chamber. The use of LES for the investigation of combustion instabilities is essential.

6. Acknowledgment

The present work was a part of the subproject A7 of the Collaborative Research Centre (CRC) 606 – "Unsteady Combustion: Transportphenomena, Chemical Reactions, Technical Systems" at the Karlsruhe Institute of Technology. The project was supported by the German Research Foundation (Deutsche Forschungsgemeinschaft - DFG). We also acknowledge support by DFG and Open Access Publishing Fund of Karlsruhe Institute of Technology. The calculations have been performed using the HP XC4000 of the Steinbuch Centre for Computing (SCC) in Karlsruhe.

7. References

Annaswamy, A.M., Ghoniem, A.F. (2002). Active Control of Combustion Instability: Theory and Practice, *IEEE Control Systems Magazine*, Vol.22, Nr.6, pp. 37-54

Arnold, G., Büchner, H. (2003). Modelling of the Transfer Function of a Helmholtz-Resonator-Type combustion chamber, *Proceedings of the European Combustion Meeting 2003 (ECM2003)*, Orléans, France

Baade, P. (1974). Selbsterregte Schwingungen in Gasbrennern, *Klima- + Kälteingenieur*, Nr. 4/74, Teil 6, pp. 167-176

Bellucci, V., Schuermans, B., Nowak, D., Flohr, P., Paschereit, C.O. (2005). Thermoacoustic Modeling of a Gas Turbine Combustor Equipped with Acoustic Dampers, *Journal of Turbomachinery*, Vol.127, Nr.2, pp. 372-379

Bockhorn, H.; Fröhlich, J.; Suntz, R. (2003). SFB606 – a German Research Initiative on Unsteady Combustion, *ERCOFTAC Bulletin*, Vol.59, pp. 40-44

Büchner, H. (2001). *Strömungs- und Verbrennungsinstabilitäten in technischen Verbrennungssystemen*, professorial dissertation, University of Karlsruhe, Germany

Büchner, H., Bockhorn, H., Hoffmann, S. (2000). Aerodynamic Suppression of Combustion-Driven Pressure Oscillations in Technical Premixed Combustors, *Proceedings of Symposium on Energy Engineering in the 21st Century (SEE 2000)*, Hong Kong, China, Ping Cheng (Ed.), Begell House, New York, Vol.4, pp.1573-1580

Büchner, H., Külsheimer, C. (1997). Untersuchungen zum frequenzabhängigen Mischungs- und Reaktionsverhalten pulsierender, vorgemischter Drallflammen, *GASWÄRME International*, Vol.46, Nr.2, pp. 122-129

Culick, F.E.C. (1971). Nonlinear Growth and Limiting Amplitude of Acoustic Oscillations in Combustion Chambers, *Combustion Science and Technology*, Vol.3, Nr.1, pp. 1-16

Deuker, E. (1994). *Ein Beitrag zur Vorausberechnung des akustischen Stabilitätsverhaltens von Gasturbinenbrennkammern mittels theoretischer und experimenteller Analyse von Brennkammerschwingungen*, PhD thesis, Universität of RWTH Aachen, Germany

Giauque, A., Selle, L., Gicquel, L., Poinsot, T., Büchner, H., Kaufmann, P., Krebs, W. (2005). System Identification of a Large-Scale Swirled Partially Premixed Combustor Using LES and Measurements, *Journal of Turbulence*, Vol.6, N 21

Guyot, D., Taticchi Mandolini Borgia, P., Paschereit, C.O. (2008). Active Control of Combustion Instability Using a Fluidic Actuator, *Proceedings of the 46th AIAA Aerospace Sciences Meeting and Exhibit*, 7-10 January 2008, Reno, Nevada, USA, AIAA 2008-1058

Gysling, D.L., Copeland, G.S., McCormick, D.C., Proscia, W.M. (2000). Combustion System Damping Augmentation with Helmholtz Resonators. *ASME Journal of Engineering for Gas Turbines and Power*, Vol.122, Nr.2, pp. 269-275

Hantschk, C.-C. (2000). *Numerische Simulation selbsterregter thermoakustischer Schwingungen*, Fortschritt-Berichte VDI, Reihe 6, Nr. 441

Heidmann, M. F., Povinelli, L. A. (1967). An Experiment on Particulate Damping in a Two-Dimensional Hydrogen-Oxygen Combustor. *NASA TM X-52359*

Higgins, B. (1802). On the Sound Produced by a Current of Hydrogen Gas Passing Through a Tube, *Journal of Natural Philosophy, Chemistry and the Arts*, Vol.1, pp. 129-131

International Energy Agency, *Key World Energy Statistics 2010*, www.iea.org

Johnson, C.E., Neumeier, Y., Lieuwen, T., Zinn, B.T. (2000). Experimental Determination of the Stability Margin of a Combustor Using Exhaust Flow and Fuel Injection Rate Modulations, *Proceedings of the Combustion Institute*, Vol.28, pp. 757-764

Joos, F. (2006). *Technische Verbrennung*, Springer, Berlin. ISBN-13 978-3-540-34333-2

Kim, K., Jones, C.M., Lee, J.G., Santavicca, D.A. (2005). Active Control of Combustion Instabilities in Lean Premixed Combustors, *Proceedings of the 6th Symposium on Smart Control of Turbulence*, Tokyo, Japan

Külsheimer, C., Büchner, H., Leuckel, W., Bockhorn, H., Hoffmann, S. (1999). Untersuchung der Entstehungsmechanismen für das Auftreten periodischer

Druck-/Flamme-nschwingungen in hochturbulenten Verbrennungssystemen, *VDI-Berichte* 1492, pp. 463

Lefebvre, A.H. (1995). The Role of Fuel Preparation in Low-Emission Combustion, *ASME Journal of Engineering for Gas Turbines and Power*, Vol.117, pp. 617-665

Lehmann, K.O. (1937). Über die Theorie der Netztöne (thermisch erregte Schallschwingungen), *Annalen der Physik*, Vol.421, Nr.6, pp. 527-555

Lenz, W. (1980). *Die dynamischen Eigenschaften von Flammen und ihr Einfluss auf die Entstehung selbsterregter Brennkammerschwingungen*, PhD thesis, University of Karlsruhe, Germany

Lieuwen, T.C. (2005). Online Combustor Stability Margin Assessment Using Dynamic Pressure Data, *Journal of Engineering for Gas Turbines and Power*, Vol.127, Nr.3, pp. 478-482

Lieuwen, T.C., Yang, V. (2005). *Combustion Instabilities in Gas Turbine Engines*, AIAA, ISBN-13 978-1-56347-669-3

Lilly, D. K. (1967). The Representation of Small-Scale Turbulence in Numerical Simulation Experiments. In *Proceedings of the IBM Scientific Computing Symposium on Environmental Sciences*, Yorktown Heights, N.Y., IBM form no. 320-1951, White Plains, New York, pp. 195-210

Lohrmann, M., Arnold, G., Büchner, H. (2001). Modelling of the Resonance Characteristics of a Helmholtz-Resonator-Type Combustion Chamber with Energy Dissipation, *Proceedings of the International Gas Research Conference (IGRC)*, Amsterdam, Netherlands

Lohrmann, M., Bender, C., Büchner, H., Zarzalis, N. (2004) Scaling of Stability Limits by Use of Universal Flame Transfer Functions, *Proceedings of the Joint Congress CFA/DAGA'04*, Vol.2, Strasbourg, France

Lohrmann, M., Büchner, H. (2005). Prediction of Stability Limits for LP and LPP Gas Turbine Combustors, *Combustion Science and Technology*, Vol.177, Nr.12, pp. 2243-2273

Lohrmann, M., Büchner, H. (2004). Scaling of Stability Limits of Lean-Premixed Gas Turbine Combustors, *Proceedings of ASME Turbo Expo*, Wien, Austria

Magagnato, F. (1998). KAPPA – Karlsruhe parallel program for aerodynamics. *TASK Quarterly*, Vol.2, pp. 215-270

Magagnato, F., Pritz, B., Büchner, H., Gabi, M. (2005). Prediction of the Resonance Characteristics of Combustion Chambers on the Basis of Large-Eddy Simulation, *International Journal of Thermal and Fluid Sciences*, Vol.14, Nr.2, pp. 156-161

Petsch, O., Pritz, B., Magagnato, F., Büchner, H. (2005). Untersuchungen zum Resonanzverhalten einer Modellbrennkammer vom Helmholtz-Resonator-Typ, *Verbrennung und Feuerungen - 22. Deutscher Flammentag*, VDI-GET, VDI-Berichte Nr. 1888, pp. 507-512

Poinsot, T., Veynant, D. (2005). *Theoretical and Numerical Combustion*, R.T. Edwards, Inc., Philadelphia, 2nd Edition, ISBN 1-930217-10-2

Priesmeier, U. (1987). *Das dynamische Verhalten von Axialstrahl-Diffusionsflammen und dessen Bedeutung für selbsterregte Brennkammerschwingungen*, PhD thesis, Universität of Karlsruhe, Germany

Pritz, B. (2010). *LES of the Pulsating, Non-Reacting Flow in Combustion Chambers*, SVH, ISBN 978-3-8381-1304-3, Saarbrücken, Germany

Pritz, B., Magagnato, F., Gabi, M. (2009). Stability Analysis of Combustion Systems by Means of Large Eddy Simulation, *Proceedings of Conference on Modelling Fluid Flow (CMFF'06), The 14th International Conference on Fluid Flow Technologies*, Budapest, Hungary

Putnam, A.A. (1971). *Combustion-Driven Oscillations in Industry*. American Elsevier, New York, USA

Rayleigh, J.W.S. (1878). *The Theory of Sound*, Volume 2. Macmillan, London, Kingdom of Great Britain

Reynst, F.H. (1961). *Pulsating Combustion*, Edited by M.W. Thring, Pergamon Press, Oxford, United Kingdom

Richards, G.A., Thornton, J.D., Robey, E.H., Arellano, L. (2007). Open-Loop Active Control of Combustion Dynamics on a Gas Turbine Engine. *Journal of Engineering for Gas Turbines and Power*, Vol.129, Nr.1, pp. 38-48

Richards, G.A., Robey, E.H. (2008). Effect of Fuel System Impedance Mismatch on Combustion Dynamics, *Journal of Engineering for Gas Turbines and Power*, Vol.130, Nr.1, 011510

Rommel, D. (1995). *Numerische Simulation des instationären, turbulenten und isothermen Strömungsfeldes in einer Modellbrennkammer*, master thesis, Engler-Bunte-Institute, University of Karlsruhe, Germany

Russ, M., Büchner, H. (2007). Berechnung des Schwingungsverhaltens gekoppelter Helmholtz-Resonatoren in technischen Verbrennungssystemen, *Verbrennung und Feuerung*, VDI-Berichte zum 23. Deutscher Flammentag

Sato, H., Nishidome, C., Kajiwara, I., Hayashi, A.K. (2007). Design of Active Control System for Combustion Instability Using H^2 Algorithm. *International Journal of Vehicle Design*, Vol.43, pp. 322-340

Sattelmayer, T., Polifke, W. (2003a). A Novel Method for the Computation of the Linear Stability of Combustors, *Combustion Science and Technology*, Vol.175, Nr.3, pp. 477-497

Sattelmayer, T., Polifke, W. (2003b). Assessment of Methods of the Computation of the Linear Stability of Combustors, *Combustion Science and Technology*, Vol.175, Nr.3, pp. 453-476

Tran, N., Ducruix, S., Schuller, T. (2009). Damping Combustion Instabilities with Perforates at the Premixer Inlet of a Swirled Burner, *Proceedings of the Combustion Institute*, Vol.32, pp. 2917-2924

Shell (2008). *Energy Scenarios to 2050*, Shell International BV, Amsterdam 2008, www.shell.com/scenarios

SFB 606 (2002). www.sfb606.kit.edu

Yi, T., Gutmark, E.J. (2008). Online Prediction of the Onset of Combustion Instability Based on the Computation of Damping Ratios, *Journal of Sound and Vibration*, Vol.310, pp. 442-447

Zinn, B.T. (1996). Pulse Combustion Applications: Past, Present and Future, In: *Unsteady Combustion*, Culick F. et al., (Ed.), NATO ASI Series, Vol.306, pp. 113-137

Zinn, B.T., Powell, E.A. (1970). Nonlinear Combustion Instabilities in Liquid Propellant Rocket Engines, *Proceedings of the Combustion Institute*, Vol.13, pp. 491-502

Zou, Z., Xu, L. (2000). Prediction of 3-D Unsteady Flow Using Implicit Dual Time Step Approach. *Acta Aeronautica et Astronautica Sinica*, Vol.21, Nr.1, pp 317-321

10

Unitary Qubit Lattice Gas Representation of 2D and 3D Quantum Turbulence

George Vahala[1], Bo Zhang[1], Jeffrey Yepez[2], Linda Vahala[3] and Min Soe[4]
[1]College of William & Mary
[2]Air Force Research Lab
[3]Old Dominion University
[4]Rogers State University
USA

1. Introduction

Turbulence is of vital interest and importance to the study of fluid dynamics Pope (1990). In classical physics, turbulence was first studied carefully for *incompressible* flows whose evolution was given by the Navier-Stokes equations. One of the most celebrated results of incompressible classical turbulence (CT) is the existence of an inertial range with the cascade of kinetic energy from large to small spatial scales until one reaches scale lengths on the order of the dissipation wave length and the eddies/vortices are destroyed. The Kolmogorov kinetic energy spectrum in this inertial range follows the power law in wave number space.

$$E(k) \sim k^{-5/3} \tag{1}$$

Independently, quantum turbulence (QT) was being studied in the low-temperature physics community on superfluid ^4He Pethick & Smith (2009); Tsubota (2008). However, as this QT dealt with a two-component fluid (an inviscid superfluid and a viscous normal fluid interacting with each other) it considered phenomena not present in CT and so no direct correspondence could be made.

With the onset of experiments in the Bose Einstein condensation (BEC) of dilute gases, we come to a many body wave function that at zero temperature reduces to a product of one-body distributions. The evolution of this one-body distribution function $\varphi(\mathbf{x}, \mathbf{t})$ is given by the Gross-Petaevskii (GP) equation Gross (1963); Pitaevskii (1961):

$$i\partial_t \varphi = -\nabla^2 \varphi + a(g|\varphi|^2 - 1)\varphi, \tag{2}$$

with the nonlinear term $|\varphi|^2 \varphi$ arising from the weak boson-boson interactions of the dilute BEC gas at temperature $T = 0$. Eq. (2) is ubiquitous in nonlinear physics: in plasma physics and astrophysics it appears as the envelope equation of the modulational instability while in nonlinear optics it is known as the Nonlinear Schrodinger (NLS) equation Kivshar & Agrawal (2003).

The quantum vortex is a topological singularity with the wave function $|\varphi| \to 0$ at the vortex core Pethick & Smith (2009). A 2π circumnavigation about the vortex core leads to an integer multiple of the fundamental circulation about the core: i.e., the circulation is quantized. Thus the quantum vortex is a more fundamental quantity than a classical vortex. In CT one has at any time instant a sea of classical vortices with continuously varying circulation and sizes, making it difficult to define what is a vortex. Thus there was the hope that QT (which consists of entangled quantum vortices Feynman (1955), while of fundamental importance in its own right, might shed some light on CT Barenghi (2008). Under the Madelung transformation

$$\varphi = \sqrt{\rho}\, e^{i\theta/2}, \qquad \text{with} \qquad \mathbf{v} = \nabla\theta \tag{3}$$

the GP Eq. (2) can be rewritten in conservation fluid form (ρ the mean density and $\mathbf{v}$ the mean velocity)

$$\partial_t \rho + \nabla \cdot (\rho \mathbf{v}) = 0 \tag{4}$$

$$\rho(\partial_t \mathbf{v} + \mathbf{v} \cdot \nabla \mathbf{v}) = -2\rho\nabla(g\rho - \frac{\nabla^2\sqrt{\rho}}{\sqrt{\rho}}). \tag{5}$$

Thus QT in a BEC gas will occur in a *compressible inviscid* quantum fluid whose pressure now has two major contributions: a barotropic pressure $\sim g\rho^2$ and a s-called quantum pressure $\sim -\sqrt{\rho}\,\nabla^2\sqrt{\rho}$.

In the seminal paper on QT in a BEC , Nore et. al. Nore et al. (1997) stress the Hamiltonian nature of the GP system and performed 3D QT simulations on a 512^3 grid using Taylor-Green vortices as initial conditions. Nore et. al. Nore et al. (1997) find a transient glimmer of a $k^{-5/3}$ incompressible energy spectrum in the low-k wave number range, but later in time this disappeared. Other groups Kobayashi & Tsubota (2007); Machida et al. (2010); Numasato et al. (2010); Tsubota (2008)] attacked QT in GP and just recently achieved a simulation on a 2048^3 grid using pseudo-spectral methods Machida et al. (2010) . However, most the algorithms of the Tsubota group introduce an ad hoc dissipative term to the GP equation. Ostensibly, this added dissipative term damps out the very large k-modes and a hyperviscosity effect is commonly used in (dissipative) Navier-Stokes turbulence to suppress numerical instabilities. However the introduction of dissipation into the GP Eq. (2) may be more severe since it destroys its Hamiltonian nature - while in CT this dissipation just augments the actual viscosity in the Navier-Stokes equation. There are a considerable number of studies of the effects of hyperviscosity on backscatter and bottlenecks in the kinetic energy spectrum. The presence of ad hoc dissipation in QT can have important effects on the energy backscatter from the large k to smaller k that is even seen in CT simulations [ref..]. They Machida et al. (2010) find a pronounced small-k region that exhibited a Kolmogorov-like incompressible energy spectrum of $k^{-5/3}$, which after some time decayed away.

We now briefly outline the sections in this chapter. In Sec. 2 the difference between 2D and 3D CT are briefly summarized. In Sec. 3 we introduce our novel unitary quantum lattice gas (QLG) algorithm to solve the GP equation. The untiary nature of our algorithm has several important features: (a) it completely respects the Hamiltonian structure of the $T = 0$ BEC dilute gas, and (b) it permits a determination of a class of initial conditions that exhibit very short Poincare recurrence times. This has not been seen before in other simulations of the GP equation because either this class of initial conditions has been missed or because the introduction of any ad hoc dissipative terms destroy the existence of any Poincare recurrence. In Sec. 4 we discuss our QLG simulations on 2D QT using a variety of initial conditions and compare our findings to the recent papers of Horng et. al. Horng et al. (2009) and Numasato et. al. Numasato et al. (2010) In Sec. 5 QLG simulations of 3D QT are presented with particular emphasis on the energy spectra. Finally in Sec. 6 we discuss future directions of QLG.

1.1 Coherence length, ζ

Before we summarize the differences between 2D and 3D CT we shall consider the concept of *coherence* length that is very commonly introduced in low-temperature physics. In particular we consider some of the ways the coherence length ζ is introduced for BECs and for QT. The first is a back-of-the-envelope definition of the coherence length Pethick & Smith (2009) : it is the length scale at which the kinetic energy is of the same order as the nonlinear interaction term in the GP Eq. (2). Approximating $\nabla \sim \zeta^{-1}$ and the nonlinear term by the asymptotic density $|\varphi| \sim \sqrt{\rho_0}$

$$\zeta = (\sqrt{ag\rho_0})^{-1} \tag{6}$$

A more quantitatively derived expression for ζ for1D vortices is given by Pethick & Smith Pethick & Smith (2009) who readily show that the exact steady state solution to the GP Eq. (2) is

$$\varphi(x) = \sqrt{\rho_0} \tanh(x/\sqrt{2}\zeta) \tag{7}$$

under the boundary conditions: $\varphi(0) = 0$ and $\varphi(x \to \infty) = \sqrt{\rho_0}$. The coherence length is thus the minimal distance from the singular core to the asymptotic value of the vortex wave function. A similar result, under the same boundary conditions, hold for 2D and 3D line vortices when one uses the Pade approximant expression of the radial part of the wave function following Berloff Berloff (2004) . Altenatively, Nore et. la. Nore et al. (1997) base their definition of coherence length on a *linear* perturbuation dispersion relation about uniform density. As Proment et. al.Proment et al. (2010) mention, the coherence length is strictly defined only for a single isolated vortex and to extend this to QT one assumes that

$$\zeta = (\sqrt{ag <\rho_0 >})^{-1} \tag{8}$$

where $< \rho_0 >$ is the spatially averaged BEC density.

The concept of coherence length becomes much more subtle for strongly nonlinear flows and when the (initial) wave function in the simulation is not a quasi-solution of the GP equation. For example, we shall consider random initial conditions or rescaled wave functions which thus are no longer solutions to the GP equation because the GP equation is nonlinear. For such problems the coherence length is at best a qualitative concept with possibly $\zeta = \zeta(t)$.

2. Incompressible Classical Turbulence: 2D and 3D CT

2.1 2D CT

In 2D incompressible CT it is convenient to introduce the scalar vorticity

$$\omega = (\nabla \times \mathbf{v}) \cdot \mathbf{e_z} \tag{9}$$

and write the 2D Navier-Stokes equation in the form

$$\partial_t \omega + \mathbf{v} \cdot \nabla \omega = -\nu \nabla^2 \omega \tag{10}$$

In the *inviscid* limit, $\nu \to 0$, there are two constants of the motion - the energy E and the enstrophy Z

$$2E = \int d\mathbf{r} \, |\mathbf{v}|^2, \qquad 2Z = \int d\mathbf{r} \, |\omega|^2 \tag{11}$$

Under the assumption of incompressibility, isotropy and self-similarity, Batchelor (1969); Kraichman (1967) have argued for the existence of a dual cascade in the inertial range:

an inverse cascade of (incompressible) kinetic energy to small k while a direct cascade of enstrophy to large k. From dimensional arguments they then predicted the spectral exponents of the energy spectrum $E(k)$, where $E = \int dk E(k)$:

$$\text{from the energy inverse cascade to small } k: \quad E(k) \propto k^{-5/3}; \tag{12}$$

$$\text{from the enstrophy direct cascade to large } k: \quad E(k) \propto k^{-3}. \tag{13}$$

2.2 3D CT

In 3D incompressible CT there is only one inviscid constant of the motion - the total energy. This leads to a direct cascade of energy from large to small k with the well-known Kolmogorov spectrum

$$E(k) \propto k^{-5/3} \tag{14}$$

When one considers *compressible* classical turbulence in real flows one is typically forced into some subgrid large eddy simulation (LES) modeling. It is interesting to note that in these models one typically requires closure schemes that will produce a Kolmogorov $k^{-5/3}$ spectrum for the subgrid *total* kinetic energy Menon & Genin (2010)

3. Unitary quantum lattice gas algorithm for the Gross-Pitaevskii equation

3.1 The unitary algorithm

We briefly outline the unitary quantum lattice gas (QLG) algorithm for the solution of the GP equation. Instead of employing a direct scheme to solve the GP equation we move to a more fundamental mesoscopic level and introduce qubits on a lattice. In particular, to recover the scalar GP equation we need to just introduce 2 qubits at each lattice site. A classical bits can take on only one of 2 possible values, which shall be denoted by $|0\rangle$ or $|1\rangle$. However a qubit can take on an arbitrary superposition of these two states: $|q\rangle = \alpha_q |0\rangle + \beta_q |1\rangle$. To take advantage of quantum entanglement we need to introduce at least 2 qubits per lattice site if the unitary collision operator is restricted to entangle only on-site qubits. To recover the GP Eq. (2), it will turn out that we will need to consider just 2 qubits per lattice site and of the basis set of the 4 posisble states $|00\rangle, |01\rangle, |10\rangle$ and $|11\rangle$, only the subset $|01\rangle, |10\rangle$ are required. We introduce the complex probability amplitudes α and β for these states $|01\rangle, |10\rangle$. Thus at each cubic lattice position x we introduce the (complex) two-spinor field

$$\psi(x,t) = \begin{pmatrix} \alpha(x,t) \\ \beta(x,t) \end{pmatrix} \tag{15}$$

and construct the evolution operator $U[\Omega]$ - consisting of an appropriate sequence of non-commuting unitary collision and streaming operators - so that in the continuum limit the two spinor equation

$$\psi(x,t+\Delta t) = U[\Omega]\,\psi(x,t). \tag{16}$$

will reduce to the GP equation for the 1-particle boson wave function φ under the projection

$$(1,1)\cdot\psi = \varphi. \tag{17}$$

Consider the unitary collision operator C that locally entangles the complex amplitudes α and β

$$C \equiv e^{i\frac{\pi}{4}\sigma_x(1-\sigma_x)} = \begin{pmatrix} \frac{1-i}{2} & \frac{1+i}{2} \\ \frac{1+i}{2} & \frac{1-i}{2} \end{pmatrix}, \tag{18}$$

where the σ are the Pauli spin matrices

$$\sigma_x = \begin{pmatrix} 0 & 1 \\ 1 & 0 \end{pmatrix} \qquad \sigma_y = \begin{pmatrix} 0 & -i \\ i & 0 \end{pmatrix} \qquad \sigma_z = \begin{pmatrix} 1 & 0 \\ 0 & -1 \end{pmatrix}. \tag{19}$$

Now C^2 turns out to be just the swap operator because of its action on the amplitudes α and β:

$$C^2 \begin{pmatrix} \alpha \\ \beta \end{pmatrix} = \begin{pmatrix} \beta \\ \alpha \end{pmatrix}. \tag{20}$$

Hence C is typically known as the square-root-of-swap gate.
The streaming operators shift just one of these amplitudes at x to a neighboring lattice point at $x + \Delta x$:

$$S_{\Delta x,0} \begin{pmatrix} \alpha(x,t) \\ \beta(x,t) \end{pmatrix} \equiv \begin{pmatrix} \alpha(x + \Delta x, t) \\ \beta(x,t) \end{pmatrix} \tag{21a}$$

$$S_{\Delta x,1} \begin{pmatrix} \alpha(x,t) \\ \beta(x,t) \end{pmatrix} \equiv \begin{pmatrix} \alpha(x,t) \\ \beta(x + \Delta x, t) \end{pmatrix}. \tag{21b}$$

The subscript $\gamma = 0$ on the streaming operator $S_{\Delta x,\gamma}$ refers to shifting just the amplitude α while the subscript $\gamma = 1$ refers to just shifting the amplitude β. These streaming operators can be expressed in an explicit unitary form by using the Paul spin matrices

$$S_{\Delta x,0} = n + e^{\Delta x \partial_x} \bar{n}, \qquad S_{\Delta x,1} = \bar{n} + e^{\Delta x \partial_x} n, \tag{22}$$

where $n = (1 - \sigma_z)/2$, $\bar{n} = (1 + \sigma_z)/2$. Note that the collision and streaming operators do not commute: $[C, S] \neq 0$.
Now consider the following interleaved sequence of unitary collision and streaming operators

$$J_{x\gamma} = S_{-\Delta x,\gamma} C S_{\Delta x,\gamma} C \tag{23}$$

Since $|\Delta x| \ll 1$ and $C^4 = I$, Eq.(23) yields $J_{x\gamma}^2 = I + O(\Delta x)$, where I is the identity operator. This is because the streaming operators are $O(\Delta x)$ deviations from the identity operator I. We first consider the effect of the evolution operator $U_\gamma[\Omega(x)]$

$$U_\gamma[\Omega(x)] = J_{x\gamma}^2 J_{y\gamma}^2 J_{z\gamma}^2 e^{-i\varepsilon^2 \Omega(x)}, \tag{24}$$

acting on the γ component of the 2-spinor ψ. Here $\varepsilon \ll 1$ is a perturbative parameter and Ω is a function to be specified later.
Using perturbation theory, it can be shown that the time advancement of ψ Yepez et al. (2009; 2010)

$$\psi(x, t + \Delta t) = U_\gamma[\Omega] \, \psi(x, t). \tag{25}$$

yields the spinor equation

$$\psi(x, t + \Delta t) = \psi(x, t) - i\varepsilon^2 \left[-\frac{1}{2}\sigma_x \nabla^2 + \Omega \right] \psi(x, t) +$$
$$\frac{(-1)^\gamma \varepsilon^3}{4} (\sigma_y + \sigma_z) \nabla^3 \psi(x, t) + \mathcal{O}(\varepsilon^4), \tag{26}$$

with $\gamma = 0$ or 1 and $\Delta x = O(\varepsilon)$. Since the order ε^3 term in Eq. (26) changes sign with γ, one can eliminate this term by introducing the symmetrized evolution operator

$$U[\Omega] = U_1\left[\frac{\Omega}{2}\right] U_0\left[\frac{\Omega}{2}\right].\qquad(27)$$

rather than just a single U_γ operator.

Under diffusion ordering, $\Delta t = O(\varepsilon^2) = \Delta x^2$, the evolution equation

$$\psi(x, t + \Delta t) = U[\Omega(x)]\,\psi(x, t)\qquad(28)$$

leads to the spinor equation

$$i\partial_t\psi(x, t) = \left[-\frac{1}{2}\sigma_x\nabla^2 + \Omega\right]\psi(x, t) + \mathcal{O}(\varepsilon^2),\qquad(29)$$

where the function Ω is still arbitrary. To recover the scalar GP Eq.2), one simply rescales the spatial grid $\nabla \to a^{-1}\nabla$, contracts the 2-component spinor field ψ to the (scalar) BEC wave function φ

$$\varphi = (1, 1)\cdot\psi = \alpha + \beta\qquad(30)$$

and chooses $\Omega = g|\varphi|^2 - 1$:

$$i\partial_t\varphi = -\nabla^2\varphi + a(g|\varphi|^2 - 1)\varphi + \mathcal{O}(\varepsilon^2).\qquad(31)$$

Thus the QLG algorithm that recovers the scalar GP Eq.(31) in the mesoscopic 2-spinor representation Eq.(28) with the evolution operator Eq.(27) and the component unitary operators defined by Eq.(24). It is important to realize that there is nothing per se in the QLG that enforces diffusion ordering - which is critical for QLG to recover the GP Eq.(2). This diffusion ordering must be recovered numerically by appropriate choices of available parameters and initial amplitudes.

Since the QLG is unitary, the norm of the spinor ψ is automatically conserved. However, one finds (small) fluctuations in the mean density due to the overlap of the components of the 2-spinor

$$\delta\bar{\rho} = \int dx(|\varphi|^2 - |\psi|^2) = \int dx(\alpha^\dagger\beta + \alpha\beta^\dagger).$$

If α is kept purely imaginary and β is kept purely real (or vice versa), the overlap between the two components vanishes and the mean density is conserved exactly. This is achieved by introducing two modifications to the QLA algorithm described above :

- initialize the 2-spinor so that $\alpha = Re[\varphi]$, $\beta = i\,Im[\varphi]$;

- replace the scalar potential function Ω by the unitary non-diagonal matrix:

$$\Omega_N = \begin{pmatrix} \cos[\Omega\Delta t] & -i\sin[\Omega\Delta t] \\ -i\sin[\Omega\Delta t] & \cos[\Omega\Delta t] \end{pmatrix},$$

such that $\sum_\gamma(\Omega_N\cdot\psi)_\gamma = e^{-i\Omega\Delta t}\varphi$, with $\Delta t = \varepsilon^2$.

Now QLG, like its somewhat distant mesoscopic cousin the lattice Boltzmann algorithm Succi (2001), are explicit second order accurate schemes in both space and time. We have found that QLG and lattice Boltzmann have similar numerical accuracy in 1D soliton simulations. Moreover, it has been shown that lattice Boltzmann approaches the accuracy of spectral methods Succi (2001). This has been attributed to the extremely small coefficient multiplying the second order error term.

All mesoscopic algorithms have to be carefully benchmarked - particularly as there can be some scaling requirements in order that the more general mesoscopic algorithm actually simulates the physics of interest. In particular, we have benchmarked our 1D QLG code against exact (theoretical) soliton solutions of the nonlinear Schrodinger equation - both scalar Vahala et al. (2003) and vector Vahala et al. (2004) soliton collisions - of nonlinear optics.

3.2 Parallelization performance on various architectures

One important property of QLG is that at this mesoscopic level the collision operator acts only on local data stored at the lattice site, while the streaming operator moves the post collision data to the nearest neighbor cell. This leads to ideal parallelization on supercomputers. Its unitary properties will permit direct encoding onto quantum computers. The scalings we report here are for recent codes that solve the coupled set of GP equations ($BEC2$) or for the $T > 0$ system which involves a coupling between the BEC ground state and the Bogoliubov modes (BdG). While the $BEC2$ code is an immediate vector generalization of the QLG of Sec. 2, the BdG code is a significant variation. Yepez has developed a 2-qubit QLG that now involves all 4-qubit states with 4×4 coupling matrices rather than the 2×2 coupling matrices in $BEC2$. The physics from these codes will be discussed elsewhere.

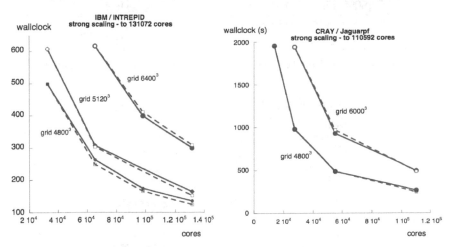

Fig. 1. Strong scaling of BEC2 code on IBM BlueGene/Intrepid and CRAY XT5/Jaguarpf for various grids. The solid (blue) line is wallclock time for the BEC2 code while the dashed (red) line is ideal timing. 100 iterations.

In strong scaling, one considers the code's performance on a fixed grid. One increases the number of processors and checks the wallclock time needed to do the same fixed number of time steps as before. Ideal scaling occurs if the wallclock time decreases by 0.5 if the number of processors are doubled. We see that the strong scaling of the $BEC2$-code is excellent. The

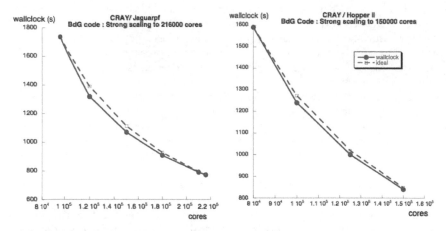

Fig. 2. Strong scaling of the $T > 0$ BdG code on CRAY XT5/Jaguarpf and CRAY XE6/Hopper II. The solid (blue) line is wallclock time for the BdG code while the dashed (red) line is ideal timing. Some superlinear scaling is apparent. Grid 8400^3, 100 iterations.

GRID	CORES	WALLCLOCK (IDEAL: 143.6s)	% PEAK
400^3	64	143.6	19.5%
800^3	512	144.5	19.4%
1600^3	4096	144.8	19.4%
3200^3	32768	150.8	18.6%

Table 1. Weak scaling of the BEC2 code on IBM/BlueGene Intrepid (100 iterations)

GRID	CORES	WALLCLOCK (IDEAL: 246.8s)
600^3	216	246.8
1200^3	1728	245.6
2400^3	13824	249.6
4800^3	110592	266.1

Table 2. Weak scaling of the BEC2 code on CRAY XT5/Jaguarpf (100 iterations)

GRID	CORES	WALLCLOCK (IDEAL: 261.0s)
600^3	216	261.0
1200^3	1728	261.8
2400^3	13824	263.9
3600^3	46656	268.4

Table 3. Weak scaling of the BdG code on CRAY XT5/Jaguarpf (100 iterations)

GRID	CORES	WALLCLOCK (IDEAL: 209.2S)	% PEAK
600^3	216	209.2	9.5%
1200^3	1728	212.7	9.4%
2400^3	13824	212.1	9.3%
3600^3	46656	212.7	9.0%
4000^3	64000	213.3	9.1%
4800^3	110592	214.0	

Table 4. Weak scaling of the BdG code on CRAYXE6/Hopper II (100 iterations)

dashed curve is the ideal scaling while the solid curve is our QLA code scaling. On the IBM BlueGene/P *Intrepid* this has been tested to 131072 cores and on larger grids (6400^3) it shows superlinear scaling (Fig. 1). Similar results are achieved on the CRAY *XT5Jaguarpf*, tested to 110592 cores and grids 6000^3. The *BdG*-code, with its 4×4 entangling matrices on the 2 qubits, shows quite strong superlinear scaling on both the CRAY *XT5* and *XE6*, testing to 216000 cores on *Jaguarpf* and to 150000 cores on *HopperII* (Fig. 2).

In weak scaling, one keeps the work done by each processor fixed. Thus if one doubles the grid in each direction, then one would need to increase the number of processors by 2^3 in a 3D simulations. For ideal scaling the wallclock time to completion should remain invariant. The weak scaling of these codes are given in Tables I - IV, and again are excellent, with fluctuations in timings typically below 5% as one moves from 216 cores to 110592 cores.

4. 2D QT

We first consider 2D quantum turbulence. From the GP Eq.2, the total energy in 2D QT is conserved

$$E_{TOT} = \int dx^2 \left(\frac{1}{2}\rho|\mathbf{v}|^2 + ag\rho^2 + 2|\nabla\sqrt{\rho}|^2 - a\rho \right) = const. \tag{32}$$

Since the GP equation also conserves the number of particles, the last term in Eq.(32) yields an ignorable constant. The other terms in Eq.(32) can be categorized as Nore et al. (1997)

$$\text{kinetic energy: } E_{kin} = \frac{1}{2} \int dx^2 \rho|\mathbf{v}|^2 \tag{33}$$

$$\text{internal energy: } E_{int} = ag \int dx^2 \rho^2 \tag{34}$$

$$\text{quantum energy: } E_{qnt} = 2 \int dx^2 |\nabla\sqrt{\rho}|^2. \tag{35}$$

To examine the effect of compressibility on the GP Eq.(2), E_K can be further decomposed into its incompressible and compressible components via Helmholtz decomposition Nore et al. (1997). Because a quantum vortex is a topological singularity, one needs to regularize the using definitions of velocity and vorticity. In particular, one defines a density weighted velocity $\mathbf{q} = \sqrt{\rho}\mathbf{v}$ and its Fourier transform to be $\tilde{\mathbf{q}}$. This differs from the standard Favre averaging used in standard *compressible* computational fluid dynamics (CFD) in that in QT one uses $\sqrt{\rho}$

rather than ρ. Unlike CFD, in QT at the vortex core the density $\rho \to 0$ with $\sqrt{\rho}\mathbf{v} \to const.$. The longitudinal component and the transverse component of $\tilde{\mathbf{q}}$ are

$$\tilde{\mathbf{q}}_l = \frac{\mathbf{k} \cdot \tilde{\mathbf{q}}}{|\mathbf{k}|^2} \tag{36}$$

$$\tilde{\mathbf{q}}_t = \tilde{\mathbf{q}} - \frac{\mathbf{k} \cdot \tilde{\mathbf{q}}}{|\mathbf{k}|^2} \tag{37}$$

so that $\mathbf{k} \times \tilde{\mathbf{q}}_l = 0$, $\mathbf{k} \cdot \tilde{\mathbf{q}}_t = 0$. Correspondingly the compressible energy and incompressible energy can be defined as

$$\text{compressible}: E_C = (2\pi)^2 \int dk^2 |\tilde{\mathbf{q}}_l|^2 \tag{38}$$

$$\text{incompressible}: E_{IC} = (2\pi)^2 \int dk^2 |\tilde{\mathbf{q}}_t|^2. \tag{39}$$

Consequently the energy densities of compressible and incompressible energy become

$$\varepsilon_{c,ic}(k) = k \int_0^{2\pi} d\theta |\tilde{\mathbf{q}}_{l,t}(k,\theta)|^2, \tag{40}$$

using polar coordinates k, θ. For 2D QT it is useful to introduce a renormalized vorticity Horng et al. (2009)

$$\omega_q = (\nabla \times [\sqrt{\rho}\mathbf{v}]) \cdot \mathbf{e}_z. \tag{41}$$

The two components of ω_q are:

- $(\nabla \sqrt{\rho}) \times \mathbf{v} \cdot \mathbf{e}_z$, with major contributions coming from density variations near vortices and from sound and shock waves;
- $\sqrt{\rho}\nabla \times \mathbf{v} \cdot \mathbf{e}_z \propto \delta(\mathbf{r} - \mathbf{r}_0)$, which pinpoints the locations of the vortices.

Consequently, the time evolution of the density weighted enstrophy, $Z_q = \int dx^2 |\omega_q|^2$, is an excellent measure of the time variation in the total number of vortices present in the 2D QT. Incompressibility and the conservation of enstrophy are crucial elements for the existence of the dual cascade in 2D CT. However, in 2D QT, one has neither incompressibility nor enstrophy conservation. Bogoliubov elementary excitations, permitted by the compressibility of the quantum fluid, can create quantum vortices. On the other hand, counter-rotating vortex pairs can annihilate each other (in CT one has the merging of like-rotating vortices). Lacking the conservation of enstrophy due to compressibility, 2D QT does not necessitate dual cascade. Another important feature of the GP Eq.(2) is that it is Hamiltonian with total energy conserved. Thus we are guaranteed that after a sufficiently long time, the system will return to a state that is very close to its initial state. For nearly all continuous Hamiltonian systems, this recurrence time T_P is effectively infinite. However Tracy et al. (1984) has demonstrated that in the 1D NLS (i..e the GP) system the Poincaré recurrence time can be unexpectedly short for certain initial manifolds. Arnold's cat map is an invertible chaotic map of a torus onto itself which is ergodic, mixing and structurally stable and it exhibits a short Poincaré recurrence time. To visualize such a recurrence under the Arnold cat map, we discretize a sample 2D square picture into $m \times m$ segments. Each segment is indexed with its 'coordinate' $[q_t, p_t]$, with $t = 1....m$. The Arnold cat map is applied to these segments:

$$q_{t+1} = Mod(2q_t + p_t, m) \tag{42}$$

$$p_{t+1} = Mod(q_t + p_t, m). \tag{43}$$

These segments are then patched together to form a new picture according to their updated coordinate $[q_{t+1}, p_{t+1}]$. We now consider the effect of pixel resolution on the Arnold cat map by considering $m = 74$ and $m = 300$. For $m = 74$ we find Poincaré recurrence after $T_P = 114$ iterations, Fig.3, while for the higher resolution $m = 300$, $T_P = 300$ iterations, Fig.4. What is quite interesting is the image after $t = T_P/2$ iterations: for the low resolution run, $m = 74$ there is a point inversion, Fig.3 (b) , while there is no point inversion for the high pixel resolution $m = 300$, Fig.4 (b). However, the Arnold cat map exhibits quite complex characteristics with pixel resolution. For example, at resolution $m = 307$, one *does* see a point inversion of the image at $t = T_P/2 = 22$.

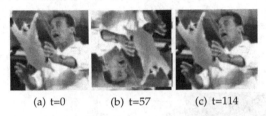

(a) t=0   (b) t=57   (c) t=114

Fig. 3. Poincaré recurrence under the Arnold cat mapping (I). Picture resolution: 74×74 pixels. Poincaré recurrence time $T_P = 114$. Notice at the semi-Poincaré time t=57, the picture is almost a point inversion of the initial condition.

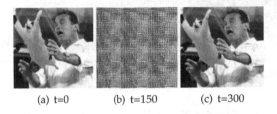

(a) t=0   (b) t=150   (c) t=300

Fig. 4. Poincaré recurrence under the Arnold cat mapping (II). Picture resolution: 300×300 pixels. Poincaré recurrence time $T_P = 300$. At the semi Poincaré time, the picture shows no strong symmetry to the initial condition.

In our simulations of 2D QT we encounter unexpected short Poincaré recurrence time provided that the ratio between the time-averaged internal energy and averaged kinetic energy is sufficiently small:

$$\gamma = \langle E_{int}(t) \rangle / \langle E_{kin}(t) \rangle \lesssim \mathcal{O}(10^{-1}) \tag{44}$$

The plots of amplitude, vorticity and phase in the following sections will adopt the "thermal" scheme: *blue* stands for low values while *red* stands for high values.

4.1 Poincaré recurrence in 2D QT
4.1.1 Four vortices with winding number 1 embedded in a Gaussian background BEC
For the first set of simulations, we study the evolution of a set of vortices embedded in a Gaussian BEC background. Periodic boundary conditions are assumed. The initial wave

function takes the form of products of individual quantum vortices, modulated by the inhomogeneous Gaussian background:

$$\psi(\mathbf{r}) = h\, e^{-a\, w_g\, r^2} \prod_i \phi_i(\mathbf{r} - \mathbf{r}_i).$$

$\phi_i = tanh(\sqrt{a}|\mathbf{r} - \mathbf{r}_i|)e^{i\, n\, Arg(\mathbf{r} - \mathbf{r}_i)}$ is the approximated wave function of a single 2D vortex with winding number n. Vortices with opposite charge, e.g. $n = 1$ and $n = -1$, are interleaved to form a vortex array that satisfies the periodic boundary conditions. Fig.5 illustrates the amplitude, density weighted enstrophy and phase of the initial wave function. The vortices can be identified by two methods: a) peaks in the vorticity plots or b) the end points of the branch cuts in the phase plots. For example, in Fig. 5(c), vortices with positive charge (phase varying from $-\pi$ to π counter-clockwise around the singularity) are encircled in black while those with negative charge (phase varying from $-\pi$ to π clockwise around the singularity) are encircled in white. In this simulation, grid size is 512 and the total number of iteration

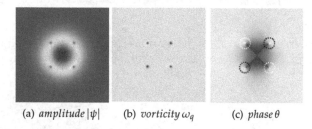

(a) amplitude $|\psi|$ (b) vorticity ω_q (c) phase θ

Fig. 5. Four vortices with winding number 1 embedded in a Gaussian background. $h = 0.05, a = 0.01, w_g = 0.01, g = 5.0$. Distance between vortices is L/4 with L being grid size 512. Since winding number is 1, the wave function's phase undergoes $\pm 2\pi$ change around the singularities.

steps is 50000. The ratio parameter $\gamma = 0.018$. A recurrence structure is clearly visible via the evolution of internal energy, $E_{int}(t)$, c.f. Fig.6. To further investigate such structure, we

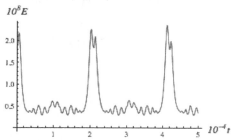

Fig. 6. Evolution of internal energy for winding number 1. The similarities exhibited by the two peaks around t=21000 and t=41900 clearly indicates a periodical structure.

plot the wave function's amplitude, vorticity and phase distribution. At Poincaré recurrence time $T_P = 41900$, the wave function recovers the initial condition except for some background noise. At semi-Poincaré recurrence time, $t = T_P/2 = 21000$, the wave function would be the same as initial condition if the origin of domain was shifted by $(L/2, L/2)$. This resembles the appearance of up-side-down point inversion pattern of the Arnold's cat mapping at a resolution of 74-pixels.

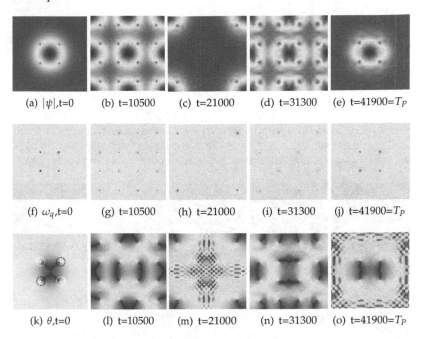

| (a) $|\psi|$,t=0 | (b) t=10500 | (c) t=21000 | (d) t=31300 | (e) t=41900=T_P |

| (f) ω_q,t=0 | (g) t=10500 | (h) t=21000 | (i) t=31300 | (j) t=41900=T_P |

| (k) θ,t=0 | (l) t=10500 | (m) t=21000 | (n) t=31300 | (o) t=41900=T_P |

Fig. 7. Poincaré recurrence for vortices with winding number $n = 1$. At Poincaré time $T_P = 41900$, all three distributions replicate the initial state except for some background noise, such as the small vortices at the boundaries appearing on the phase plot (o). At the semi-Poincaré time, due to periodic boundary conditions, the system is almost identical to its initial state provided the origins shift by $(-L/2, -L/2)$, i.e., a point inversion. Grid length $L = 512$.

4.1.2 Four vortices with winding number $n = 2$ embedded in a Gaussian BEC background
Vortices with winding number $n = 2$ are energetically unstable and will rapidly split into two $n = 1$ vortices. The Poincaré recurrence for winding number $n = 2$ case should be viewed as reproducing the state immediately following the vortex splitting. In this simulation, the energy ratio $\gamma = 0.0036$, which is much smaller than for the winding number $n = 1$ case due to the rotation induced by more vortices. The recurrence structure is most clearly visible via the evolution of the internal energy, Fig.8. With more vortices present in the system, there is more energy exchange between vortices and sound waves. Therefore when the Poincaré recurrence occurs, more background noise is to be expected, and is seen in Fig.9.
One interesting observation in this simulation is that Kelvin-Helmholtz instability is important for vortices generation as suggested in Blaauwgeers et al. (2002); Henn et al. (2009). When counter-rotating vortices approaches each other, in the regions between the vortex cores the velocity gradient can reach a certain critical value and trigger the Kelvin-Helmholtz instability. This instability will create pairs of counter-rotating vortices. As the number

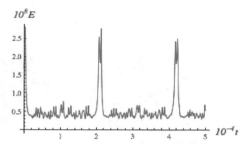

Fig. 8. Evolution of internal energy for winding number 2. Around $t \sim 21000$ and $t \sim 41900$, internal energy reaches peaks. Comparing with winding number 1 case, more fluctuation is observed and the peaks are much narrower. This is caused by the high density of number of vortices and the frequent annihilation and creation of counter-rotating vortex pairs.

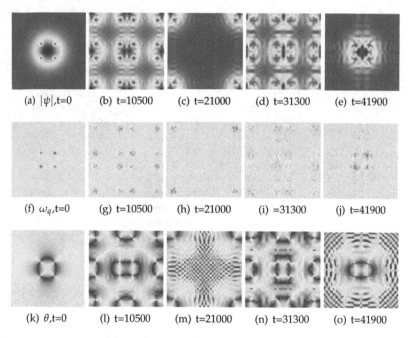

| (a) $|\psi|$,t=0 | (b) t=10500 | (c) t=21000 | (d) t=31300 | (e) t=41900 |

| (f) ω_q,t=0 | (g) t=10500 | (h) t=21000 | (i) =31300 | (j) t=41900 |

| (k) θ,t=0 | (l) t=10500 | (m) t=21000 | (n) t=31300 | (o) t=41900 |

Fig. 9. Poincaré recurrence for winding number $n =2$. At the semi-Poincaré time, the wave function approximates the initial condition with shift in the origins. At $t = 41900$, the wave function bears a similar structure as at $t = 0$, but with more noise.

of vortices increases in the system, the probability for Kelvin-Helmholtz triggered vortex generation increases. Fig.10 demonstrates such a process.

4.1.3 Random phase initial condition

In this simulation, the initial wave function features a constant amplitude and random phase $\psi(\mathbf{r}) = h\,e^{i\theta(\mathbf{r})}$. Bicubic interpolation is adopted to produce random phase which satisfies periodic boundary condition, c.f. Keys (1981). Under this interpolation, a 2D function

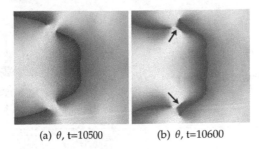

(a) θ, t=10500   (b) θ, t=10600

Fig. 10. Vortices generation via the Kelvin-Helmholtz instability. The region depicted in these plots is a blow-up of $[-256, -128] \times [-128, 128]$. (a) phase θ at t=10500; (b) θ at t=10600. At $t = 10600$, a new pair of counter-rotating vortices (pointed out by the black arrows) are generated between neighboring counter-rotating vortices.

defined on a unit square is approximated by polynomials: $p(x,y) = \sum_{i=0}^{3}\sum_{j=0}^{3} a_{i,j}x^i y^j$. The coefficients $a_{i,j}$ are determined by the enforced continuity at the corners. Since there are 16 unknown coefficients, 16 equations are needed to determine these $a_{i,j}$. Usually one can enforce continuity for $p(x,y), \partial_x p(x,y), \partial_y p(x,y), \partial_{x,y} p(x,y)$. To generate random phase in the domain L^2, the following procedure is followed:

- Discretize the domain into $m \times m$ squares. m here is the level of randomness;

- Generate 4 pseudo-random numbers at the 4 corners of each sub-square. These random numbers correspond to $p, \partial_x p, \partial_y p, \partial_{x,y} p$ at the corners of this sub-square;

- Enforce periodicity at the domain boundaries;

- Solve $a_{i,j}$ belonging to each single sub-square.

For a grid of 512^2 we consider a randomness level $m = 8$. The random phase wave function is dynamically unstable. Sound waves are immediately emitted and create quantum vortices. Typically these vortices will decay away and the GP system tends to a thermal equilibrium, as demonstrated by Numasato et al. (2010). However, if the initial condition is chosen such that energy ratio $\gamma \ll 1$, a Poincaré recurrence emerges. In our simulation, $\gamma = 0.00287$ and number of iteration is 100000. From the energy evolution plot, Fig.11, the Poincaré recurrence time can be clearly identified by the abrupt energy exchanges (i.e., the spikes).
The first spike in Fig.11 appears around $t = 21000$, with the second at $t = 41900$. These are just the semi-Poincaré and Poincaré times. One thus expects the phase distribution of the wave function at $t = 0, t = 10500, t = 21000$ and $t = 41900$ to illustrate the recurrence, c.f. Fig.12. What is remarkable is that the randomly distributed vortices suddenly disappear from the system at $T_P/2$ and T_P. At $t = 41900$, the phase distribution is very close to the initial state despite a constant shift in the central region. In Fig.13 we have plotted the density-weighted vorticity at various times: there are no vortices at $t = 0$ or at $T_P/2$, (e), although a considerable amount of sound waves.
As energy ratio γ increases, the strength of the Poincaré recurrence is weakened by noise. Fig.14 demonstrate how the Poincaré recurrence is lost as γ increases. When $\gamma = 0.0567$, one still can observe the depletion of vortices from the system at $T_P/2 = 21000$, however, at $T_P = 41900$, the initial condition, which is vortex free, can not be reproduced. For $\gamma = 0.133$,

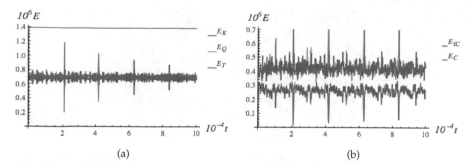

Fig. 11. Energy evolution of the GP system with random phase initial condition. (a): evolution of energies (E_K, E_Q, E_T). Internal energy is negligible compared to the kinetic energy E_K. (b) evolution of incompressible (*red*), E_{IC}, and compressible kinetic energy (*blue*), E_C.

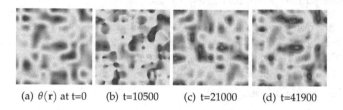

(a) $\theta(\mathbf{r})$ at t=0   (b) t=10500   (c) t=21000   (d) t=41900

Fig. 12. Poincaré recurrence with random phase initial condition. At $t = 10500$, many randomly distributed vortices can be identified via the branch cuts. At $t = 21000$ and $t = 41900$, no vortices exist in the system since there are no branch cuts. There is an induced phase shift seen in the color scheme of the phase plots at $t = 0$ and $t = 41900$, but the geometric patterns are the same.

no trace of a short Poincaré recurrence can now be found.

It needs to be pointed out that in our simulations, the Poincaré recurrence is characterized by abrupt energy exchange E_K and E_Q as well as among the compressible and incompressible components of the kinetic energy, E_C and E_{IC}. Therefore such phenomena can not be analyzed via standard turbulence theories invoking things like inverse cascades....

4.2 Energy cascade in 2D QT

For the simulations with vortices initially embedded in an Guassian BEC background, a k^{-3} power law is found ubiquitously in the compressible, incompressible and quantum energy spectra whenever vortices are present in the system. Fig.15 describes the time evolution of the incompressible energy spectrum $k^{s_{IC}}$. The linear regression fit is over $k \in [50, 100]$. Simulation grid 512^2.

In the time interval around $t \sim 24500$ for winding number $n = 1$ embedded vortices, we examine the sudden drop in the spectral exponent s_{IC}. In Fig.16, the linear regression fit for the incompressible kinetic energy spectrum is made over the wave number interval $k \in [50, 100]$. At $t = 24400$ and $t = 24600$, vortices are present in the system and the spectrum exponent $s_{IC} \sim -3$. At $t = 24500$, when all the vortices are depleted, the incompressible kinetic energy spectral exponent decreases to $s_{IC} = -5.828$. This could well indicate that the existence in

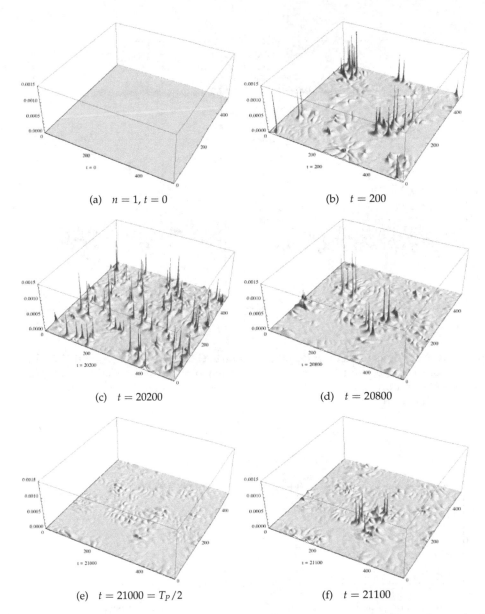

(a) $n = 1, t = 0$

(b) $t = 200$

(c) $t = 20200$

(d) $t = 20800$

(e) $t = 21000 = T_P/2$

(f) $t = 21100$

Fig. 13. The evolution of the vorticity $|\omega(\mathbf{x}, \mathbf{t})|$. At a quantum vortex $\omega(\mathbf{x}) \sim \mathbf{ffi}(\mathbf{x} - \mathbf{x_i})$. At $t = 0$, the initial conditions for the wave function $\varphi = \sqrt{\rho}\, exp(i\theta)$ are $\rho = const$ and θ random. Thus there are no quantum vortices at $t = 0$. Very rapidly vortices are born. The vortices are annihilated at $t = T_P/2$. Grid 512^2. T_P scales as L^2, diffusion ordering, independent of whether 2D or 3D.

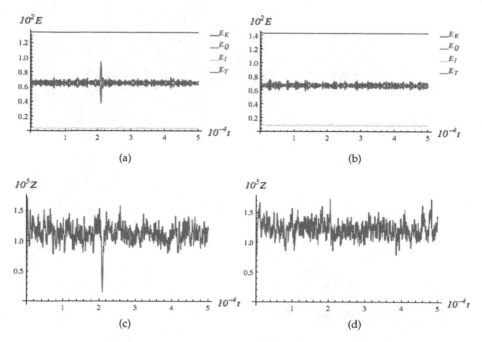

Fig. 14. Loss of Poincaré recurrence. (a) evolution of E_T, E_K, E_Q and E_I, $\gamma = 0.0567$; (b) evolution of E_T, E_K, E_Q and E_I, $\gamma = 0.133$; (c) evolution of enstrophy Z_q, $\gamma = 0.0567$; (d) evolution of enstrophy Z_q, $\gamma = 0.133$. Depletion of vortices can be identified from the sharp decrease of enstrophy.

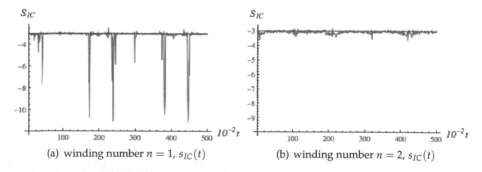

Fig. 15. Time evolution of the incompressible kinetic energy spectrum s_{IC}. The red horizontal line indicates the k^{-3} power law. For winding number $n = 1$, there are spikes in the slope with $s_{IC} \gg 3$ in many instances. While for winding number $n = 2$ the variation in s_{IC} is greatly reduced.

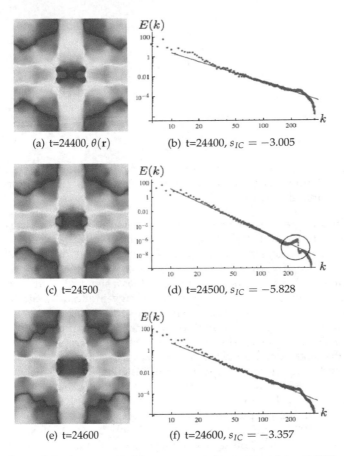

(a) t=24400, $\theta(\mathbf{r})$ (b) t=24400, $s_{IC} = -3.005$

(c) t=24500   (d) t=24500, $s_{IC} = -5.828$

(e) t=24600   (f) t=24600, $s_{IC} = -3.357$

Fig. 16. Phase plot and the incompressible energy spectrum round $t \sim 24500$. At $t = 24400$ and $t = 24600$ there are branch cuts and vortices in the BEC with spectral exponent $s_{IC} \sim -3$. But at $t = 24500$, no branch cuts exist in the phase plot, indicative of no vortices in the system. The incompressible spectrum exhibits a discontinuity in the high-k region with a strong decrease in the exponent, $s_{IC} \sim -5.8$.

the incompressible kinetic energy spectrum of k^{-3} power law in the high-k region could be the by-product of the spectrum of a topological singularity - at least in 2D QT. It should be remembered that in 2D QT there can be no quantum Kelvin wave cascade since the quantum vortex core is just a point singularity, unlike 3D QT where the vortex core is a line or loop.

To examine the spectral exponents of the compressible and incompressible kinetic energies in more detail we now discuss some high grid resolution runs: (a) grid 32768^2 with random phase initial conditions, and (b) grid 8192^2 with winding number $n = 6$ linear vortices in a uniform BEC background.

(a) For the 32768^2 run, we choose initial conditions similar to Numasato et al. (2010), with parameters yielding a 'coherence length' $\xi = (a\,g\,|\psi_0|^2)^{-1/2} = 33.33$ - even though, of course, initially there are no vortices in the system because of the random phase initial condition for the wave function.

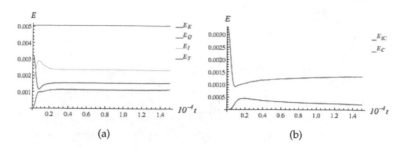

(a)   (b)

Fig. 17. (a) The time evolution of the kinetic, quantum, internal and total energies on a 32768^2 grid with random phase initial condition. (b) Evolution of theincompressible energy (red) and compressible energy (blue).

On this 32768^2 grid we performed a relatively short run to $t_{max} = 15000$. The evolution of the kinetic, quantum, internal and total energies are shown in Fig.17. Based on the time evolution of the energies, the dynamics can be broadly categorized into two stages: (I) generation of vortices and (II) decay of vortices. In stage (I), the compressible energy decreases rapidly while the incompressible energy increases rapidly. Thus a significant amount of energy in the sound waves is transformed into incompressible energy induced by the rotational motion of vortices. In stage (II), the randomly distributed vortices disappear. The energy of the vortices is transferred into sound waves through vortex-vortex annihilation. Note that in this stage, the only major energy exchange occurs between the incompressible and compressible energies while the quantum and internal energies remain almost constants. The spectra for incompressible and compressible energy at $t = 8000$ is given in Fig.18.

At large-k region ($k > 3000$), a k^{-3} power law is present which can be interpreted as result of FFT of quantum vortices. It is interesting to notice that at $k \sim k_\xi$, e.g. $k \in [700, 1200]$, a k^{-4} power law is observed. We sampled the incompressible energy spectra every 50 iteration steps between $6000 < t < 10000$ within a wave number window $k \in [800, 1200]$. The time averaged slope $\langle s_{IC} \rangle = -4.145 \pm 0.066$. This is in good agreement with the results obtained in Horng et al. (2009). This k^{-4} power law can be interpreted as the result of dissipation of randomly distributed vortices, as suggested in Horng et al. (2009). However, in low-k region (region I and II in Fig.18(b) and Fig. 18(c)) where semi-classical Kolmogorov cascade is expected, we did not observe the $k^{-5/3}$ power law. This could be attributed to the compressibility of quantum fluid.

Finally, we consider the case of 12 vortices of winding number $n = 6$ on an 8192^2 grid. The wave function is rescaled so that it is not a quasi-eigenstate of the GP Eq. (2). This will lead more quickly to turbulence. The *compressible* kinetic energy spectrum is shown in Fig.19(a) and clearly exhibits a triple cascade: a small-k region with spectral exponent $s_C = -1.83$, an intermediate-k region with exponent $s_C = -8.1$, and a large-k region with exponent $s_C = -2.85$. We will comment on this triple cascade spectrum more when we discuss QT in 3D. In Fig.19(b) we plot the *incompressible* kinetic energy spectrum and notice the dual cascade region: for high-k we see the ubiquitous exponent $s_{IC} = -2.93$ while for lower-k, the exponent becomes similar to the Saffman exponent, $s_{IC} = -4.0$. The k_ξ at which we have this dual spectrum meet is around the join of the steep intermediate range spectrum with the k^{-3} spectral tail. We shall see this also in 3D QT.

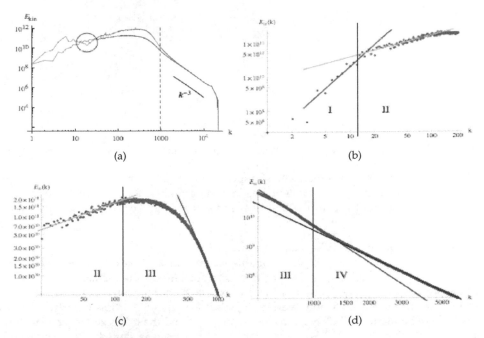

(a)   (b)

(c)   (d)

Fig. 18. (a): The incompressible spectrum $E_{inc}(k)$ (*red*) and compressible spectrum $E_{comp}(k)$ (*blue*) on a 32768^2 grid at $t = 8000$. The dashed vertical line indicates the location of k_ξ, based on the qualitative notation of the coherence length ξ. The encircled dip in the compressible energy propagates towards the lower-k region, resembling a backward propagating pulse. (b) *Incompressible* kinetic energy spectrum ain Region I ($k \lesssim 0.01k_\xi$) and Region II ($0.01k_\xi \lesssim k \lesssim 0.1k_\xi$). The spectral exponents are $s_{IC} = +2.34$ (red line) and $s_{IC} = +0.65$ (green line). (c) *Incompressible* energy spectrum in Region II and Region III ($0.1k_\xi \lesssim k \lesssim k_\xi$). Spectral exponents are: $s_{IC} = 0.65$ (green line); $s_{IC} = -4.17$ (purple line). (d) *Incompressible* energy spectra in Region III and Region IV ($k_\xi \lesssim k$). Spectral exponents are: $s_{IC} = -4.17$ (purple line) and $s_{IC} = -3.03$ (black line).

5. 3D QT

The major differences between 2D and 3D CT is in the behavior of the vorticity vector. In 2D, the vorticity is always perpendicular to the plane of motion while in 3D the vorticity vector can have arbitrary orientation. Here, in 3D we will employ variants of a set of linear vortices following the Pade approximant methods of Berloff Berloff (2004). For winding number $n = 1$, using cylindrical polar coordinates (r, ϕ, z), a linear vortex that lies along the z-axis (and centered at the origin) is given by

$$\varphi(r) = g^{-1/2} e^{i\phi} \sqrt{\frac{11a\, r^2(12 + a\, r^2)}{384 + a\, r^2(128 + 11a\, r^2)}} = g^{-1/2}\, \varphi_0(r)\, e^{i\phi}, \qquad (45)$$

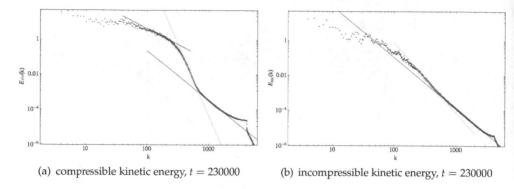

(a) compressible kinetic energy, $t = 230000$ (b) incompressible kinetic energy, $t = 230000$

Fig. 19. The (a) compressible kinetic energy spectrum and (b) incompressible kinetic energy spectrum for winding number $n = 6$ vortices in a uniform BEC gas. Grid 8192^2. Notice the triple cascade region for the compressible energy spectrum and the dual cascade spectrum for the incompressible energy.

with $|\varphi| \rightarrow 1/\sqrt{g}$, and $|\varphi_0| \rightarrow 1$ as r $\rightarrow \infty$, and $|\varphi| \sim r\sqrt{a/g}$ as r $\rightarrow 0$. Eq.(45) is an asymptotic solution of the GP Eq.(2). For this isolated linear vortex, the coherence length, from Eq. (6), $\xi \sim 4/\sqrt{a}$. This is one of the reasons for introducing the factor a into the GP Eq.(2): a small a permits resolution of the vortex core in the simulations. If one starts with a periodic set of well-spaced non-overlapping Pade asymptotic vortices (clearly, of course, this will be dependent on the choice of the parameter a and the grid size L of the lattice) an asymptotic solution of the GP Eq.(2) is simply a product of the shifted φ_0's , weighted by $g^{-1/2}$. The system will evolve slowly into turbulence because this initial state is very weakly unstable. For these wave functions the coherence length is initially fairly well defined. On the other hand, in most of our runs, we just simply rescaled the asymptotic basis vortex function $\varphi \rightarrow g^{\sigma}\varphi$, for some σ. Because the GP Eq.((2) is nonlinear, $g^{\sigma}\varphi$ is no longer an asymptotic solution and the definition of coherence length becomes fuzzy.

In Fig. 20 we show a somewhat complex initial vortex core situation. The initial wave function has winding number $n = 5$ and the positions of the initial line vortices are chosen so that there is considerable overlap of the wave functions around the center of the domain. In this plot we show not only the phase information on the vortex core isosurfaces but also the phase information on the boundary walls. On the vortex isosuface at $t = 0$ one can distinguish the 5 periods around the core. On the boundary walls, the intersection of the cores with the walls gives the location of the 4 branch-point like topological singularities. Emanating from each of these singularities are 5 branch cuts because of the chosen winding number $n = 5$. These branch cuts then join the branch points. Because the $n = 5$ state is energetically unfavorable, the initial state rapidly decays into $5\times$ winding number $n = 1$ vortices, Fig. 20(b). It is tempting to identify the wave structures on the vortices as quantum Kelvin waves. Sound waves can also be identified on the boundaries. Near the center of the lattice, where there was initially considerable overlap of the vortices, many vortex loops have now formed.

5.1 Poincare recurrence for certain classes of initial conditions
As in 2D, a class of initial conditions will also be found for which the Poincare recurrence time is very short. The definitions of the incompressible and compressible kinetic energy, the quantum energy and the internal energy are immediate generalizations of those given in

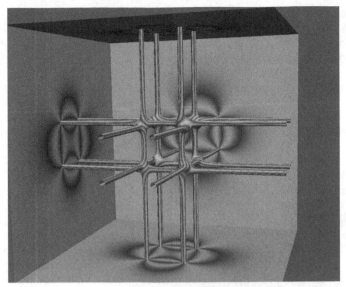

(a) $n = 5, t = 0$

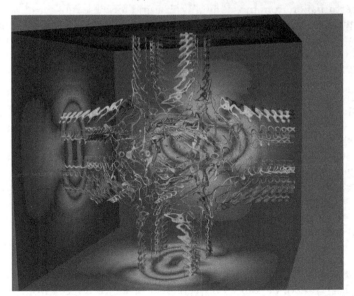

(b) $n = 5, t = 3000$

Fig. 20. The $|\varphi|$ isosurfaces very near the vortex core singularity for winding number $n = 5$: (a) t = 0, (b) t = 3000 . Phase information is given on both the vortices and the boundary walls. The winding number $n = 5$ is evident from both the 5-fold periodicity around the each vortex as well as the 5 branch cuts emanating from each branch point on the boundary. By $t = 3000$, (b), the 5-fold degeneracy is removed with what seem like quantum Kelvin waves on the $n = 1$ cores. Basic phase coding : $\phi = 0$ in blue, $\phi = 2\pi$ in red. Grid 2048^3

$2D$. As in the 2D GP case, short Poincare recurrence will be found for initial conditions such that $E_{int}(0), E_{qu}(0) << E_{kin}(0)$ with $E_{comp}(0) << E_{incomp}(0)$. These conditions are readily satisfied in 3D by considering localized quantum line vortices so that $\rho \sim const.$ with $\nabla \sqrt{\rho} \sim 0$ throughout much of the lattice. The evolution of a set of 3×16 linear vortices in the 3 planes and are examined for winding numbers $n = 1$ and $n = 2$, Fig. 21. Phase information is shown on the boundary walls: $\phi = 0$ - $blue$, $\phi = 2\pi$ - red. First consider the case of winding number $n = 1$. At $t = 0$, the phase information on the straight line vortex cores are clearly identified on their intersection with the boundaries. The corresponding branch cuts join the 48 branch points. A snapshot of the vortex isosurfaces is shown at $t = 84000$ and shows strong vortex entanglement with many vortex loops - basically a snapshot of a quantum turbulence state. However by $t = 115000$ we see a point inversion of the Poincare recurrence of the initial line vortices at $t = 0$, as was also seen in 2D GP flows. The full Poincare recurrence occurs around $t = 230000$, (d). The kinks along the the vortex cores may be quantum Kelvin waves: since one outputs at discrete time intervals, one is not at the exact T_P.

The robustness of the Poincare recurrence time is further exhibited by considering the evolution of 48 quantum vortices with winding number $n = 2$. The two-fold degeneracy translates into a more complex phase on the boundary walls, as can be seen on comparing (e) with (a) in Fig. 21. At $t = T_P = 230000$, one sees considerable very small scale vortex loops that have arisen from the splitting of the confluent degeneracy although the overall structure of the line vortices can be seen globally seen. The phase information on the boundaries are only a slight perturbation from those initially. The grid for these runs was 1200^3.

For winding number $n = 2$ we show the details of the isosurfaces at the semi-Poincare recurrence time $t = 115000$ with phase information on the boundaries, and from a slightly different perspective. Also we show a detailed zoomed-in isoruface plot of the vortex cores at the Poincare recurrence time $t = 230000$ with phase information on the vortices themselves, Fig.22.

As in 2D, the signature of the occurrence of the Poincare recurrence can be seen in the evolution of the kinetic and quantum energies as a function of time, Fig. 23. The total energy is very well conserved throughout this run, $t_{max} = 250000$, by our unitary algorithm on a grid 1200^3: $E_{TOT} = const.$ For the parameters chosen here the internal energy is negligible. Note that the peaks in the kinetic energy are well preserved for vortices with winding number $n = 1$:$E_{kin}(0) \approx E_{kin}(t = 115000) \approx E_{kin}(t = 230000) \approx \cdots$. However, for line vortices with winding number $n = 2$, there is a gradual decay in the peak in the kinetic energy $E_{kin}(0) \geq E_{kin}(t = 115000) \geq E_{kin}(t = 230000) \geq \cdots$. This also explains why the Poincare recurrence in the isosurfaces for winding number $n = 2$ is not as clean as for winding number $n = 1$. Also it can be seen that in the evolution of E_{kin} the vortex motion is much more turbulent for winding number $n = 2$. Since the internal energy for these runs is so low, the quantum energy evolution is the complement of kinetic energy $(E_{TOT} = const. = E_{kin}(t) + E_{qu}(t) + E_{int}(t))$. It should be noted that the time evolution of $E_{kin}(t)$ and $E_{qu}(t)$ (and, of course, $E_{int}(t)$) are determined directly from their definitions, Eq.(33) and their sum then gives us the (conserved) total energy.

We note the loss of the semi-Poincaré time as the pixel resolution of the Arnold cat is increased from 74 to 300×300 yet find the persistence of the semi-Poincaré time for grid resolution from 512^3 to 1200^3. Presumably this is because the QLG algorithm strictly obeys diffusion ordering so that T_P on a 512^3 grid occurs at $T_P = 41775$ and at $T_P = 230000$ on a 1200^3 grid. (Diffusion ordering would give t = 229477). There is no such physics scaling laws in the Arnold cat map.

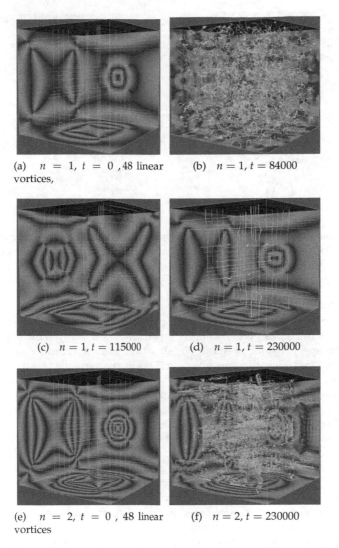

(a) $n = 1, t = 0$,48 linear vortices,

(b) $n = 1, t = 84000$

(c) $n = 1, t = 115000$

(d) $n = 1, t = 230000$

(e) $n = 2, t = 0$, 48 linear vortices

(f) $n = 2, t = 230000$

Fig. 21. The evolution of quantum core singularities from an initial set of 48 straight line vortices. (a) winding number $n = 1$ and the corresponding wall phase information at $t = 0$, (b) winding number $n = 1$ isosurface cores at $t = 84000$. (c) winding number $n = 1$ isosurface cores at $t = 115000 = 0.5T_{Poin}$. The 2π phase changes at the core singularity intersections at the walls is very evident. (d) winding number $n = 1$ isosurface cores at $t = 230000 = T_{Poin}$ showing only small perturbative changes from the initial state given in (a). (e) winding number $n = 2$ and the corresponding wall phase information at $t = 0$ showing the confluence degeneracy. (f) The corresponding isosurface cores at $t = 230000 = T_{Poin}$ for winding number $n = 2$. The wall phase information is a simple perturbative change to that at $t = 0$, (e) - but there is much small scale vortex loops that has evolved due to the initial confluent degeneracy . Phase coding : $\phi = 0$ *blue*, $\phi = 2\pi$ *red*. Grid 1200^3

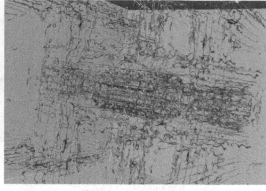

(a) $n = 2, t = 115000$ (b) $n = 2, t = 230000$

Fig. 22. The vortex isosurfaces for winding number $n = 2$ at the semi- and full Poincare times: (a) $t = 115000$ with phase information on the boundaries and at a different perspective, (b) $t = 230000$ with phase information on the isosurfaces $|\varphi| = const..$ Phase coding : $\phi = 0$ *blue*, $\phi = 2\pi$ *red*. Grid 1200^3

5.2 Energy spectra on 1200^3 grid

We first determine the incompressible and compressible kinetic energy spectrum for the initial profiles considered in Fig. 24 . Nearly all the kinetic energy initially is incompressible.

Very quickly the spectra tend to quasi-steady state, with a typical $E_{comp}(k)$ and $E_{inc}(k)$ spectrum as in Fig. 25 For winding number $n = 1$ the incompressible spectrum $E_{inc}(k)$ exhibits two spectral domains $k^{-\alpha}$: for very large k ($k > 100$) the spectral exponent $\alpha \sim 3.05$, while in the intermediate k range ($15 < k < 50$) one finds $\alpha \sim 5.0$. The compressible kinetic energy spectrum $E_{comp}(k)$ exhibits three spectral regions: a very fuzzy large k region, preceded by a steep spectral region which then merges into the small k region. It is interesting to note that around the wave number $k_\xi \sim 70$ at which the steep compressible spectral exponent (typically $\alpha > 7$) joins to the large k spectrum, we find the switch over in the incompressible spectral exponents. We have noticed this in basically every simulation we have performed (and grids up to 4096^3). For vortices with winding number $n = 2$, the kinetic energy spectra are much cleaner, presumably because the vortex entanglements are stronger and hence the QT is stronger. For the incompressible spectrum $E_{inc}(k)$ one again sees two spectral cascade regions: for $k > k_\xi$ the spectral exponent is $\alpha \sim 3.07$ while for $k < k_\xi$, the exponent $\alpha \sim 3.90$. For the *compressible* spectrum, we find three spectral energy cacades: for very low k ($5 < k < 30$) a Kolmogorov-like cascade with exponent $\alpha \sim 1.67$, with a steep spectra decay followed for $k > k_\xi$ a compressible kinetic energy with exponent $\alpha \sim 3.28$. (The *total* kinetic energy spectrum has the exponents $\alpha \sim 1.64$ for low k, and $\alpha \sim 3.17$ for large k). The crossover $k_\xi \sim 70 - 90$.

Somewhat surprising, we find a time interval during which we loose the incompressible kinetic energy spectrum $E_{inc}(k) \sim k^{-3}$ for winding number $n = 1$ vortices. In particular, in the time interval $81400 < t < 84300$ - except for very brief transient reestablishment of the k^{-3} spectrum - we find spectra as shown in Fig. 26. In Figs. 26 (b)-(d) there is a sharp drop in the incompressible energy spectrum for wave numbers $k > 100$, except for a very brief transient recovery around $t \sim 83000$. There is also a sharp cutoff in the compressible spectrum for $k > 500$. Around these intermittencies, the incompressible kinetic energy spectrum also

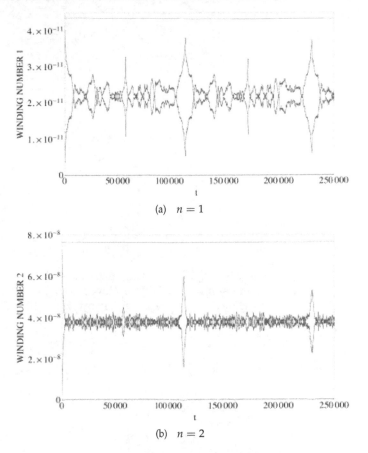

(a) $n = 1$

(b) $n = 2$

Fig. 23. The time evolution of the $E_{\text{kin}}(t)$ (in blue) and $E_{\text{qu}}(t)$ (in red) for $0 \le t \le 250000$ for (a) Winding Number $n = 1$, and (b) Winding Number $n = 2$. Grid 1200^3.

exhibits a triple cascade $k^{-\alpha}$ with $\alpha \sim 3.7$ for small k, an $\alpha \sim 6$ for the intermediate cascade, and $\alpha \sim 3.0$ for the large-k cascade. At the intermittency, the large-k exponent increases to a noisy $\alpha \sim 5.2$ as well as a steeped intermediate wave number exponent.

To investigate the cause of this intermittent loss of the incompressible k^{-3} spectrum, we then examined the vortex isosurfaces around this time interval, Fig.27.

One notices that the loss of the k^{-3} corresponds to the apparent loss of vortex loops, i.e., of vortices. This would be consistent with the assumption that the *incompressible* kinetic energy spectrum of k^{-3} in the very large-k regime is due to the Fourier transform of an isolated vortex Nore et al. (1997). As the vortex loops are reestablished, so is the incompressible k^{-3} kinetic energy spectrum. An alternative but somewhat more speculative explanation rests on the assumption that the incompressible k^{-3} spectrum is due to the quantum Kelvin wave cascade on the quantum vortices. As the quantum vortex loop shrink topologically, the Kelvin waves are inhibited and hence the loss of the k^{-3} spectrum. Moreover, if one looks at the time evolution of the mean kinetic $E_{kin}(t)$ and quantum $E_{qu}(t)$ energies one notices that this loss of the vortex loops occurs around the $t \sim 82000$ around which the $E_{kin}(t)$ has a secondary

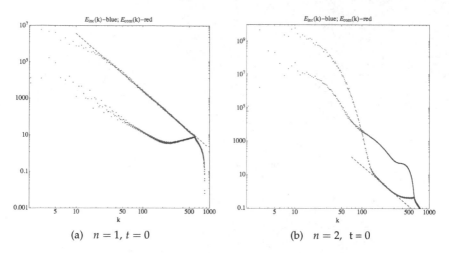

Fig. 24. The initial incompressible (*blue*) $E_{inc}(k,0)$ and compressible (*red*) $E_{comp}(k,0)$ kinetic energy spectra for (a) Winding Number $n = 1$, and (b) Winding Number $n = 2$. The linear regression (*blue dashed line*) fit to the incompressible kinetic energy, $E_{inc}(k,0) \sim k^{-\alpha}$ is: (a) $\alpha = 3.15$, (b) $\alpha = 3.30$. Grid 1200^3.

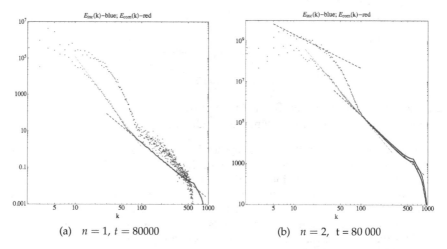

Fig. 25. At t = 80 000, the incompressible (*blue dots*) $E_{inc}(k)$ and compressible (*red dots*) $E_{comp}(k)$ kinetic energy spectra for (a) Winding Number $n = 1$, and (b) Winding Number $n = 2$. The linear regression (*blue dashed line*) fit to the incompressible kinetic energy, $E_{inc}(k,0) \sim k^{-\alpha}$ is: (a) $\alpha = 3.05$, (b) $\alpha = 3.07$. The intermediate k range has incompressible kinetic energy exponents (a) $\alpha \sim 5.0$, and (b) $\alpha \sim 3.9$ - (*green dashed line*). The small k-range for the compressible energy exhibits a weak Kolmogorov-like spectrum for winding number $n = 2$, (b), of $\alpha \sim 1.67$ - (*read dashed line*). Grid 1200^3.

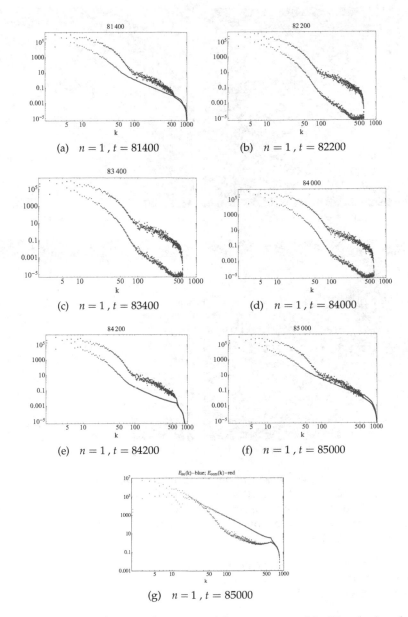

Fig. 26. 7 snapshots for winding number $n = 1$ of the incompressible (*blue dots*) and compressible (*red dots*) kinetic energy spectrum at times (a) t = 81400, (b) t = 82200, (c) t = 83400, (d) t = 84000, (e) t = 84200, (f) t = 85000, and (g) $t = 115000 = T_P/2$. There is a very brief transient recovery of the k^{-3} spectrum around $t \sim 83000$. Also shown in s the spectrum at the $t = T_P/2$ and should be compared to the initial spectrum in Fig.24(a) . Grid 1200^3

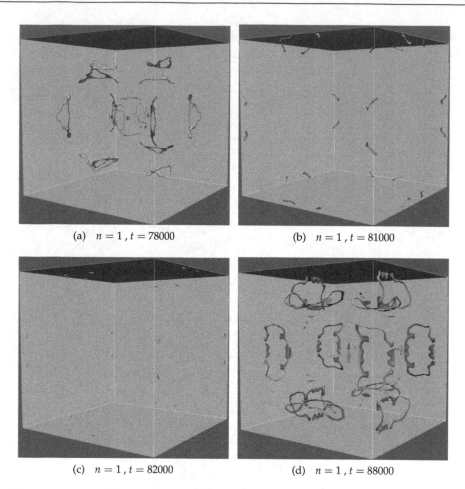

(a) $n = 1$, $t = 78000$ (b) $n = 1$, $t = 81000$

(c) $n = 1$, $t = 82000$ (d) $n = 1$, $t = 88000$

Fig. 27. 4 snapshots of the winding number $n = 1$ singular vortex core isosurfaces at times (a) t = 78000, (b) t = 81000, (c) t = 82000, and (d) t = 88000. The phase information (blue: $\phi = 0$, red: $\phi = 2\pi$ on the vortex core singularities clearly shows the 2π phase change in circumnavigating the vortex core loops. At $t = 82000$ there is a different morphology in the $|\varphi|$ - isosurfaces. Grid 1200^3

peak. As there is another secondary peak in $E_{kin}(t)$ around $t \sim (82000 + T_P/2)$ one expects another transient loss in the k^{-3} spectrum and in the vortex loops. This is indeed found to occur around $196400 < t < 199300$. For winding number $n = 2$ vortices, we do not find such intermittent loss of vortex loops or any intermittent loss of the k^{-3} spectrum. These results are in agreement with those found earlier in 2D QT. Moreover, there is not a similar intermittent loss of vortex loops for 48 linear vortices with winding number $n = 1$ (c.f., Fig. 21(b))

5.3 Total kinetic energy spectrum for large grid simulations on 5760^3

We have performed simulations on 5760^3 grid using winding number $n = 6$ straight line vortices as initial conditions. By $t = 40000$ one obtains the following *total* kinetic and quantum energy spectra, Fig. 28. Due to an oversight, we did not retain data for the incompressible

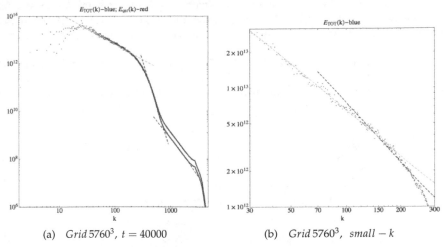

(a) $Grid\,5760^3$, $t = 40000$ (b) $Grid\,5760^3$, $small - k$

Fig. 28. (a) Total kinetic energy spectrum, $E_{TOT}(k)$ — (*blue dots*), and quantum energy spectrum, $E_{qu}(k)$ — (*red dots*), at $t = 40000$ for winding number $n = 6$ on a large 5760^3 grid. 3 energy cascade regions can be readily distinguished, with both the total and quantum energy spectra being very similar. Dashed curves – linear regression fits. (b) Linear regression fits for different k-bands in the small-k region: for $30 < k < 200$ (*green dashed line*), the spectra exponent $\alpha \sim 1.30$, while for $100 < k < 250$ (*red dashed line*), the spectral exponent $\alpha \sim 1.68$. The standard deviation error is 0.06 in both cases.

and compressible components of the kinetic energy spectrum. Both $E_{TOT}(k)$ and $E_{qu}(k)$ have basically the same spectral properties. One sees 3 distinct energy cascade regions $k^{-\alpha}$: a small-k band with $\alpha \sim 1.30$, an intermediate-k band with steep slope $\alpha \sim 7.76$ and a large-k band with $\alpha \sim 3.00$.

In Yepez et al. (2009), we tried to identify these 3 regions as the small-k classical Kolmogorov casacde, followed by a semi-classical intermediate-k band (with non-universal exponent α) which is then followed a quantum Kelvin wave cascade for the very large-k band. Objections were raised against this interpretation Krstulovic & Brachet (2010); L'vov & Nazarenko (2010) based on (a) the kinetic energy spectrum of a single isolated vortex is k^{-3} for all k (for a straight line vortex all the kinetic energy is incompressible); and (b) using the standard definition of the coherence length ξ for the parameters chosen in our simulations, $\xi > 2000$ – i.e., it is claimed that we are investigating the physics of very strong vortex-vortex core overlapping wave functions. We counter that the definition of ξ is based on a boundary value solution of the GP equation for an isolated single vortex under the condition that the wave function asymptotes to the background value as one moves away from the vortex core. Our simulations are with periodic boundary conditions and we have no pointwise convergence of our wave function to some nice smooth 'background' value. While it can be argued that one should simply replace the usual background density ρ_0 by its spatial average $< \rho >$, the definition now of ξ becomes qualitative and does not handle large fluctuations about $< \rho >$. It is clear that we cannot categorically claim that the ubiquitous k^{-3} spectrum for the large-k band is due to quantum Kelvin wave cascade on single vortices - especially as this k^{-3} spectrum is also seen in our 2D QT simulations and in 2D there are no quantum Kelvin waves since the vortex core is just a point singularity. It is possible, however, that with the co-existence of this triple cascade region in the kinetic energy spectrum, and with the small-k region exhibiting a quasi-classical Kolmogorov $k^{-5/3}$ spectrum, that this large-k band k^{-3} spectrum could be dominated by the spectrum of a single vortex as we are now at very small scales in the problem.

We have investigated in some detail the 3D QT for winding number $n = 2$ on a 3076^3 grid. One again finds the triple cascade region in the *compressible* kinetic energy spectrum, Fig. 29, (*red squares*), as well as in the quantum energy spectrum (*gold diamonds*). However the

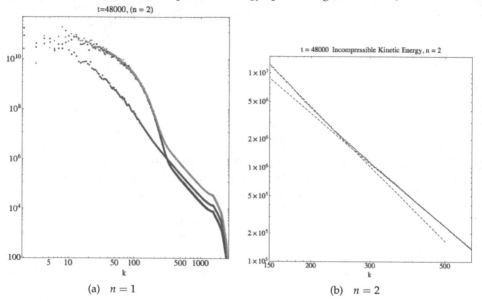

(a) $n = 1$ (b) $n = 2$

Fig. 29. (a) Energy spectra at $t = 48000$ for winding number $n = 2$. Blue circles - incompressible kinetic energy, red squares - compressible kinetic energy, gold diamonds - quantum energy. A triple cascade is quite evident in both the quantum and compressible kinetic energy spectra. These two spectra only deviate around the transition from the medium k to large k cascade, i.e., around $k \sim 300$. (b) A blow-up of the transition in the *incompressible* kinetic energy pectrum from $k^{-3.6}$ to $k^{-3.0}$ around $k \sim 300$. Grid 3072^3

incompressible kinetic energy (*blue circles*) has a slight bend in its spectral exponent around the wave number $k_\xi \sim 300$ from the large-k exponent of $\alpha \sim 3.0$ for $k > k_\xi$ to $\alpha \sim 3.6$ for $k < k_\xi$. This bend in the incompressible kinetic energy spectrum occurs at the k_ξ where the compressible and quantum energy spectra make their transition from the intermediate-k band large spectral exponent to the large-k band spectral exponent of $\alpha \sim 3$. It is interesting to note similar behavior in 2D QT, Fig.18(d) and Fig.19(b). The spectral exponents for $E_{TOT}(k)$ are: a Kolmogorov $\alpha \sim 1.66$ for the small-k band $15 < k < 90$, a steep $\alpha \sim 8.53$ for the intermediate-k band $180 < k < 280$ and exponent $\alpha \sim 3.04$ in the large-k band.

6. Conclusion

Here we have discussed a novel unitary qubit algorithm for a mesoscopic study of quantum turbulence in a BEC dilute gas. Since it requires just 2 qubits/lattice site with unitary collision operators that entangle the qubit probabilities and unitary streaming operators that propagate this entanglement throughout the lattice, QLG has a small memory imprint. This permits production runs on relatively few processors: e.g., production runs on 5760^3 grids using just over 10000 cores. Using standard pseudo-spectral codes, the largest grids achieved so far have been 2048^3 by Machida et al. (2010) - and these codes required the ad hoc addition of dissipative terms to damp out high-k modes and for numerical stability. For BEC turbulence this destroys the Hamiltonian structure of the GP Eq. (2) and destroys any Poincare recurrence phenomena.

In its current formulation, it is critical that parameters are so chosen that the mesoscopic QLG algorithm yields diffusion ordering at the macroscopic GP level. This does restrict the choice in the values of kinetic, quantum and internal energies. If parameter choices are made that violate the diffusion ordering then the QLG algorithm will not be simulating the GP Eq. (2). There are various tests for the validity of our QLG solution of the GP equation: the conservation of total energy, and the fact the the Poincare recurrence time scales as $grid^2$ - whether in 2D or 3D GP. This replaces the naive thought that the time for QLG phenomena would necessarily scale as $grid^D$, where D is the dimensionality of the macroscopic problem. On the other hand, standard CFD algorithms have the spatial and time step independent which, of course, is quite beneficial.

We have presented significant spectral results for both 2D and 3D QT - although their interpretation is not straightforward and much still needs to be done in this area. Much of the controversy surrounds the significance of the coherence length ζ and the k^{-3} spectrum in the high-k region. We believe there is much new physics occurring in our QT simulations even if the k^{-3} spectrum is attributed to the dominance of the spectrum of an isolated vortex: (a) there is clear evidence in both 2D and 3D QLG of a dual cascade in the incompressible kinetic energy spectrum, with a spectrum of k^{-4} followed by the k^{-3}. This k^{-4} spectrum has also been seen by Horng et al. (2009) and connection implied with the Saffman spectrum arising from vorticity discontinuities; (b) the triple cascade in the *total* kinetic energy spectrum with the small-k regime yielding a quasi-Kolmogorov $k^{-5/3}$ spectrum. It is a bit strange that the QT community is that concerned with achieving the $k^{-5/3}$ energy spectrum in the *incompressible* energy $E_{inc}(k)$ in the small k-region where the quantization of the vortex core circulation becomes unimportant. The dynamics of the GP Eq.(2) is fundamentally compressible. In *compressible* CFD simulations the emphasis changes to the *total* kinetic energy spectrum and that its power law is $k^{-5/3}$. However much work remains to be done on clarifying the role of quantum Kelvin waves on the energy spectrum.

Finally, we comment on future directions of QLG. In this chapter we have restricted ourselves to the *scalar* GP equation. This is appropriate for a BEC gas with spin f confined in a magnetic well. The spin of the atom is aligned to the magnetic field and so the BEC dynamics is given by just one scalar GP equation. However if this BEC gas is confined in an optical lattice, the spin is no longer constrained and one now must work with $2f + 1$ GP equations. These so-called spinor BECs yield an enormous field of future research Ueda & Kawaguchi (2010). Moreover, the quantum vortices of a scalar BEC are necessarily Abelian vortices, but the vortices of spinor BECs can be non-Abelian in structure. Since QT is driven by vortex-vortex interactions, research needs to be performed to ascertain the role played by the non-Abelian in the energy cascades. (These non-Abelian vortices have non-integer multiples of the base circulation - not dissimilar to fractional quantum electron charge in the fractional quantum Hall effect). Other interesting vortices are skyrmions - used by high energy physicists to model baryons.

7. Acknowledgments

The computations were performed at both DoE and DoD supercomputer facilities. Much help was given by Sean Ziegeler (HPTi) for the graphics. The authors were supported by grants from the AFOSR and DoE.

8. References

Barenghi, C. (2008). *Physica D* 237: 2195.

Batchelor, G. (1969). *Physics of Fluids* 12(12): II–223.

Berloff, N. G. (2004). *J. Phys. A* 37: 1617.

Blaauwgeers, R., Eltsov, V. B., Eska, G., Finne, A. P., Haley, R. P., Krusius, M., Ruohio, J. J., Skrbek, L. & Volovik, G. E. (2002). *Phys. Rev. Lett.* 89(15): 155301.

Feynman, R. (1955). *Prog. Low Temp. Phys.* 1.

Gross, E. (1963). *J. Math. Phys.* 4: 195.

Henn, E. A. L., Seman, J. A., Roati, G., Magalhães, K. M. F. & Bagnato, V. S. (2009). *Phys. Rev. Lett.* 103(4): 045301.

Horng, T.-L., Hsueh, C.-H., Su, S.-W., Kao, Y.-M. & Gou, S.-C. (2009). *Phys. Rev. A* 80(2): 023618.

Keys, R. (1981). *IEEE transactions on Signal Processing, Acoustics, Speech, and Signal Processing* 29(1153).

Kivshar, Y. & Agrawal, G. (2003). *Optical Solitons*, Elsevier.

Kobayashi, M. & Tsubota, M. (2007). *Phys. Rev. A* 76: 045603.

Kraichman, R. (1967). *Physics of Fluids* 10(7): 1417.

Krstulovic, G. & Brachet, M. (2010). *Phys. Rev. Lett.* 105: 129401.

L'vov, V. & Nazarenko, S. (2010). *Phys. Rev. Lett.* 104: 219401.

Machida, M., Sasa, N., L'vov, V., Rudenko, O. & Tsubota, M. (2010). *ArXiv e-prints* (1008.3050v1).

Menon, S. & Genin, F. (2010). *J. Turbulence* 11: 1.

Nore, C., Abid, M. & Brachet, M. E. (1997). *Physics of Fluids* 9(9): 2644–2669.

Numasato, R., Tsubota, M. & L'vov, V. S. (2010). *Phys. Rev. A* 81: 063630.

Pethick, C. & Smith, H. (2009). *Bose-Einstein condensation in dilute gases*, Cambridge University Press.

Pitaevskii, L. P. (1961). *JETP* 13: 451.

Pope, S. (1990). *Turbulent Flows*, Cambridge University Press.

Proment, D., Nazarenko, S. & Onorato, M. (2010). *ArXiv e-prints* (1008.0096v1).

Succi, S. (2001). *The Lattice Boltzmann Equation for Fluid Dynamics and Beyond*, Oxford University Press.

Tracy, E., Chen, H. & Lee, Y. (1984). *Phys. Rev. Lett.* 53: 218.

Tsubota, M. (2008). *J. Phy. Soc. Japan* 77: 111006.

Ueda, M. & Kawaguchi, Y. (2010). *ArXiv e-prints* (1001.2072v2).

Vahala, G., Vahala, L. & Yepez, J. (2003). *Phys. Lett. A.* 310: 187.

Vahala, G., Vahala, L. & Yepez, J. (2004). *Phil. Trans. Roy. Soc. London A* 362: 215.

Yepez, J., Vahala, G., Vahala, L. & Soe, M. (2009). *Phys. Rev. Lett.* 103: 0845001.

Yepez, J., Vahala, G., Vahala, L. & Soe, M. (2010). *Phys. Rev. Lett.* 105: 129402.

11

Turbulent Boundary Layer Models: Theory and Applications

José Simão Antunes do Carmo

University of Coimbra

Portugal

1. Introduction

Shallow coastal areas are extremely dynamic regions where the fluid motions associated with both surface waves and currents interact with the bottom sediments. The prediction of the wave effects on sediment transport in shallow water conditions and in intermediate depth is still frequently restricted to monochromatic and unidirectional wave models. However, in real shallow water conditions, the nonlinear process of sediment transport responds in a rather different way to the idealized regular wave case. Therefore, in these regions, both the wave non-linearity and the wave-current interaction become important factors to be considered. Forecasts of morphological changes are invariably dependent on the correct prediction of the sand transport rate under the action of waves and currents, which requires accurate estimation of the friction at bed level, considering all resulting complex interactions effects in its entirety. A major consequence of the fluid dynamics resulting from the combined wave and current motions is the response of the movable seabed, which is significantly altered from that expected for a linear superposition of a pure wave motion with a pure current. In recent years, various attempts have been made to improve the state of knowledge of the flow in the bottom boundary layer regarding the wave non-linearity and complex wave-current effects on the sand-transport rate, using theoretical models. The erosion and sediment transport estimation around usual structures in the fluvial and coastal environment, like bridge piers, groynes and breakwaters, are of a major concern for designing these structures and for considering preventive measures. After a brief discussion on turbulence, the following sections present mathematical and numerical approaches of different complexity. Starting by the fundamental equations of the Fluid Mechanics, a complex unresolved formulation without further assumptions is obtained. Afterwards, considering some physical hypotheses, practical models of different complexity are shown, followed by simple parametric approaches and applications.

2. Turbulence

Turbulence has been a long standing challenge for human mind. Five centuries after the first studies of Leonardo da Vinci, understanding turbulence continues to attract a great deal of attention. This may be due to its fascinating complexity and ubiquitous presence in a variety of flows in nature and engineering. The first turbulence references by Leonardo da Vinci are based on visual observations. In 1883, Osborne Reynolds introduced the concept of averages, which became the base of great theoretical-experimental studies. In 20th century, Taylor by the thirties presented the first statistical theory for isotropic turbulence, Kolmogorov by the year 1941 formulated theoretical developments for local turbulence, Batchelor by the year 1953 distinguish himself for theoretical and experimental studies about free turbulence of waves and jets. Then, much more other studies were presented, mainly about wall turbulence, boundary layer and air models. Several resumes can be found in Monin and Yaglom (1971), Tennekes and Lumley (1972), Launder and Spalding (1972), Hinze (1975), Schiestel (1993), Nezu and Nakagawa (1993), Rodi (1980, 1993), Mohammadi and Pironneau (1994), Lumley (1996), Chen et al. (1996) and Lesieur (1997), among others.

The detailed accurate computation of large scale turbulent flows has become increasingly important and considerable effort has been devoted to the development of models for the simulation of complex turbulent flows in several applications over the last decades. The description of turbulence flows is based on the assumption that instantaneous flow variables satisfy the Navier-Stokes equations, which contain a full description of turbulence, given that they describe the motion of every Newtonian incompressible fluid based on conservation principles without further assumptions. Analysing the applicability of continuum concepts to the description of turbulence, Moulden et al. (1978) conclude that if the Newtonian constitutive relation is valid, then it is plausible to accept that turbulent flows instantaneously satisfy the same dynamical equations as laminar flows. For laminar flows, analytical or numerical solutions can be directly compared to experimental results in some cases. Moser (2006) declared that despite the increasing range of turbulence spatial scales as the Reynolds number increases, in turbulence, the continuum assumption and the Navier-Stokes equations are an increasingly good approximation.

The aforementioned assumption seems to be well supported as DNS "Direct Numerical Simulation", in which all scales of the motion are simulated using solely the Navier-Stokes equations. It is the most natural approach to the numerical simulation of turbulent flows but, since by Kolmogorov's theory, small scales exist down to O. ($Re^{-3/4}$), in order to capture them on a mesh, a meshsize $h \approx Re^{-3/4}$ and consequently (in 3D) $N \approx Re^{9/4}$ mesh points are necessary. Thus, it only could be applied for simple and low-Reynolds number turbulent flows (Kaneda & Ishihara, 2006; McComb, 2011). Even if DNS were feasible for hydraulic practical interest, it is not possible to define, with the precision required by the smallest scales of the motion, proper initial and boundary conditions. This fact is of significant importance due to non-linear character of the advection terms, which results in the production and maintenance of instabilities which in turn excite small scales in the motion. The presence of non-linear terms also precludes the existence, in the most general case, of unique solutions for a given set of initial and boundary conditions. Thus, as a large Reynolds number turbulent flow is inherently unstable, even small boundary perturbations may excite the already existing small scales, with possible corresponding perturbation amplifications. The lack of solution uniqueness and the infeasibility of defining precise initial and boundary conditions combine themselves in a way that the resultant flow appears random in character. Indeed, the uncontrollable nature of the boundary conditions (in terms of wall roughness size and distribution, wall vibration, etc.) forces the analyst to characterize them as "random forcings" which, consequently, produce random responses (Aldama, 1990). The Navier-Stokes equations can then exhibit great sensitivity to initial and boundary conditions leading to unpredictable chaotic behaviour. Although the fundamental laws behind the Navier-Stokes equations are purely deterministic, these equations, similar to other simpler deterministic equations, can often behave chaotically under certain conditions. Due to the randomness in turbulent flows, it is hopeless to track instantaneous behaviour. Instead, the goal is to measure this behaviour in the temporal or spatial mean.

Most researchers in the turbulence field accept that instantaneous flow variables satisfy the Navier-Stokes equations as an axiom and use it as the basis for the development of models for numerical simulation. Assuming that details of motion at the level small and intermediate scales, which tend to exhibit high randomness levels and peculiar characteristics such as isotropy, are not required in most applications of interest in engineering and geophysics, the establishment of two approaches, which have the potential for being applied to problems of engineering interest, can be defined. The first approach is based on the use of filters for the flow variables of interest, Large Eddy Simulation (LES). The second one relies on the use of statistical averages on the same variables, Reynolds-averaged Navier-Stokes equations. Although the former is formally superior to the latter, its use implies paying a computational price which is too high for applications of practical interest. LES requires less computational effort than direct numerical simulation (DNS), but more effort than those methods that solve the Reynolds-averaged Navier-Stokes equations (RANS). These equations, derived by Osborne Reynolds in 1985, describe the dynamics of the "mean flow" in terms of a time average, and later defined as average in the probability space "ensemble average". The Reynolds stresses produced by advection terms, which are second order correlations in statistical terms, are determined by exact transport equations for the Reynolds stresses derived from the Navier-Stokes equations. However, third-order correlations appear in such expressions and four-order correlations will appear in the exact transport equations for the third-order correlations. This is called the problem of closure of the statistical treatment. The approach of neglecting correlations of higher order has proved to be unsuccessful because the turbulent flows are not completely random. Experimental investigations have made it possible to identify, through the use of conditional sampling techniques, "coherent structures" such as shear layers imbedded in turbulent flows, and that the degree of coherence is scale dependent. In the solution of complicated sets of nonlinear partial differential equations, the interaction between physics and numerical approach is very strong, and the use of second approach in question makes it possible to have a better understanding of that interaction and, as a consequence, to control it. Four main approaches have been followed to find ways to close the Reynolds equations by introducing hypotheses based on physical insight and observational evidence: 1-transport; 2- mean velocity field; 3- turbulent field, and 4- invariant models. The resulting model equations contain a number of empirical constants which, in general, increase with their complexity. These models have the base on important concepts and hypotheses as the eddy viscosity concept by Boussinesq, in 1877, Prandtl's mixing length concept, in 1925, Kolmogorov's isotropic dissipation assumption, in 1941, and Rotta's energy redistribution hypothesis, in 1951 (Monin & Yaglom, 1971; Rodi, 1984).

3. Governing equations

The fundamental equations of the Fluid Mechanics applied to a three-dimensional flow of an incompressible and viscous fluid, with sediment in suspension, are written:

a) $\dfrac{\partial u_i}{\partial x_i} = 0$

b) $\dfrac{\partial u_i}{\partial t} + u_j \dfrac{\partial u_i}{\partial x_j} = -\dfrac{1}{\rho_0}\dfrac{\partial p}{\partial x_i} + v_l \dfrac{\partial^2 u_i}{\partial x_j^2} + g_i \dfrac{\rho - \rho_0}{\rho_0}$ (1)

c) $\dfrac{\partial C}{\partial t} + \left(u_i + w_{s_i}\right)\dfrac{\partial C}{\partial x_i} = \gamma_m \dfrac{\partial^2 C}{\partial x_i^2}$

d) $\rho = \rho_s C + \left(1 - C\right)\rho_0$

where u is the instantaneous velocity of the flow; C is the volumetric concentration of the sediment; p is the pressure; v_l is the kinematic viscosity; g_i is the acceleration due to gravity; ρ is the density; ρ_0 is the density of the fluid; ρ_s is the density of the sediment; w_s is the sediment settling velocity, and γ_m is the molecular diffusivity.

3.1 Turbulence closure model with sediment in suspension

Following the classical Osborne Reynolds procedure, and assuming that the fluid is in a randomly unsteady turbulent state and applying time averaging to the basic equations of motion, the fundamental equations of incompressible turbulent motion are obtained. These are known as the Reynolds equations, and involve both mean and fluctuating quantities – the turbulent inertia tensor components. We consider only incompressible turbulent flow with constant transport properties but with possible significant fluctuations in velocity, pressure, and concentration, i.e.:

$$u_i = \bar{u}_i + u_i' \ ; \ p_i = \bar{p}_i + p_i' \ ; \ C_i = \bar{C}_i + C_i'$$

Substituting these functions into the basic equations (1), and taking the time average of each entire equation, we obtain (2) (Rodi, 1984):

a) $\dfrac{\partial \bar{u}_i}{\partial x_i} = 0$

b) $\dfrac{\partial \bar{u}_i}{\partial t} + \bar{u}_j \dfrac{\partial \bar{u}_i}{\partial x_j} = -\dfrac{1}{\rho_0}\dfrac{\partial \bar{p}}{\partial x_i} + \dfrac{\partial}{\partial x_j}\left[v_l \dfrac{\partial \bar{u}_i}{\partial x_j} - \overline{u_i' u_j'}\right] + g_i \dfrac{\bar{p} - \rho_0}{\rho_0}$ (2)

c) $\dfrac{\partial \bar{C}}{\partial t} + \left(\bar{u}_i + w_{s_i}\right)\dfrac{\partial \bar{C}}{\partial x_i} = \dfrac{\partial}{\partial x_i}\left[\gamma_m \dfrac{\partial \bar{C}}{\partial x_i} - \overline{u_i' \rho'}\right]$

d) $\bar{p} = \rho_s \bar{C} + \left(1 - \bar{C}\right)\rho_0$

where $-\overline{u_i' u_j'}$ are the tensor components of the Reynolds stresses, and $-\overline{u_i' \rho'}$ are the tensor components of density-velocity correlations. Thus the mean momentum equation and the equation for the concentration are complicated by new terms involving the turbulent inertia tensor $u_i' u_j'$ and density fluctuations $u_i' \rho'$. The new terms are never negligible in any turbulent flow with sediment in suspension, and can be defined only through knowledge of the detailed turbulent structure, which is, in its turn, unavailable. These turbulent quantities are related not only to the fluid physical properties but also to local flow conditions. As no physical laws are available, most attempts have been made to resolve this dilemma. Many attempts have been made to add turbulence conservation relations to the time-averaged equations above.

3.2 Boussinesq hypothesis (first order turbulence closure model)

According to the Boussinesq hypothesis, the turbulent shear stresses $\overline{u_i' u_j'}$ are modelled in terms of the gradients of the mean flow velocities through (3),

$$-\overline{u_i' u_j'} = v_t \left[\frac{\partial \overline{u}_i}{\partial x_j} + \frac{\partial \overline{u}_j}{\partial x_i} \right] - \frac{2}{3} \delta_{ij} K \; ; \quad -\overline{u_i' \rho'} = \gamma_t \frac{\partial \overline{\rho}}{\partial x_i} \; ; \quad i, j = 1, 2, 3 \tag{3}$$

where $K = \overline{u_i' u_j'}/2 = \left(\overline{u_1'}^2 + \overline{u_2'}^2 + \overline{u_3'}^2 \right)/2$ is the turbulent kinetic energy, per mass unit; v_t is the turbulent viscosity, and γ_t is the turbulent diffusivity. In contrast to the molecular viscosity v_l, the turbulent viscosity v_t is not a fluid property, but depends strongly on the state of the turbulence and may vary considerably over the flow field. A turbulence model thus usually has the task of determining the distribution of v_t over the flow field, by relating the turbulence correlations to the averaged dependent variables. As a first order turbulence closure, the turbulent viscosity v_t is obtained through the *mixing-length* theory of Prandtl (1925), who, by analogy with kinetic theory, proposed that each turbulent fluctuation could be related to a length l_m scale and a velocity gradient,

$$v_t = l_m^2 \left| \frac{\partial u_i}{\partial z} \right| = l_m^2 \sqrt{\left(\frac{\partial u}{\partial z} \right)^2 + \left(\frac{\partial v}{\partial z} \right)^2} \tag{4}$$

For the l_m scale different relations have been proposed. We suggest $l_m = kz\sqrt{1 - z/z_\delta}$, where $k \approx 0.4$ is the von Kármán constant and z_δ is the boundary layer thickness.

3.3 Second order turbulence closure model

A derivation of the turbulent shear stresses, where $i \neq j$, involves subtracting the above time-averaged equation (2-b) from its instantaneous value (1-b), for both the x_i and x_j directions. The ith result is then multiplied by u_j' and added to the jth result multiplied by u_i'. This relation is then time-averaged to yield the following *Reynolds stress equation* $\overline{u_i' u_j'}$:

$$\frac{\partial}{\partial t} \left(\overline{u_i' u_j'} \right) + \overline{u}_k \frac{\partial}{\partial x_k} \left(\overline{u_i' u_j'} \right) = -\overline{u_i' u_k'} \frac{\partial \overline{u}_j}{\partial x_k} - \overline{u_j' u_k'} \frac{\partial \overline{u}_i}{\partial x_k} + \frac{1}{\rho_0} \left(\overline{g_i u_j' \rho'} + \overline{g_j u_i' \rho'} \right)$$

$$-\frac{\partial}{\partial x_k} \left(\overline{u_i' u_j' u_k'} \right) - \frac{1}{\rho_0} \left(\overline{u_i' \frac{\partial p}{\partial x_j}} + \overline{u_j' \frac{\partial p}{\partial x_i}} \right) + v_l \frac{\partial^2}{\partial x_k^2} \left(\overline{u_i' u_j'} \right) - 2 v_l \overline{\frac{\partial u_i'}{\partial x_k} \frac{\partial u_j'}{\partial x_k}} \tag{5}$$

In equation (5), the three terms of different nature $-\dfrac{\partial}{\partial x_k} \left(\overline{u_i' u_j' u_k'} \right)$, $-\dfrac{1}{\rho_0} \left(\overline{u_i' \dfrac{\partial p}{\partial x_j}} + \overline{u_j' \dfrac{\partial p}{\partial x_i}} \right)$ and

$v_l \dfrac{\partial^2}{\partial x_k^2} \left(\overline{u_i' u_j'} \right) - 2 v_l \overline{\dfrac{\partial u_i'}{\partial x_k} \dfrac{\partial u_j'}{\partial x_k}}$ are to be either neglected or related to other variables. Let us consider these terms in some detail.

- The third-order velocity correlations $-\dfrac{\partial}{\partial x_k}\left(\overline{u_i'\,u_j'\,u_k'}\right)$ express the process as the Reynolds stresses are conservatively transmitted from one region of the flow to another; they are usually obtained through $C_t\dfrac{\partial}{\partial x_k}\left[\overline{q}L\dfrac{\partial}{\partial x_k}\left(\overline{u_i'\,u_j'}\right)\right]$ (Lewellen, 1977), where $\overline{q}$ is the root-mean-square value of the total velocity fluctuation, L is the macroscale of the eddies, and C_t is a constant;

- The pressure-velocity correlations $-\dfrac{1}{\rho_0}\left(\overline{u_i'\dfrac{\partial p'}{\partial x_j}}+\overline{u_j'\dfrac{\partial p'}{\partial x_i}}\right)$, which redistribute to the mean flow the turbulent energy produced, are commonly approximated by $-C_p\dfrac{\sqrt{\overline{q^2}}}{L}\left(\overline{u_i'\,u_j'}-\dfrac{1}{3}\delta_{i,j}\overline{q^2}\right)$, where $\delta_{i,j}$ is the Kronecker symbol and C_p is a constant;

- The dissipation terms $\nu_l\dfrac{\partial^2}{\partial x_k^2}\left(\overline{u_i'\,u_j'}\right)$ and $-2\nu_l\overline{\dfrac{\partial u_i'}{\partial x_k}\dfrac{\partial u_j'}{\partial x_k}}$, which represent the destruction of the mean flow energy by viscous effects, are jointly modelled through $-C_v\dfrac{1}{L}\delta_{i,j}\left(\overline{q^2}\right)^{3/2}$, where C_v is a constant. Inserting these approximations in (5), the following equation (6) for the $\overline{u_i'\,u_j'}$ correlations is obtained:

$$\frac{\partial}{\partial t}\left(\overline{u_i'\,u_j'}\right)+\overline{u}_k\frac{\partial}{\partial x_k}\left(\overline{u_i'\,u_j'}\right)=-\overline{u_i'\,u_k'}\frac{\partial \overline{u}_j}{\partial x_k}-\overline{u_j'\,u_k'}\frac{\partial \overline{u}_i}{\partial x_k}+\frac{1}{\rho_0}\left(g_i\overline{u_j'\,\rho'}+g_j\overline{u_i'\,\rho'}\right)$$

$$+C_t\frac{\partial}{\partial x_k}\left[\overline{q}L\frac{\partial}{\partial x_k}\left(\overline{u_i'\,u_j'}\right)\right]-C_p\frac{\sqrt{\overline{q^2}}}{L}\left(\overline{u_i'\,u_j'}-\frac{1}{3}\delta_{i,j}\overline{q^2}\right)-C_v\frac{1}{L}\delta_{i,j}\left(\overline{q^2}\right)^{3/2} \qquad (6)$$

where $\overline{q^2}=2K=\overline{u_i'\,u_i'}$, and the constants have the following values: $C_t\approx 0.30$, $C_p\approx 1.0$ and $C_v\approx 1/12$. By analogy to equation (6), a density-velocity correlations tensor $\overline{u_i'\,\rho'}$ is obtained:

$$\frac{\partial}{\partial t}\left(\overline{u_i'\,\rho'}\right)+\overline{u}_j\frac{\partial}{\partial x_j}\left(\overline{u_i'\,\rho'}\right)=-\overline{u_i'\,u_j'}\frac{\partial \overline{\rho}}{\partial x_j}-\overline{u_j'\,\rho'}\frac{\partial \overline{u}_i}{\partial x_j}$$

$$+g_i\frac{\overline{\rho'^2}}{\rho_0}+C_t\frac{\partial}{\partial x_j}\left[\overline{q}L\frac{\partial}{\partial x_j}\left(\overline{u_i'\,\rho'}\right)\right]-C_q\frac{\overline{q}}{L}\overline{u_i'\,\rho'} \qquad (7)$$

with the quadratic term $\overline{\rho'^2}$ calculated through (8),

$$\frac{\partial}{\partial t}\left(\overline{\rho'^2}\right)+\overline{u}_j\frac{\partial}{\partial x_j}\left(\overline{\rho'^2}\right)=-2\overline{u_j'\,\rho'}\frac{\partial \overline{\rho}}{\partial x_j}$$

$$+C_t\frac{\partial}{\partial x_j}\left[\overline{q}L\frac{\partial}{\partial x_j}\left(\overline{\rho'^2}\right)\right]-C_r\frac{\overline{q}}{L}\left(\overline{\rho'^2}\right) \qquad (8)$$

An equation for the turbulent length scale (or macroscale of the eddies), L, is written:

$$\frac{\partial L}{\partial t}+\overline{u}_i\frac{\partial L}{\partial x_i}=C_l\frac{L}{q^2}\overline{u'_i u'_j}\frac{\partial \overline{u}_i}{\partial x_j}+C_s\overline{q}$$

$$+C_t\frac{\partial}{\partial x_i}\left[\overline{q}L\frac{\partial L}{\partial x_i}\right]-C_q\frac{1}{2\overline{q}}\left(\frac{\partial \overline{q}L}{\partial x_i}\right)^2+C_z\frac{L}{q^2}g_i\frac{\overline{u'_i \rho'}}{\rho_0} \tag{9}$$

where $C_q \approx 0.75$, $C_r \approx 0.1125$, $C_l \approx 0.35$, $C_s \approx 0.075$ and $C_z \approx 0.80$. As can be easily seen, an equation for the turbulent length scale L is, like all other approximations, of the form:

$$\frac{\partial L}{\partial t}+\overline{u}_i\frac{\partial L}{\partial x_i}= \text{PRODUCTION } + \text{ DISSIPATION}+ \text{ DIFFUSION } + \text{ BUOYANCY} \tag{10}$$

Modelling the production and dissipation terms by $C_l\frac{L}{q^2}\overline{u'_i u'_j}\frac{\partial u_i}{\partial x_j}$ and $C_s\overline{q}$, respectively,

the diffusion terms by $C_t\frac{\partial}{\partial x_i}\left[\overline{q}L\frac{\partial L}{\partial x_i}\right]-C_q\frac{1}{2\overline{q}}\left(\frac{\partial \overline{q}L}{\partial x_i}\right)^2$, as suggested by Lewellen (1977), and

adding the buoyancy term, the approximation (9) above is newly obtained.
In summary, equations (2), along with equations (6), (7), (8) and (9) for turbulence closure, constitute a complete 3D turbulent boundary layer model with sediment in suspension.

3.4 Simplified turbulent boundary layer models
Proceeding with a non-dimensional analysis of the mean flow equations, without sediment in suspension, and considering:
1.   A sinusoidal wave $\left(\hat{U}_w, T, L_w\right)$.
2.   The following boundary layer approximations:
  -   Small boundary layer thickness $(z_\delta \ll \lambda/2\pi)$, λ being the wave length;
  -   Nikuradse equivalent bottom rugosity much inferior to the boundary layer thickness ($k_N \ll z_\delta$).
3.   Small wave amplitude and Stokes hypothesis, which assumes that:
  -   The maximum wave velocity amplitude is much inferior to the celerity $(\hat{U}_w \ll \sqrt{gh}$).
4.   Local equilibrium turbulence, along with the turbulent kinetic energy is equivalent to the viscous dissipation. Assuming local equilibrium there is no time evolution or spatial diffusion of the correlations, and the Reynolds stress equation $\overline{u'_i u'_j}$ can be reduced.

In summary, assuming these hypotheses we can: i) consider a horizontal flow ($u, v, w = 0$); ii) neglect the convective and horizontal diffusion transport, and iii) simplify the turbulent transport equations, cancelling the remaining time variation terms and the diffusion terms of the velocity correlations.
Considering the above hypotheses in the pure hydrodynamic Reynolds equations (2-b), without stratification, the following approximations (11) result:

$$\frac{\partial u}{\partial t}=-\frac{1}{\rho_0}\frac{\partial P}{\partial x}-\frac{\partial}{\partial z}\left(\overline{u'w'}\right); \;\frac{\partial v}{\partial t}=-\frac{1}{\rho_0}\frac{\partial P}{\partial y}-\frac{\partial}{\partial z}\left(\overline{v'w'}\right) \tag{11}$$

On the other hand, under the same assumptions, the *Reynolds stress equation* $\overline{u_i\,u_j}$ (6) can be written explicitly:

$$\frac{\partial \overline{u'w'}}{\partial t} = -\overline{w'^2}\frac{\partial u}{\partial z} + C_t\frac{\partial}{\partial z}\left(\overline{q}L\frac{\partial \overline{u'w'}}{\partial z}\right) - C_p\frac{q}{L}\overline{u'w'}$$

$$\frac{\partial \overline{v'w'}}{\partial t} = -\overline{w'^2}\frac{\partial v}{\partial z} + C_t\frac{\partial}{\partial z}\left(\overline{q}L\frac{\partial \overline{v'w'}}{\partial z}\right) - C_p\frac{q}{L}\overline{v'w'}$$

$$\frac{\partial \overline{u'^2}}{\partial t} = -2\overline{u'w'}\frac{\partial u}{\partial z} + C_t\frac{\partial}{\partial z}\left(\overline{q}L\frac{\partial \overline{u'^2}}{\partial z}\right) - C_p\frac{q}{L}\left(\overline{u'^2} - \frac{q^2}{3}\right) - C_v\frac{q^3}{L} \qquad (12)$$

$$\frac{\partial \overline{v'^2}}{\partial t} = -2\overline{v'w'}\frac{\partial v}{\partial z} + C_t\frac{\partial}{\partial z}\left(\overline{q}L\frac{\partial \overline{v'^2}}{\partial z}\right) - C_p\frac{q}{L}\left(\overline{v'^2} - \frac{q^2}{3}\right) - C_v\frac{q^3}{L}$$

$$\frac{\partial \overline{w'^2}}{\partial t} = \qquad\quad + C_t\frac{\partial}{\partial z}\left(\overline{q}L\frac{\partial \overline{w'^2}}{\partial z}\right) - C_p\frac{q}{L}\left(\overline{w'^2} - \frac{q^2}{3}\right) - C_v\frac{q^3}{L}$$

Adding the last three equations we get (13) for q^2 :

$$\frac{\partial q^2}{\partial t} = -2\overline{u'w'}\frac{\partial u}{\partial z} - 2\overline{v'w'}\frac{\partial v}{\partial z} + C_t\frac{\partial}{\partial z}\left(\overline{q}L\frac{\partial q^2}{\partial z}\right) - 3C_v\frac{q^3}{L} \qquad (13)$$

Taking now into account local equilibrium turbulence (Sheng, 1984), which can be assumed when the scale L/q is much smaller than the time scale of the mean flow and when the turbulent quantities have a small variation on the macroscale of the eddies L. In addition, neglecting both variations in time and diffusive transport terms, from equations system (12) the following equations (14) are obtained:

$$\overline{w'^2}\frac{\partial u}{\partial z} + \frac{q}{L}\overline{u'w'} = 0 \;\; ; \;\; \overline{w'^2}\frac{\partial v}{\partial z} + \frac{q}{L}\overline{v'w'} = 0$$

$$2\overline{u'w'}\frac{\partial u}{\partial z} + \frac{q}{L}\left(\overline{u'^2} - \frac{q^2}{3}\right) + C_v\frac{q^3}{L} = 0$$

$$2\overline{v'w'}\frac{\partial v}{\partial z} + \frac{q}{L}\left(\overline{v'^2} - \frac{q^2}{3}\right) + C_v\frac{q^3}{L} = 0 \qquad (14)$$

$$\frac{q}{L}\left(\overline{w'^2} - \frac{q^2}{3}\right) + C_v\frac{q^3}{L} = 0$$

This system of equations allows us to obtain (15):

$$\overline{u'^2} = 6\,C_v L^2\left(\frac{\partial u}{\partial z}\right)^2 + 3\,C_v\,q^2 \;\; ; \;\; \overline{v'^2} = 6\,C_v\,L^2\left(\frac{\partial v}{\partial z}\right)^2 + 3C_v\,q^2 ;$$

$$\overline{w'^2} = 3\,C_v\,q^2 \;\; ; \;\; -\overline{u'w'} = 3\,C_v\,q\,L\frac{\partial u}{\partial z} \;\; ; \;\; -\overline{v'w'} = 3\,C_v\,q\,L\frac{\partial v}{\partial z} \qquad (15)$$

Comparing the last two equations of (15) with $-\overline{u'w'} = v_t \dfrac{\partial u}{\partial z}$ and $-\overline{v'w'} = v_t \dfrac{\partial v}{\partial z}$ it is clear that:

$$v_t = 3\,C_v\,qL = 3\,C_v\sqrt{2K}\,L = \frac{\sqrt{2K}\,L}{4} \tag{16}$$

In addition, it can be seen from the first three equations of (15) that:

$$q^2 = 2K = 24\,C_v\,L^2\left[\left(\frac{\partial u}{\partial z}\right)^2 + \left(\frac{\partial v}{\partial z}\right)^2\right] = 2L^2\left[\left(\frac{\partial u}{\partial z}\right)^2 + \left(\frac{\partial v}{\partial z}\right)^2\right] \tag{17}$$

3.5 1DV turbulent boundary layer models

Considering now a horizontal flow along x-direction (u, $v = 0$, $w = 0$) with sediment in suspension, so with the buoyancy terms, and local equilibrium turbulence, the system (12) is written in the following form (18):

$$
\begin{aligned}
\frac{\partial \overline{u'w'}}{\partial t} &= 0 = -\overline{w'^2}\frac{\partial u}{\partial z} & -\frac{g}{\rho_0}\overline{u'\rho'} - C_p\frac{q}{L}\overline{u'w'} \\[2mm]
\frac{\partial \overline{w'^2}}{\partial t} &= 0 = & -\frac{2g}{\rho_0}\overline{w'\rho'} - C_p\frac{q}{L}\left(\overline{w'^2} - \frac{q^2}{3}\right) - C_v\frac{q^3}{L} \\[2mm]
\frac{\partial \overline{u'\rho'}}{\partial t} &= 0 = -\overline{w'\rho'}\frac{\partial u}{\partial z} - \overline{u'w'}\frac{\partial \rho}{\partial z} & -C_q\frac{q}{L}\overline{u'\rho'} \\[2mm]
\frac{\partial \overline{w'\rho'}}{\partial t} &= 0 = -\overline{w'^2}\frac{\partial \rho}{\partial z} & -\frac{g}{\rho_0}\overline{\rho'^2} -C_q\frac{q}{L}\overline{w'\rho'} \\[2mm]
\frac{\partial \overline{\rho'^2}}{\partial t} &= 0 = -2\overline{w'\rho'}\frac{\partial \rho}{\partial z} & -C_r\frac{q}{L}\overline{\rho'^2}
\end{aligned}
\tag{18}
$$

Solving this equation system for $\overline{u'w'}$ and $\overline{w'\rho'}$ we obtain (19):

$$\overline{u'w'} = -\frac{(1-16.444\,\Omega)}{(1-19.778\,\Omega)(1-\Omega)}\frac{qL}{4}\frac{\partial u}{\partial z}\;;\quad \overline{w'\rho'} = -\frac{1}{(1-19.778\,\Omega)}\frac{qL}{3}\frac{\partial \rho}{\partial z} \tag{19}$$

where $\Omega = \dfrac{4}{3}\dfrac{g}{\rho_0}\dfrac{\partial \rho}{\partial z}\dfrac{L^2}{q^2}$. As shown before and from (3) $-\overline{u'w'} = v_t\dfrac{\partial u}{\partial z}$ and $-\overline{w'\rho'} = \gamma_t\dfrac{\partial \rho}{\partial z}$; comparing with the expressions (19) above we can write (20) and (21):

$$v_t = \frac{(1-16.444\,\Omega)}{(1-19.778\,\Omega)(1-\Omega)}\frac{qL}{4} = \frac{(1-16.444\,\Omega)}{(1-19.778\,\Omega)(1-\Omega)}\frac{\sqrt{2K}L}{4} \tag{20}$$

$$\gamma_t = \frac{1}{(1-19.778\,\Omega)}\frac{qL}{3} = \frac{(1-\Omega)}{(1-16.444\,\Omega)}\frac{4}{3}v_t \tag{21}$$

3.5.1 Two-equation K-L 1DV boundary layer model

Taking into account the assumptions stated before, a complete set of governing equations (22) for the two-equation $K-L$ model is written (Tran-Thu & Temperville, 1994):

$$\frac{\partial u}{\partial t} = -\frac{1}{\rho_0}\frac{\partial P}{\partial x} + \frac{\partial}{\partial z}\left(v_t \frac{\partial u}{\partial z}\right); \quad \frac{\partial v}{\partial t} = -\frac{1}{\rho_0}\frac{\partial P}{\partial y} + \frac{\partial}{\partial z}\left(v_t \frac{\partial v}{\partial z}\right)$$

$$\underbrace{\frac{\partial K}{\partial t}}_{\substack{\text{rate of}\\\text{change}}} = \underbrace{v_t\left[\left(\frac{\partial u}{\partial z}\right)^2 + \left(\frac{\partial v}{\partial z}\right)^2\right]}_{\text{production}} - \underbrace{\frac{\sqrt{2K}}{4L}K}_{\text{dissipation}} + \underbrace{0.30\frac{\partial}{\partial z}\left(\sqrt{2K}L\frac{\partial K}{\partial z}\right)}_{\text{diffusion}} + \underbrace{\frac{g}{\rho_0}\gamma_t\frac{\partial \rho}{\partial z}}_{\text{buoyancy}}$$

$$\underbrace{\frac{\partial L}{\partial t}}_{\substack{\text{rate of}\\\text{change}}} = \underbrace{-0.35\frac{v_t}{2K}\left[\left(\frac{\partial u}{\partial z}\right)^2 + \left(\frac{\partial v}{\partial z}\right)^2\right]L}_{\text{production}} + \underbrace{0.075\sqrt{2K}}_{\text{dissipation}}$$

$$\underbrace{+0.30\frac{\partial}{\partial z}\left(\sqrt{2K}L\frac{\partial L}{\partial z}\right) - \frac{0.375}{\sqrt{2K}}\left[\frac{\partial}{\partial z}\left(\sqrt{2K}L\right)\right]^2}_{\text{diffusion}} + \underbrace{0.80\frac{L}{2K}\frac{g}{\rho_0}\gamma_t\frac{\partial \rho}{\partial z}}_{\text{buoyancy}}$$

$$\frac{\partial C}{\partial t} = \frac{\partial(w_s C)}{\partial z} + \frac{\partial}{\partial z}\left(\gamma_t\frac{\partial C}{\partial z}\right)$$

(22)

where u and v are horizontal components of flow velocity in the boundary layer; C is the volumetric concentration; w_s is the sediment settling velocity; K is the turbulent kinetic energy, and L is the length scale of the large vortices.

The turbulent viscosity v_t and the turbulent diffusivity γ_t are given by equations (20) and (21), respectively. The hydrodynamic equations and the concentration equation are coupled through the equation (23) for the density:

$$\rho = \rho_0 + (\rho_s - \rho_0)C$$

(23)

where ρ_0 and ρ_s are the densities of the fluid and sediment, respectively.

3.5.2 One-equation K-L 1DV boundary layer model

With $L = f(k,z,K)$, a complete one-equation $K-L$ turbulence closure model is simply written:

$$\frac{\partial u}{\partial t} = -\frac{1}{\rho_0}\frac{\partial P}{\partial x} + \frac{\partial}{\partial z}\left(v_t \frac{\partial u}{\partial z}\right) \quad ; \quad \frac{\partial v}{\partial t} = -\frac{1}{\rho_0}\frac{\partial P}{\partial y} + \frac{\partial}{\partial z}\left(v_t \frac{\partial v}{\partial z}\right)$$

$$\frac{\partial K}{\partial t} = v_t\left[\left(\frac{\partial u}{\partial z}\right)^2 + \left(\frac{\partial v}{\partial z}\right)^2\right] - \frac{\sqrt{2K}}{4L}K + 0.30\frac{\partial}{\partial z}\left(\sqrt{2K}L\frac{\partial K}{\partial z}\right) + \frac{g}{\rho_0}\gamma_t\frac{\partial \rho}{\partial z}$$

$$L = f(k,z,K) \quad ; \quad \frac{\partial C}{\partial t} = \frac{\partial(w_s C)}{\partial z} + \frac{\partial}{\partial z}\left(\gamma_t\frac{\partial C}{\partial z}\right)$$

(24)

where (20), (21) and (23) apply.

A number of empirical equations for the length scale L could be found in the literature; some examples are (with $k = 0.4$):

- $L = k \sqrt[4]{c_l} \, z$, where $c_l = 0.08$.

- $L = k \, z \sqrt{1 - z/z_\delta}$, z_δ being the boundary layer thickness.

- $L = k \sqrt{K} \left\{ \int_{z_0}^{z} K^{-1/2} dz + z_0 K_0^{-1/2} \right\}$, $K_0 = K(z_0, t)$.

- $L = k \, z_w \left(1 - e^{z_w^+ / A} \right)$, where $A \approx 26$, z_w is the distance to the wall and $z_w^+ = (z_w \, u_T)/v_l$, u_T being the friction velocity.

The influence of a stable stratification on L can be taken into account through (25),

$$\left(\frac{L}{L_o} \right)^2 = \left(1 + \beta R_i \right)^n , \text{ with } \beta \approx (10, \ 14) \text{ and } n \approx (-0.5, \ -1.5) \tag{25}$$

where $R_i = -\dfrac{g}{\rho} \dfrac{\partial \rho}{\partial z} \bigg/ \left[\left(\dfrac{\partial u}{\partial z} \right)^2 + \left(\dfrac{\partial v}{\partial z} \right)^2 \right]$ is the Richardson number and L_o is the length scale L

value without stratification.

3.5.3 Zero-equation boundary layer model

Defining the mixing length as $l_m = k \, z \sqrt{1 - z/z_\delta}$, where $k \approx 0.4$ and z_δ is the boundary layer thickness, equations (26) for the u and v variables in the boundary layer are obtained:

$$\frac{\partial u}{\partial t} = -\frac{1}{\rho_0} \frac{\partial P}{\partial x} + \frac{\partial}{\partial z} \left(l_m^2 \left| \frac{\partial u}{\partial z} \right| \frac{\partial u}{\partial z} \right) \ ; \ \frac{\partial v}{\partial t} = -\frac{1}{\rho_0} \frac{\partial P}{\partial y} + \frac{\partial}{\partial z} \left(l_m^2 \left| \frac{\partial v}{\partial z} \right| \frac{\partial v}{\partial z} \right) \tag{26}$$

Stable stratification effects on l_m could be taken into account through the relation $(l_m / l_{mo})^2 = (1 + 10 R_i)^{-0.5}$, where l_{m0} is the mixing length l_m value without stratification, and R_i is the Richardson number, as defined above. We now assume in (26):

$$-\frac{1}{\rho_0} \frac{\partial P}{\partial x} = \frac{\partial U}{\partial t} \text{ and } -\frac{1}{\rho_0} \frac{\partial P}{\partial y} = \frac{\partial V}{\partial t} \tag{27}$$

where U and V are the velocity components outside of the boundary layer. Defining the deficit velocity components (u_d, v_d) as (28),

$$u_d(z,t) = u(z,t) - U(t) \ ; \ v_d(z,t) = v(z,t) - V(t) \tag{28}$$

and substituting in (26) the following equations (29) are obtained,

$$\frac{\partial u_d}{\partial t} = \frac{\partial}{\partial z} \left(l_m^2 \left| \frac{\partial u_d}{\partial z} \right| \frac{\partial u_d}{\partial z} \right) \ ; \ \frac{\partial v_d}{\partial t} = \frac{\partial}{\partial z} \left(l_m^2 \left| \frac{\partial v_d}{\partial z} \right| \frac{\partial v_d}{\partial z} \right) \tag{29}$$

These equations are non-linear and no analytical solutions are available, so they have to be solved numerically, as will be shown later.

3.6 Boundary conditions for 1DV turbulent boundary layer models
3.6.1 One- and two-equation boundary layer models of the K-L type

- At the lower limit of the boundary layer, $z = z_0$

  - $u(z_0) = v(z_0) = 0$; $\partial K/\partial z = 0$; $L(z_0) = \alpha z_0$, with $\alpha = 0.67$ (empirical constant).

  - At the hydraulic rough regime, the level z_0 is taken to be $k_N/30$, with $k_N = 2.5\,d$ the Nikuradse equivalent roughness of a bed of sand with diameter d. In the transitory regime, k_N and z_0 are calculated following Sleath (1984) (Tran Thu and Temperville, 1994).

  - For the reference concentration at the bottom, C_b, the following relations may be used: $C = C_b \approx 0.63$, or $C = C(\psi)$, where $\psi = \tau_b(t)/(\rho(s-1)gd)$.

- At the upper limit of the boundary layer, $z = z_\delta$

  Assuming that the instantaneous velocity $\vec{U}(t)$ is given at a level $z = z_\delta$ outside the boundary layer, the boundary conditions are:

  - $\vec{u}(z_\delta) = \vec{U}(t)$, $\vec{U}(t)$ may contain a component of the mean current U_c as well as oscillatory components of the wave;

  - $K(z_\delta) = 0$ (pure wave), or $\partial K/\partial z|_{z_\delta} = 0$ (combined wave and current);

  - $L(z_\delta) = 0$ (pure wave), or $L(z_\delta) = \alpha z_\delta$ (combined wave and current);

  - Depending on the problem, the condition $\partial L/\partial z|_{z_\delta} = 0$ may be also adequate;

  - $C(z_\delta) = 0$ (pure wave), or $w_q C + \gamma_t \partial C/\partial t|_{z_\delta} = 0$ (combined wave and current).

  Initial values for u, v, K and L are the solution for the initial field current velocities (U_c, V_c).

- Estimation of the boundary layer thickness, z_δ

  Considering a pure current ($\hat{U}_w = 0$) in a channel with a water column h, the boundary layer thickness is $z_\delta = h$.

  Assuming now a pure wave ($U_c = V_c = 0$) propagating in a channel, the boundary layer thickness reaches its minimum value and can be approximated by $z_\delta/k_N = 0.246\,(\hat{a}/k_N)^{0.81}$ (Huynh-Thanh, 1990), where the orbital amplitude is given by $\hat{a} = \hat{U}_{weq} T_{ch}/(2\pi)$ for an equivalent sinusoidal wave with $\hat{U}_{weq}$, and during a characteristic signal period T_{ch} (Antunes do Carmo et al., 1996). The relation proposed for z_δ/k_N corresponds to the thickness beyond which K is zero.

A general rough estimation for z_δ can be obtained by (30):

$$z_\delta = \frac{|\vec{U}_c|\,h + |\hat{U}_w|\,z_{\delta w}}{|\vec{U}_c| + |\hat{U}_w|} \;=> \; z_\delta = \frac{|\vec{U}_c|\,h + 0.246\,k_N\,(\hat{a}/k_N)^{0.81}|\hat{U}_w|}{|\vec{U}_c| + |\hat{U}_w|} \tag{30}$$

3.6.2 Zero-equation model

The following conditions (31) are imposed at the lower limit $z = z_0$ and at the upper limit $z = z_\delta$ of the boundary layer:

$$u_d(z_0,t) = -U \;;\; u_d(z_\delta,t) = 0 \;;\; v_d(z_0,t) = -V \;;\; v_d(z_\delta,t) = 0 \tag{31}$$

3.7 2DV turbulent boundary layer model

Over movable beds, the interaction of flow and sediment transport creates a variety of bed forms such as ripples, dunes, antidunes or other irregular shapes and obstacles. Their presence, in general, causes flow separation and recirculation, which can alter the overall flow resistance and, consequently, can affect sediment transport within the water mass and bottom erosion. For dunes, in particular, the flow is characterized by an attached flow on their windward side, separation at their crest and formation of a recirculation eddy in their leeside (Fourniotis et al., 2006). A detailed description of the flow over a dune is then of fundamental interest because the pressure and friction (shear-stress) distributions on the bed determine the total resistance on the bottom and the rate of sediment transport. Over bed forms a 1DV version of the turbulent boundary layer is no able to describe the main processes that occur above and close to the bed surface. Consequently, a 2DV turbulent boundary layer model is developed herein.

Considering a two-dimensional mean non-stratified flow in the vertical plane $(u, v = 0, w)$, only non-zero y-derivatives are present. The physical problem is outlined in figure 1 below, under the action of a wave. Knowing that the wavelength is always greater than the length of the ripples, i.e. $L_w \gg L_r$, we can restrict the domain of calculation, instead of investigate all the domain over of the whole wavelength.

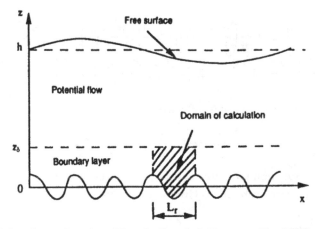

Fig. 1. Scheme of the physical system (Huynh-Thanh & Temperville, 1991)

The basic equations of the model are derived from the previous ones (2). In order to simplify the numerical resolution of the equations we make use of the stream function (Ψ) and vorticity (ξ) variables, instead of the velocities u and v, and a transformation of the physical domain into a rectangular one. Considering that only two-independent spatial derivatives are involved in the flow, in the xz-plane, i.e., a flow with only velocity components $u(x,z,t)$ and $w(x,z,t)$, the equations of motion are restricted to the continuity equation and the two components of the Reynolds equations. Under these assumptions, from (2) the two components (32) and (33) of the pure hydrodynamic momentum equation are written:

$$\frac{\partial u}{\partial t} + u\frac{\partial u}{\partial x} + w\frac{\partial u}{\partial z} = -\frac{1}{\rho}\frac{\partial p}{\partial x} + \frac{\partial}{\partial x}\left(-\overline{u'^2}\right) + \frac{\partial}{\partial z}\left(-\overline{u'w'}\right) \tag{32}$$

$$\frac{\partial w}{\partial t}+u\frac{\partial w}{\partial x}+w\frac{\partial w}{\partial z}=-\frac{1}{\rho}\frac{\partial p}{\partial z}+\frac{\partial}{\partial x}\left(-\overline{u'w'}\right)+\frac{\partial}{\partial z}\left(-\overline{w'^2}\right) \tag{33}$$

Substituting in (32) and (33) the approximations (34),

$$-\overline{u'^2}=2\,\nu_t\frac{\partial u}{\partial x}\;;\;\;-\overline{u'w'}=\nu_t\left(\frac{\partial u}{\partial z}+\frac{\partial w}{\partial x}\right)\;\text{and}\;-\overline{w'^2}=2\,\nu_t\frac{\partial w}{\partial z} \tag{34}$$

the governing equations (35) and (36) result:

$$\frac{\partial u}{\partial t}+u\frac{\partial u}{\partial x}+w\frac{\partial u}{\partial z}=-\frac{1}{\rho}\frac{\partial p}{\partial x}+2\frac{\partial}{\partial x}\left(\nu_t\frac{\partial u}{\partial x}\right)+\frac{\partial}{\partial z}\left[\nu_t\left(\frac{\partial u}{\partial z}+\frac{\partial w}{\partial x}\right)\right] \tag{35}$$

$$\frac{\partial w}{\partial t}+u\frac{\partial w}{\partial x}+w\frac{\partial w}{\partial z}=-\frac{1}{\rho}\frac{\partial p}{\partial z}+\frac{\partial}{\partial x}\left[\nu_t\left(\frac{\partial u}{\partial z}+\frac{\partial w}{\partial x}\right)\right]+2\frac{\partial}{\partial z}\left(\nu_t\frac{\partial w}{\partial z}\right) \tag{36}$$

The unknown pressure gradient due to the bed forms can now be eliminated from equations (35) and (36) by cross-differentiation, i.e., taking the curl of the two-dimensional vector momentum equations. The result reads:

$$\frac{\partial}{\partial t}\left(\frac{\partial u}{\partial z}-\frac{\partial w}{\partial x}\right)+u\frac{\partial}{\partial x}\left(\frac{\partial u}{\partial z}-\frac{\partial w}{\partial x}\right)+w\frac{\partial}{\partial z}\left(\frac{\partial u}{\partial z}-\frac{\partial w}{\partial x}\right)=$$
$$-\frac{\partial^2}{\partial x^2}\left[\nu_t\left(\frac{\partial u}{\partial z}+\frac{\partial w}{\partial x}\right)\right]+2\frac{\partial^2}{\partial x\partial z}\left[\nu_t\left(\frac{\partial u}{\partial x}-\frac{\partial w}{\partial z}\right)\right]+\frac{\partial^2}{\partial z^2}\left[\nu_t\left(\frac{\partial u}{\partial z}+\frac{\partial w}{\partial x}\right)\right] \tag{37}$$

By definition, the following relations (38) account:

$$u=\frac{\partial\Psi}{\partial z}\;;\;\;w=-\frac{\partial\Psi}{\partial x}\;;\;\;\xi=\frac{\partial u}{\partial z}-\frac{\partial w}{\partial x} \tag{38}$$

Inserting the stream function (Ψ) and vorticity (ξ) variables in equation (37) the following result (39) for the vorticity is obtained (Huynh-Thanh, 1990; Tran-Thu, 1995):

$$\frac{\partial\xi}{\partial t}-\frac{\partial(\Psi,\xi)}{\partial(x,z)}=\nabla^2\left(\nu_t\xi\right)-2\left(\frac{\partial^2\nu_t}{\partial x^2}\frac{\partial^2\Psi}{\partial z^2}-2\frac{\partial^2\nu_t}{\partial x\partial z}\frac{\partial^2\Psi}{\partial x\partial z}+\frac{\partial^2\nu_t}{\partial z^2}\frac{\partial^2\Psi}{\partial x^2}\right) \tag{39}$$

where $\dfrac{\partial(\Psi,\xi)}{\partial(x,z)}=\dfrac{\partial\Psi}{\partial x}\dfrac{\partial\xi}{\partial z}-\dfrac{\partial\Psi}{\partial z}\dfrac{\partial\xi}{\partial x}$ and $\nabla^2=\dfrac{\partial^2}{\partial x^2}+\dfrac{\partial^2}{\partial z^2}$.

An equation for the stream function is obtained through the definitions (38), substituting u and v in ξ:

$$\nabla^2\Psi=\xi \tag{40}$$

which is known as the Poisson equation. The turbulent viscosity ν_t is obtained assuming local equilibrium turbulence. Once more in the vertical plane $(u,v=0,w)$, the following equations system (41) can be written:

$$\frac{\partial \overline{u'w'}}{\partial t} = 0 = -\left(\overline{u'^2}\frac{\partial w}{\partial x} + \overline{w'^2}\frac{\partial u}{\partial z}\right) \quad -C_p\frac{q}{L}\overline{u'w'}$$

$$\frac{\partial \overline{u'^2}}{\partial t} = 0 = -2\left(\overline{u'^2}\frac{\partial u}{\partial x} + \overline{u'w'}\frac{\partial u}{\partial z}\right) \quad -C_p\frac{q}{L}\left(\overline{u'^2} - \frac{q^2}{3}\right) - C_v\frac{q^3}{L}$$

$$\frac{\partial \overline{v'^2}}{\partial t} = 0 = \qquad\qquad\qquad\qquad\qquad -C_p\frac{q}{L}\left(\overline{v'^2} - \frac{q^2}{3}\right) - C_v\frac{q^3}{L}$$

$$\frac{\partial \overline{w'^2}}{\partial t} = 0 = -2\left(\overline{u'w'}\frac{\partial w}{\partial x} + \overline{w'^2}\frac{\partial w}{\partial z}\right) - C_p\frac{q}{L}\left(\overline{w'^2} - \frac{q^2}{3}\right) - C_v\frac{q^3}{L}$$

(41)

The third equation of this system allows us to obtain $\overline{v'^2} = q^2/4 = K/2$, with $C_p = 1.0$ and $C_v = 1/12$.

Assuming identical production along both x- and z-directions, from the second and fourth equations we find that $\overline{u'^2} = \overline{w'^2}$. This hypothesis is supported by laboratory experiments over a bottom with ripples conducted by Sato *et al.* (1984), among others. Therefore, as $q^2 = 2K = \overline{u'^2} + \overline{v'^2} + \overline{w'^2}$, the above results show that $\overline{u'^2} = \overline{w'^2} = 3\,q^2/8 = 3K/4$. On the other hand, from the first equation of the system (41) we find that:

$$-\overline{u'w'} = \frac{L}{q}\left(\overline{u'^2}\frac{\partial w}{\partial x} + \overline{w'^2}\frac{\partial u}{\partial z}\right) = \frac{L}{q}\frac{3q^2}{8}\left(\frac{\partial w}{\partial x} + \frac{\partial u}{\partial z}\right) = \frac{3}{8}\sqrt{2K}L\left(\frac{\partial w}{\partial x} + \frac{\partial u}{\partial z}\right)$$

(42)

Therefore,

$$v_t = \frac{3}{8}\sqrt{2K}L$$

(43)

The equation for the turbulent kinetic energy, K, is obtained through the earlier already presented in two-dimensions in the vertical plane:

$$\frac{\partial K}{\partial t} + u\frac{\partial K}{\partial x} + w\frac{\partial K}{\partial z} = v_t\left[2\left(\frac{\partial u}{\partial x}\right)^2 + \left(\frac{\partial u}{\partial z} + \frac{\partial w}{\partial x}\right)^2 + 2\left(\frac{\partial w}{\partial z}\right)^2\right]$$

$$- \frac{K\sqrt{2K}}{4L} + 0.30\frac{\partial}{\partial x}\left(\sqrt{2K}L\frac{\partial K}{\partial x}\right) + 0.30\frac{\partial}{\partial z}\left(\sqrt{2K}L\frac{\partial K}{\partial z}\right)$$

(44)

Inserting the stream function (Ψ) in equation (44), we find (45):

$$\frac{\partial K}{\partial t} - \underbrace{\frac{\partial(\Psi,K)}{\partial(x,z)}}_{\text{advection}} = \underbrace{v_t\left[4\left(\frac{\partial^2\Psi}{\partial x\partial z}\right)^2 + \left(\frac{\partial^2\Psi}{\partial z^2} - \frac{\partial^2\Psi}{\partial x^2}\right)^2\right]}_{\text{production}}$$

$$\underbrace{-\frac{2}{3}v_t\frac{K}{L^2}}_{\text{dissipation}} + \underbrace{0.80\frac{\partial}{\partial x}\left(v_t\frac{\partial K}{\partial x}\right) + 0.80\frac{\partial}{\partial z}\left(v_t\frac{\partial K}{\partial z}\right)}_{\text{diffusion}}$$

(45)

The length scale L is directly imposed by the analytical solution (46):

$$L = 0.67\, z\sqrt{1 - z/z_\delta} \tag{46}$$

In order to describe the space-time distribution of the sediments concentration over a bottom with ripples, an equation for C is included, considering in it the advection and diffusion terms in both x-horizontal and z-vertical directions:

$$\frac{\partial C}{\partial t} + \frac{\partial}{\partial x}(uC) + \frac{\partial}{\partial z}\left[(w - w_s)C\right] = \frac{\partial}{\partial x}\left(\gamma_t \frac{\partial C}{\partial x}\right) + \frac{\partial}{\partial z}\left(\gamma_t \frac{\partial C}{\partial z}\right) \tag{47}$$

In order to simplify the numerical resolution of the equations, as well as the description of the boundary conditions at the ripples surface, the physical domain in coordinates (x, z) is transformed into a rectangular one (the computation domain) utilizing orthogonal curvilinear coordinates (X, Z) (Figure 2), using the following transformations (48) (Sato et al., 1984; Huynh-Thanh, 1990; Tran-Thu, 1995; Silva, 2001):

$$X = x + \sum_{n=1}^{N} a_n \exp\left(-n\frac{2\pi}{L_r}Z\right) \sin\left(n\frac{2\pi}{L_r}X - \theta_n\right)$$
$$Z = z - \sum_{n=1}^{N} a_n \exp\left(-n\frac{2\pi}{L_r}Z\right) \cos\left(n\frac{2\pi}{L_r}X - \theta_n\right) \tag{48}$$

where N, a_n and θ_n are coefficients to be determined in such a way that the curve $Z = 0$ represents the real ripple.

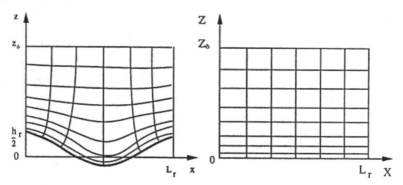

Fig. 2. Physical and computational domains. Transformation of coordinates $(x,z) \rightarrow (X,Z)$

The Jacobian of the transformation is defined by (49):

$$J = \frac{\partial(X,Z)}{\partial(x,z)} = \frac{\partial X}{\partial x}\frac{\partial Z}{\partial z} - \frac{\partial X}{\partial z}\frac{\partial Z}{\partial x} = \left(\frac{\partial X}{\partial x}\right)^2 + \left(\frac{\partial X}{\partial z}\right)^2 \tag{49}$$

which is calculated from the inverse transformation of the Jacobian J_0 $\left(J = J_0^{-1}\right)$. After carried out the transformation of coordinates $(x,z) \rightarrow (X,Z)$, the above equations (39), (40), (43), (45) and (47) are written and solved iteratively as will be shown later (see Huynh-Thanh, 1990, Tran-Thu, 1995, and Silva, 2001, for details):

$$\frac{\partial \xi}{\partial t} - J\frac{\partial(\Psi,\xi)}{\partial(X,Z)} = J\nabla^2(v_t\xi) - 2\left(\frac{\partial^2 v_t}{\partial z^2}\frac{\partial^2\Psi}{\partial x^2} - 2\frac{\partial^2 v_t}{\partial x\partial z}\frac{\partial^2\Psi}{\partial x\partial z} + \frac{\partial^2 v_t}{\partial x^2}\frac{\partial^2\Psi}{\partial z^2}\right) \tag{50}$$

$$J\nabla_{xz}^2\Psi = \xi \tag{51}$$

$$v_t = \frac{3}{8}\sqrt{2KL} \tag{52}$$

where an algebraic equation for L is used, $L = 0.67\, Z\sqrt{1 - Z/z_\delta}$. For K we get (53),

$$\frac{\partial K}{\partial t} - J\frac{\partial(\Psi,K)}{\partial(X,Z)} = 0.80\, J\left[\frac{\partial}{\partial X}\left(v_t\frac{\partial K}{\partial X}\right) + \frac{\partial}{\partial Z}\left(v_t\frac{\partial K}{\partial Z}\right)\right] - \frac{2}{3}v_t\frac{K}{L^2} + P \tag{53}$$

where $P = v_t\left[4\left(\frac{\partial^2\Psi}{\partial x\partial z}\right)^2 + \left(\frac{\partial^2\Psi}{\partial z^2} - \frac{\partial^2\Psi}{\partial x^2}\right)^2\right]$ represents the production of K, and for C:

$$\frac{\partial C}{\partial t} + J\frac{\partial}{\partial X}\left[\left(\frac{\partial\psi}{\partial Z} + w_s\frac{\partial x}{\partial Z}\right)C - \gamma_t\frac{\partial C}{\partial X}\right] - J\frac{\partial}{\partial Z}\left[\left(\frac{\partial\psi}{\partial X} + w_s\frac{\partial x}{\partial X}\right)C + \gamma_t\frac{\partial C}{\partial Z}\right] = 0 \tag{54}$$

3.8 Boundary conditions for a 2DV turbulent boundary layer model

- At the lower limit of the boundary layer, $z = z_0 = k_N/30$
  - conditions for the stream current: $\partial\Psi/\partial X = \partial\Psi/\partial Z = 0$; $\Psi = 0$.
  - condition for the turbulent kinetic energy: $\partial K/\partial Z = 0$.
  - condition for the vorticity: $\xi_0 = 2J\Psi_1/(Z_1 - Z_0)^2$, where Ψ_1 is the stream function value at height Z_1. Value for ξ_0 can be also obtained from the one obtained at the time precedent through $\xi = J\nabla_{xz}^2\Psi = J\partial^2\Psi/\partial Z^2$.
- At the upper limit of the boundary layer, $z = z_\delta$
  - condition for the stream current: $\partial\Psi/\partial Z = U(t)$, where $U(t) = U_c + U_w\sin(\omega t)$, or $\Psi(z_\delta, t) = Q(t)$ if the flow is known at the level $z = z_\delta$.
  - condition for the turbulent kinetic energy. $K = 0$ (pure current), or $\partial K/\partial Z = 0$ (combined wave and current).
  - condition for the vorticity: $\xi = 0$ (it is assumed non-rotational flow outside of the boundary layer).

At the lateral boundaries ($X = 0$ and $X = L$), a spatially periodic condition for Ψ, ξ and K is assumed.

3.9 Other simplified two-equation turbulence closure models

A relation for the turbulent viscosity, equivalent to (16), can be written as $v_t = C_\mu K^2/\varepsilon$,

where ε is the turbulent dissipation rate defined by $\varepsilon = v_1\dfrac{\overline{\partial u_i'}}{\partial x_k}\dfrac{\partial u_j'}{\partial x_k}$. Comparing this

definition of the eddy viscosity v_t with (16), a relation between L and ε is found

$\varepsilon = C_K K^{3/2}/L$. Any other combination of the form $K^m L^n$ can be utilized, for example the specific dissipation rate $\omega = C_{\omega L} K^{1/2}/L = \varepsilon/(C_{\omega \varepsilon} K)$. This suggests the use of different variables, other than the macroscale of the eddies L, with all approximations of the form (10). One of these turbulence closure schemes, possibly the best known, is the two-equation $K - \varepsilon$ model; its governing equations are written:

$$\frac{\partial K}{\partial t} + \bar{u}_j \frac{\partial K}{\partial x_j} = \frac{\partial}{\partial x_j}\left(\frac{v_t}{\sigma_K}\frac{\partial K}{\partial x_j}\right) + v_t\left(\frac{\partial \bar{u}_i}{\partial x_j} + \frac{\partial \bar{u}_j}{\partial x_i}\right)\frac{\partial \bar{u}_i}{\partial x_j} - \varepsilon \; ; \; i,j = 1,2,3 \qquad (55)$$

$$\frac{\partial \varepsilon}{\partial t} + \bar{u}_j \frac{\partial \varepsilon}{\partial x_j} = \frac{\partial}{\partial x_j}\left(\frac{v_t}{\sigma_\varepsilon}\frac{\partial \varepsilon}{\partial x_j}\right) + C_{1\varepsilon}\frac{\varepsilon}{K}v_t\left(\frac{\partial \bar{u}_i}{\partial x_j} + \frac{\partial \bar{u}_j}{\partial x_i}\right)\frac{\partial \bar{u}_i}{\partial x_j} - C_{2\varepsilon}\frac{\varepsilon^2}{K}; \; i,j = 1,2,3 \qquad (56)$$

where, $C_K \approx C_{\omega \varepsilon} = 0.08 - 0.09$, $C_{\omega L} \approx 1.0$, $\sigma_K = 1.0$, $\sigma_\varepsilon = 1.30$, $C_{1\varepsilon} = 1.44$ and $C_{2\varepsilon} = 1.92$. The turbulent viscosity is calculated by $v_t = C_\mu K^2/\varepsilon$, where $C_\mu = 0.09$.
The Wilcox (1993) model is a two-equation $K - \omega$ turbulence closure scheme. The K and ω equations are determined through (57) and (58):

$$\frac{\partial K}{\partial t} + \bar{u}_j \frac{\partial K}{\partial x_j} = \frac{\partial}{\partial x_j}\left[(v_l + C_{1K}v_t)\frac{\partial K}{\partial x_j}\right] + v_t\left(\frac{\partial \bar{u}_i}{\partial x_j} + \frac{\partial \bar{u}_j}{\partial x_i}\right)\frac{\partial \bar{u}_i}{\partial x_j} - C_{2K}K\omega \; ; \; i,j = 1,2,3 \qquad (57)$$

$$\frac{\partial \omega}{\partial t} + \bar{u}_j \frac{\partial \omega}{\partial x_j} = \frac{\partial}{\partial x_j}\left[(v_l + C_{1\omega}v_t)\frac{\partial \omega}{\partial x_j}\right] + C_{2\omega}\frac{\omega}{K}v_t\left(\frac{\partial \bar{u}_i}{\partial x_j} + \frac{\partial \bar{u}_j}{\partial x_i}\right)\frac{\partial \bar{u}_i}{\partial x_j} - C_{3\omega}\omega^2; \; i,j = 1,2,3 \qquad (58)$$

where $C_{1K} = 0.50$, $C_{2K} = 0.09$, $C_{1\omega} = 0.50$, $C_{2\omega} = 5/9$ and $C_{3\omega} = 3/40$. The turbulent viscosity is calculated by $v_t = K/\omega$.

4. Numerical approaches

4.1 1DV boundary layer models
4.1.1 One- and two-equation models of the K-L type
Equations system (22) can be easily solved applying an implicit finite-difference approach in the raw unknowns (u, v, K, L, C, v_t, and γ_t) of five differential equations, both in space and time, and two algebraic ones.
Final solution for the vertical profiles of the horizontal components of the velocity (u, v), turbulent kinetic energy (K), macroscale of the eddies (L), concentration (C) and turbulent viscosity (v_t), is obtained iteratively during the time-period T of the signal introduced at the upper limit of the boundary layer. A flowchart representing the numerical solution implemented is presented in figure 3.

4.1.2 Zero-equation model
The model equations (29) are to be solved in this section. Considering the u_d, we note that the non-linear term should be linearized in time using Taylor series. With $u_t = l_m^2 \, \partial u_d/\partial z$, the following form of the u_d-equation show how the solution could be obtained:

$$\frac{\partial u_t}{\partial t} = \frac{\partial}{\partial z}\left(|u_t|u_t\right) \;=>\; \begin{cases} \dfrac{\partial}{\partial z}(u_t) & \text{if } u_t > 0 \\[2mm] -\dfrac{\partial}{\partial z}(u_t) & \text{if } u_t < 0 \end{cases}$$

Considering the case $u_d > 0$, a discretized form of this equation reads:

$$A_j u_{d\,j-1}^{n+1} + B_j u_{d\,j}^{n+1} + C_j u_{d\,j+1}^{n+1} = D_j \;\; ; \;\; 2 \le j \le J-1$$

where the coefficients A_j, B_j, C_j and D_j are:

$$A_j = -2\frac{\Delta t}{\Delta z_j}\left(\frac{l_{m\,j-1/2}}{\Delta z_{j-1/2}}\right)^2 \left|\Delta u_{d\,j-1/2}^n\right| ; \;\; C_j = -2\frac{\Delta t}{\Delta z_j}\left(\frac{l_{m\,j+1/2}}{\Delta z_{j+1/2}}\right)^2 \left|\Delta u_{d\,j+1/2}^n\right| ;$$

$$B_j = 1 - A_j - C_j ; \;\; D_j = u_{d\,j}^n - \frac{1}{2}A_j \Delta u_{d\,j-1/2}^n - \frac{1}{2}C_j \Delta u_{d\,j+1/2}^n$$

with $\Delta u_{d\,j-1/2}^n = u_{d\,j}^n - u_{d\,j-1}^n$ and $\Delta u_{d\,j+1/2}^n = u_{d\,j+1}^n - u_{d\,j}^n$.

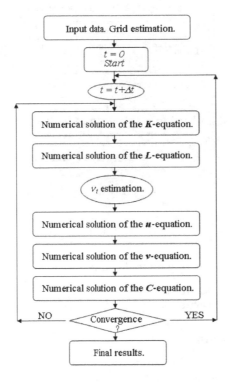

Fig. 3. Flowchart for the 1DV two-equation $K-L$ boundary layer model

Applications of 1DV boundary layer models are presented later, in this chapter.

4.2 2DV boundary layer model

Equations (50) to (54) are easily solved applying an implicit finite-difference approach centred in space and forward in time. The alternating direction implicit (ADI) method is used to solve the equations for ξ and K. The Poisson equation for Ψ is solved by the bloc-cyclic reduction method (Roache, 1976), which allows a huge saving in calculation time compared to the Gauss-Seidel iteration method (Huynh-Thanh & Temperville, 1991). Final solution is obtained iteratively during the time-period T of the signal introduced at the upper limit of the boundary layer. A flowchart representing the numerical solution implemented is presented in figure 4.

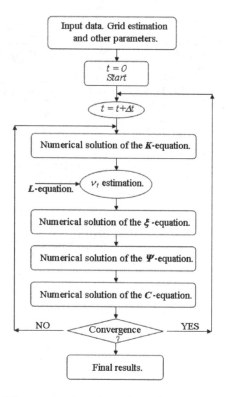

Fig. 4. Flowchart for the 2DV one-equation $K-L$ boundary layer model

Comparisons of laboratory experiments with numerical results of the 2DV boundary layer model are presented later, in section 6.

5. Parametric formulations

Following we show how different parametric approaches are derived, and tested with experimental data, using the two-equation $K-L$ boundary layer model (22). Using this model, Huynh Thanh (1990) proposed formula (59) below for the wave friction coefficient, $f_{w(r)}$, in the rough turbulent flow case:

$$f_{w(r)} = c_1 \exp \left[c_2 \left(\frac{A}{K_N} \right)^{n_1} \right]$$

(59)

where A is the wave excursion amplitude, and with the empirical coefficients c_1, c_2 and n_1 determined by Huynh Thanh, and presented in table 1 (formula HT_{fwr}). Using the same boundary layer model (22), considering the best overall fit with a large number of the model results, in the interval $6.4 \times 10^{-1} \leq A/k_N \leq 3.4 \times 10^3$, Antunes do Carmo et al. (2003) proposed formula (59) with the empirical coefficients determined in that study, listed in table 1 as formula CT_{fwr}.

Coeff. Formula	c_1	c_2	n_1
HT_{fwr}	0.00278	4.6500	-0.2200
CT_{fwr}	0.00140	4.5840	-0.1340

Table 1. Fitting coefficients c_1, c_2 and n_1, for model of Huynh Thanh (1990) (= HT_{fwr}) and proposed by Antunes do Carmo et al. (2003) (= CT_{fwr})

In the case of a current alone, Huynh Thanh found that the friction coefficient $f_{c(r)}$ coincides with the value obtained by the theoretical formula (60):

$$f_{c(r)} = 2 \left[\frac{k}{Ln \ (h/z_0) - 1} \right]^2$$

(60)

5.1 Sinusoidal wave alone

Considering rough turbulent flows, for values of the wave friction coefficient, $f_{w(r)}$, Antunes do Carmo et al. (2003) proposed formula (59) with CT_{fwr} coefficients (table 1); Tanaka & Thu (1994) suggested formula (61), Swart (1974) formula (62) and Soulsby et al. (1994) formula (63):

$$f_{w(r)} = \exp \left(-7.53 + 8.07 \ (A/z_0)^{-0.10} \right)$$

(61)

$$f_{w(r)} = 0.00251 \exp \left(5.21 \ (A/K_N)^{-0.19} \right)$$

(62)

$$f_{w(r)} = 1.39 \ (A/z_0)^{-0.52}$$

(63)

A comparison between formulae (59), with HT_{fwr} and CT_{fwr} coefficients, (61), (62) and (63) is shown in Antunes do Carmo et al. (2003). The same figure also shows experimental measurements of Sleath (1987), Kamphuis (1975), Jensen et al. (1989), Sumer et al. (1987) and Jonsson & Carlsen (1976).

According to Sleath (1991), bottom shear stress may be split into two components:

$$\hat{\tau}_{wp} = \hat{\tau}_w + \hat{\tau}_p$$

(64)

The shear stress in the fluid, $\hat{\tau}_w$, is taken into account by the model, but the value of $\hat{\tau}_p$, due to the mean pressure gradient acting on the bed roughness, is not. Using Sleath's experiments, it can be seen that a global friction coefficient may be split into the following two components:

$$f_{wp} = f_w + f_p \tag{65}$$

where f_w represents the friction coefficient obtained by the K-L model, and f_p represents the pressure gradient contribution. Assuming $K_N = 2.5\, d_{50}$, Sleath (1991) presented the formula (66):

$$f_p = 0.48 \left(A/K_N \right)^{-1} \tag{66}$$

The pressure gradient was not taken into account in experiments conducted by Sleath, Sumer, Jensen and Jonsson. Therefore, results of their experimental data are compared with model (59) considering CT_{fwr} coefficients. Excluding a small part of the Sleath's experiments, all other cases show a close agreement model (Antunes do Carmo et al., 2003). Discrepancies are explained as a consequence of some of Sleath's experiments being in the smooth-laminar transition regime. The pressure gradient is taken into account in Kamphuis' experiments, so this data should be compared with values for the following expression (67):

$$f_{wp} = f_w + f_p = 0.0014 \, \exp\left[4.584 \left(\frac{A}{K_N} \right)^{-0.134} \right] + 0.48 \left(\frac{A}{K_N} \right)^{-1} \tag{67}$$

For values of $A/K_N > 100$, the f_p term is negligible and expression (59) with CT_{fwr} coefficients (table 1) is in close agreement with results (Antunes dio Carmo et al., 2003).

5.2 Time-dependent shear stress

For the purpose of calculating time-dependent shear stress $\tau(t)$ in the case of an irregular wave whose instantaneous velocity is given by $U(t)$, Soulsby et al. (1994) propose calculating the value of the friction coefficient f_w for the equivalent sinusoidal wave with orbital velocity amplitude equal to $\sqrt{2}\, U_{rms}$ and period T_p. It can therefore be deduced (Antunes do Carmo et al., 2003):

$$f_w = 1.39 \left(\frac{A}{z_0} \right)^{-0.52} \qquad A = \frac{\sqrt{2}\, U_{rms}}{2\,\pi} \tag{68}$$

where U_{rms} = root-mean-square of orbital velocities. For a sinusoidal wave, this formulation correctly represents, in parametric form, the bottom shear stress obtained using $K - L$ model (22), but does not take into account the phase shift between $\tau(t)$ and $U(t)$. For an asymmetric wave, or an irregular wave, more important differences appear between this parametric formulation and the results calculated directly by the $K - L$ model.

To illustrate these phenomena, we consider the instantaneous velocity records presented in figure 5 for three cases (Antunes do Carmo et al., 2003): a) a sinusoidal wave, with orbital velocity amplitude 0.225 m/s and period 3.6 sec; b) a cnoidal wave, with a total velocity amplitude 1.107 m/s and period 9 sec, and c) an irregular wave obtained by the non-linear propagation of a sinusoidal wave, making use of a numerical Boussinesq-type model (Antunes do Carmo *et al.*, 1993), with a 3.0 sec period in a channel 0.30 m depth.

The values of the friction coefficient for a sinusoidal wave are shown in figure 6. Close agreement is evident between results 1 and 2. The instantaneous bottom shear stresses $\tau(t)$ have been calculated using model (59) with CT_{fwr} coefficients. In figure 7, results given by the $K-L$ model (22) (result 2) are compared both with those of model (59) (result 1) and with those obtained by a constant friction coefficient without the phase shift (result 3). Computed shear stresses for the sinusoidal wave case are presented in figure 7-a). Results of the model (59) with CT_{fwr} coefficients (result 1) are in close agreement with those of the $K-L$ model (22).

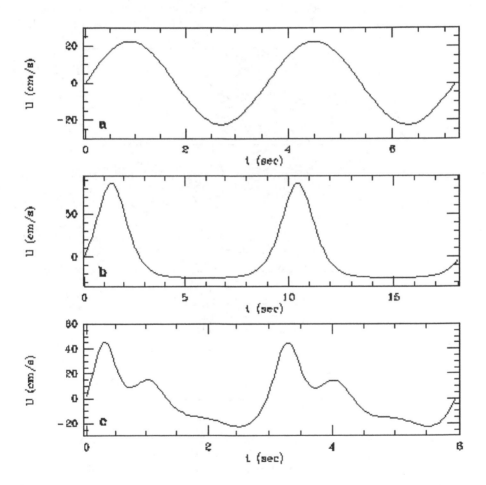

Fig. 5. Instantaneous velocity records: a – Sinusoidal wave (orbital velocity amplitude = 0.225 m/s, period = 3.6 sec); b – Cnoidal wave (total velocity amplitude = 1.107 m/s, period = 9.0 sec); c – Irregular wave (resulting from the non-linear propagation of a sinusoidal wave with a period = 3.0 sec in a channel 0.30 m depth) (Antunes do Carmo *et al.*, 2003)

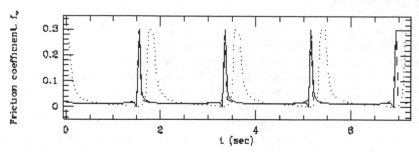

Fig. 6. Comparisons between the parameterized friction coefficient and the $K-L$ model result for a sinusoidal wave. Model (59) with CT_{fwr} coefficients (result 1: $----$; result 3: $\cdots\cdots$) and that obtained by $K-L$ model (result 2: ——) (Antunes do Carmo *et al.*, 2003)

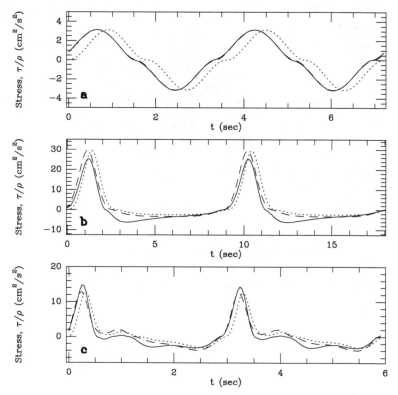

Fig. 7. Comparisons between the parameterized shear stress and the $K-L$ model result: a) Sinusoidal wave; b) Cnoidal wave; c) Irregular wave. Model (59) with CT_{fwr} coefficients (result 1: $----$; result 3: $\cdots\cdots$) and that obtained with $K-L$ model (result 2: ——) (Antunes do Carmo *et al.*, 2003)

A phase error between result 3 and result 2 ($K-L$ model) is evident. In the cnoidal wave case, the bottom shear stress calculated by the numerical boundary layer model is represented in figure 7-b) by the continuous line. As can be seen, for this case (figure 7-b), result 1 is closer to result 2 than it is to result 3, for both phase and negative values. However, asymmetries are not reproduced and a discrepancy can be seen for the maximum value. Several observations can be made concerning these results (Antunes do Carmo *et al.*, 2003): *i*) The representative curve $\tau(t)$ does not present the symmetry of velocities $U(t)$. The negative values of $\tau(t)$ are more important after the main positive peak than before it. It may be assumed that a *"turbulence memory"* created for this main peak influences what happens afterwards; *ii*) If the maximum velocity value is considered to be U_1 and the minimum velocity value U_2, it follows that:

$$\frac{\tau_2}{\tau_1} = \left(\frac{U_2}{U_1}\right)^2 = 0.08 \tag{69}$$

Figure 7-b) shows that the relation $\tau_2/\tau_1 = 0.24$ is greater than the value calculated by (69). Therefore, as in the case of a sinusoidal wave, the friction coefficient does not remain constant when velocity changes, assuming increasing values with decreasing velocity. Antunes do Carmo *et al.* (2003) propose calculating a time-dependent friction coefficient by replacing the maximum velocity with the instantaneous velocity $U(t+\theta)$, which takes into account the phase shift. The coefficient $f(t)$ will accordingly be calculated using expression (59), with CT_{fwr} coefficients (table 1), where A is given by (70):

$$A = \frac{\sqrt{2}U_{rms}\,T_p}{2\pi}\frac{U(t+\theta)}{U_{max}} \tag{70}$$

and $\tau(t)$ is defined by (71):

$$\tau(t) = \frac{f(t)}{2}U(t+\theta)\left|U(t+\theta)\right| \tag{71}$$

θ represents the phase lag between $U(t)$ and the bottom shear stress $\tau(t)$ at the upper limit of the boundary layer. Computed shear stresses for the more complex velocity case (irregular wave obtained by the non-linear propagation of an input sinusoidal wave) is presented in figure 7-c). A comparison of results 1 and 3 with result 2 shows that result 1 is still closer to that of the $K-L$ model (22) than to result 3. Also, a slight discrepancy can be seen for the maximum value. Despite the *"turbulence memory effects"*, the model (59) with CT_{fwr} coefficients fits closely with the boundary layer model results for the three cases analysed. Comparisons were made, however, assuming that results given by the $K-L$ model correctly represent the real conditions. Moreover, some discrepancies occur, especially for the maximum values.

6. Applications
6.1 K-L 1DV boundary layer model
Following closely Antunes do Carmo *et al.* (1996), an application of the $K-L$ turbulence model is presented, which corresponds to a sinusoidal mass oscillation where the velocity at the top of the bottom boundary layer is a pure sinusoidal wave with amplitude $\bar{u} = 170$ cm/s and period 7.2 sec. The following values were considered: $w_s = 2.6$ cm/s, $d_{50} = 0.021$ cm, $z_0 = 0.175 \times 10^{-2}$ cm, $z_a = 2d_{50} = 0.042$ cm and $z_\delta = 16.2$ cm. Figure 8-a) to d) show the

time series of sediment concentration computed at different levels above z_a (z = 0.10, 1.62, 2.08 and 4.54 cm). In figure 8-e) the vertical profiles of sediment concentration with phase shift of 60° are plotted (full lines), as well as the mean values over a wave period (dash lines). In each case the numerical solutions are compared to experimental data obtained by Ribberink & Al-Salem (Tran-Thu, 1995). Finally, in figure 8-f) the eddy diffusivity vertical profile averaged over a wave period is plotted.

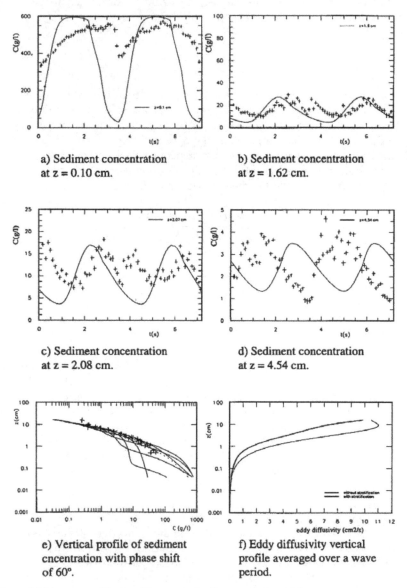

a) Sediment concentration
at z = 0.10 cm.

b) Sediment concentration
at z = 1.62 cm.

c) Sediment concentration
at z = 2.08 cm.

d) Sediment concentration
at z = 4.54 cm.

e) Vertical profile of sediment
cncentration with phase shift
of 60°.

f) Eddy diffusivity vertical
profile averaged over a wave
period.

Fig. 8. Sinusoidal mass oscillation (Antunes do Carmo *et al.*, 1996)

The analyses of results show that:

i.   the vertical distribution of sediment agrees well with experimental data;
ii. the pick concentration in the time series occurs with larger and larger phase the further away the level is located from the bed;
iii. at the upper levels, a time phase shift between the computed values of concentration and the experimental ones is observed;
iv. in the vicinity of the bottom (figure 8-a)) the time series of concentration shows the intermittence phenomena;
v.   the maximum values of sediment concentration agree well with data at all levels.

6.2 2DV boundary layer model

The flow in the bottom boundary layer established over a rippled bed was investigated through experiments and numerical calculations with a 2DV model. Experiments were conducted in an oscillatory flow tunnel illustrated in figure 9. This device was built from an existing wave flume at the Department of Civil Engineering of the University of Coimbra, Portugal. The wave tunnel has a rectangular cross section with 0.30 m width and 0.20 m high. The total length of the tunnel is 7.5 m.

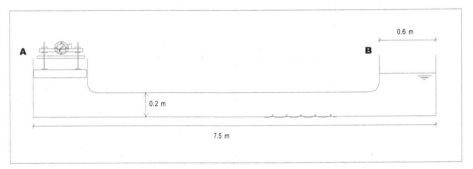

Fig. 9. Wave tunnel

At the left end (A) the vertical motion of a wave paddle produces an oscillatory flow within the tunnel. Five artificial symmetrical ripples have been placed on the tunnel's bed: each of the ripples has a length (L_r) of 7 cm and height (H_r) equal to 1.2 cm. The ripples were made in aluminium with the following profile (72):

$$z = \frac{4H_r}{L_r^2}x^2 - \frac{4H_r}{L_r}x + H_r \; ; \; 0 \leq x \leq L_r / 2 \qquad (72)$$

Sediment with a median grain diameter of 0.27 mm was glue to the surface of the ripples in order to simulate the skin roughness. Velocities were measured with an acoustic Doppler system (ADV) under sinusoidal oscillations at the wave paddle, over one ripple crest and one trough. Table 2 presents the experimental conditions considered in one of the tests made, being z_1 the height above the crest where the measurements were done. With the configuration of the ADV used, the measurements could only be done for heights above 4 cm from the bed. Figure 10 represents the mean values of the measured values of u and w at different levels during the wave cycle: u and w represent, respectively, the horizontal velocity in wave's tunnel direction and the vertical velocity.

Serie	Nr	Crest/Trough	T (s)	Z_i (cm)
S1	1	Cr	3.60	3.9
-	2	Cr	3.60	4.0
-	3	Cr	3.60	4.8
-	4	Cr	3.60	5.7
-	5	Cr	3.60	6.8
-	6	Cr	3.60	7.9
-	7	Cr	3.60	8.9
-	8	T	3.60	2.6
-	9	T	3.60	2.8
-	10	T	3.60	3.1
-	11	T	3.60	3.6
-	12	T	3.60	4.1
-	13	T	3.60	4.6
-	14	T	3.60	5.1
-	15	T	3.60	6.1
-	16	T	3.60	8.1
-	17	T	3.60	9.9

Table 2. Experimental conditions

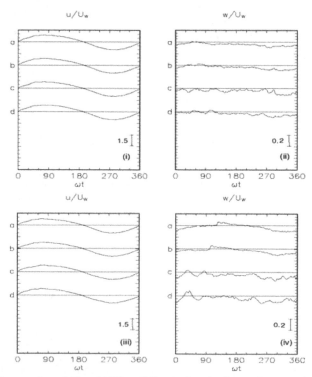

Fig. 10. Measured time series of U and W for different levels above the crest (i, ii) - (z_1 (a) = 8.9 cm, z_1 (b) = 7.9 cm, z_1 (c) = 4.0 cm, z_1 (d) = 3.9 cm) and above the trough (iii, iv) – (z_1 (a) = 9.9 cm, z_1 (b) = 8.1 cm, z_1 (c) = 2.8 cm, z_1 (d) = 2.6 cm) (Silva, 2001)

These values were divided by the amplitude of the horizontal velocity outside the boundary layer, U_w, and were obtained by averaging *equi*-phase data over approximately 20 wave periods. The analysis of figure 10 shows that: (1) The oscillatory flow in the wave tunnel does not correspond to a sinusoidal oscillation as we can observe from the velocities measured at the highest levels (a), and (2) At the lower levels (c, d) the measured values of w show oscillations with a time scale inferior to the ones observed in the highest levels: this suggest that the flow at those levels is perturbed by the lee vortex developed during the wave cycle and that are ejected from the bottom after flow reversal (0° and 180°).

A numerical simulation of the flow in the bottom boundary layer was done with the 2DV model (50) to (54). In figure 11 the numerical results are compared with the experimental data. The results are only plotted for the lower level of measurements.

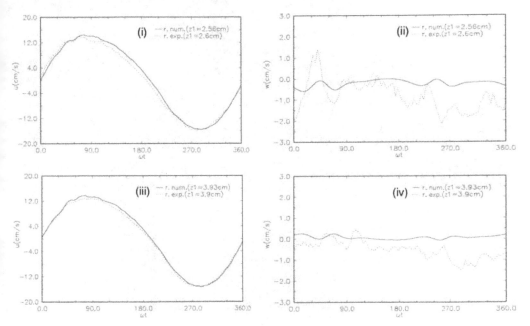

Fig. 11. Numerical results *vs* experimental data (Silva, 2001)

It is seen that there is a good agreement between the computed and measured horizontal velocity. The computed vertical velocity shows small oscillations after flow reversal, between 0°-90° and 210°-300°. The shape of these oscillations is similar to the observed one, although there is a phase shift between them. The amplitude of these oscillations is also lower than the amplitude of the measured values of w: this means that the model dissipates the kinetic energy of the ejected vortex at a rate that is superior to what it is observed. This feature has also been noted in other comparisons. To analyse with more detail the flow in the bottom boundary layer, namely the vortex paths during the wave cycle, we have plotted in figure 12 the vorticity field at different wave phases. It is seen that the lowest level of measurements over the ripple crest is above the track of the vortex that is carried by the flow after it is ejected. This justifies the poor agreement between the numerical and experimental results in the figure 11 (iv).

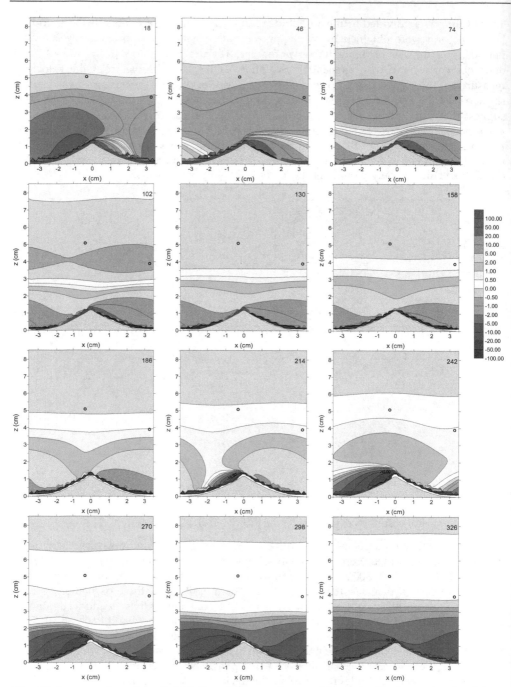

Fig. 12. Computed vorticity field (s^{-1}) at different phases of the flow. The lower levels of the measurements over the ripple crest and trough are marked •

7. Conclusion

Assuming that the fluid is in a randomly unsteady turbulent state and applying time averaging to the basic equations of motion, the fundamental equations of incompressible turbulent motion are obtained. A three-dimensional form of conservation equations for a single Reynolds stress and for the turbulent kinetic energy is derived. However, as the full three-dimensional form of equations is very complex and not easy to solve, with many unknown correlations to model, other much simpler one- and two-dimensional boundary layer forms of these relations are derived. A brief discussion about numerical models based on control volumes and finite difference approximations is presented to solve 1DV versions of the one- and two-equation rough turbulent bottom boundary layer model of the K-L type, and of the 2DV boundary layer model. These numerical models are then used to calibrate general parametric formulations for the instantaneous bottom shear stress due to both a wave and a wave-current interaction cases. They are still used to discuss some important aspects, like the *phase shift* and the *turbulence memory effects*. Mathematical formulations and parametric approaches are extended to include the effect of suspended non cohesive sediments. Comparisons with experimental results show that both 1DV and 2DV boundary layer models are able to predict quite well the complex flow properties. However, these models are strictly valid for permanent flows in the fully developed turbulent regime at high Reynolds numbers. When the flow is oscillatory, the condition of local equilibrium of the turbulence is no longer completely satisfied, particularly at the time when the velocity of the potential flow is small. Therefore, improvements are necessary to obtain more precise results for moderate Reynolds numbers.

8. References

Aldama, A.A., 1990. Filtering techniques for turbulent flows simulation, *Lecture Notes in Engineering*, Ed. C.A. Brebbia and S.A. Orszag, 56, Springer-Verlag, ISBN 3-540-52137-2.

Antunes do Carmo, J.S., Seabra-Santos, F.J. & Barthélemy, E., 1993. Surface waves propagation in shallow-water: a finite element model, *Int. J. Num. Meth. in Fluids*, *Vol. 16, No. 6, 447-459*.

Antunes do Carmo J.S., Silva P. & Seabra-Santos F.J., 1996. The K-L turbulence model *vs* parametric formulations, *Hydraulic Engineering Software VI*, 323–334, Ed. W.R. Blain, WITPRESS/CMP. ISBN 1-85312-405-2.

Antunes do Carmo, J.S., Temperville, A. & Seabra-Santos F.J., 2003. Bottom friction and time-dependent shear stress for wave-current interaction, *Journal of Hydraulic Research*, IAHR, Vol. 41, N° 1, 27-37.

Batchelor, G.K., 1953. *The Theory of Homogeneous Turbulence*. Cambridge, UK: Cambridge, Univ. Press.

Boussinesq, 1877. Essai sur la théorie des eaux courantes. Institute de France, Académie des Sciences, *Memoires présentées par divers savants*, Vol. 23, No. 1 (in French).

Chen, C.J., Shih, C., Lienau, J. & Kung, R.J., 1996. *Flow modeling and turbulence measurements*, A.A. Balkema, Rotterdam / Brookfield, ISBN: 9054108266.

Fourniotis, N.Th., Dimas, A.A & Demetracopoulos, A.C., 2006. Spatial development of turbulent open-channel flow over bottom with multiple consecutive dunes. *Proceedings of* the International Conference River Flow 2006, Lisbon, Portugal, 1023-1032.

Hinze, J., 1975. *Turbulence* (2nd Edition), McGraw-Hill Classic Textbook Reissue Series.

Huynh-Thanh, S., 1990. *Modélisation de la couche limite turbulente oscillatoire générée par l'interaction houle-courant en zone côtière*, Ph.D thesis, INP – Grenoble, France (in French).

Huynh-Thanh, S. & A. Temperville, A.,1991. A numerical model of the rough turbulent boundary layer in combined wave and current interaction. In: R.L. Soulsby and R. Betess (Editors), *Sand Transport in Rivers, Estuaries and the Sea*, Balkema, Rotterdam, 93-100.

Jensen, B.L., Sumer, B.M. & Fredsøe, J., 1989. Turbulent oscillatory boundary layers at high Reynolds numbers, *Journal of Fluid Mechanics*, 206, 265-297.

Jonsson, I.J. & Carlsen, N.A., 1976. Experimental and theoretical investigations in an oscillaory turbulent boundary layer, *Journal of Hydraulics Research*, 14(1), 45-60.

Kamphuis, J.W., 1975. Friction factor under oscillatory waves, *J. Waterw. Port Coastal Ocean Eng.*, 101 (WW2), 135-144.

Kaneda, Y. & Ishihara, T., 2006. High-resolution direct numerical simulation of turbulence, *Journal of Turbulence*, Vol. 7, No. 20.

Kolmogorov, A.N., 1941. *The local structure of turbulence in incompressible viscous fluid for very large Reynolds numbers*. Proceedings of the USSR Academy of Sciences 30: 299–303. (Russian), translated into English by Kolmogorov, Andrey Nikolaevich (July 8, 1991). *The local structure of turbulence in incompressible viscous fluid for very large Reynolds numbers*. Proceedings of the Royal Society of London, Series A: Mathematical and Physical Sciences 434 (1991): 9–13.

Launder, B.E. & Spalding, D.B., 1972. *Lectures in mathematical models of turbulence*, Academic Press London and New York, ISBN:0-12-438050-6.

Lesieur, M., 1997. Turbulence in fluids: Third revised and enlarged edition, *Fluid Mechanics and its Applications*, Vol. 40, Kluwert Academic Publishers, ISBN 0-7923-4415-4.

Lewellen, W.S., 1977. Use of the invariant modelling, in *Handbook of turbulence*, Plenum Publishing Corp., Vol. 1, 237-280.

Lumley, J.L., 1996. Fundamental aspects of incompressible and compressible turbulent flows, *Simulation and Modeling of Turbulent Flows*, Cap. 1. ICASE/LaRC Series in Computacional Science and Engineering. Edited by Gatski, Hussaini e Lumley.

McComb, W.D.: http://www.ph.ed.ac.uk/acoustics/turbulence/ (last acces May 2011).

Mohammadi, B. & Pironneau, 1994. *Analysis of the k-epsilon turbulence model*, John Wiley e Sons, ISSN:0298-3168.

Monin, A.S. & Yaglom, A.M., 1971. *Statistical Fluid Mechanics: Mechanics of Turbulence*, Vol. 1, The MIT Press, Edited by John L. Lumley, ISBN: 0-262-13062-9.

Moser, R.D., 2006. On the validity of the continuum approximation in high Reynolds number turbulence, *Phys. Fluids* 18, 078105.

Moulden, T.H., W. Frost & Garner, A.H., 1978. The complexity of turbulent fluid motion, *Handbook of turbulence*, W. Frost and T.H. Moulden, Eds, Plenum Press

Nezu, I.E. & Nakagawa, H., 1993. Turbulence in open-channel flows. IARH/AIRH Monograph Series, 3ᵃ Ed. Balkema.

Prandtl, L., 1925. Über die ausgebildete Turbulenz. *Z. Angew, Math. Mech.*, Vol. 5, 136-139.

Reynolds, O., 1883. On the experimental investigation of the circumstances which determine whether the motion of water shall be direct or sinuous, and the law of resistance in parallel channels *Phil. Trans. Roy. Soc. London Ser. A*, vol. 174, 935-982.

Reynolds, O., 1895. On the dynamical theory of incompressible viscous fluids and the determination of the criterion, *Phil. Trans. Roy. Soc. London Ser. A*, vol. 186, 123-164.

Roache, P.J., 1976. *Computational Fluid Dynamics.* Eds. Hermosa Publishers, New Mexico.

Rodi, W., 1980. *Turbulence Models and their Application in Hydraulics - A State-of-the-Art Review*, I.A.H.R – Publication.

Rodi, W., 1984. Turbulent models and their applications in *Hydraulics – A state of the art review*, 2ⁿᵈ Edition, Bookfield Publishing, 1-36.

Rodi, W., 1993. *Turbulence models and their application in hHydraulics*, Balkema.

Rotta, J.C., 1951. Statistische theorie nichthamagener turbulenz, *Zeitschrift f. Physik*, Bd. 192, pp. 547-572, and Bd. 131, 51-77.

Sato, S., Mimura, N. & Watanabe, A., 1984. Oscilatory boundary layer flow over rippled beds. Proc 19ᵗʰ *Conf. Coastal Engineering*, 2293-2309.

Sheng, Y.P., 1984. A turbulent transport model of coastal processes. Proc. 19ᵗʰ *Conf. Coastal Eng.*, 2380-2396.

Schiestel, R., 1993. *Modélisation et simulation des écoulements turbulents*, Hermes, Paris, ISBN: 2-86601-371-9.

Silva, P.M.C.A., 2001. *Contribution for the study of sedimentary dynamics in coastal regions*, Ph.D thesis, University of Aveiro, Portugal (in Portuguese).

Sleath, J.F.A., 1984. *Sea bed Mechanics.* Eds. Wiley-Interscience.

Sleath, J.F.A., 1987. Turbulent oscillatory flow over rough beds, *Journal of Fluid Mechanics*, 182, 369-409.

Sleath, J.F.A., 1991. Velocities and shear stresses in wave-current flows, *Journal of Geophysical Research*, Vol. 96, No. C8, 15, 237-15, 244.

Soulsby, R.L., Hamm, L., Klopman, G., Myrhaug, D., Simons, R.R. & Thomas, G.P., 1994. Wave-current interaction within and outside the bottom boundary layer, *Coastal Eng.*, 21, 41-69.

Sumer, B.M., Jensen, B.L. & L. Fredsøe, L., 1987. Turbulence in oscillatory boundary layers. In *Advances in Turbulence*, Springer, Heidelberg, 556-567.

Swart, D.H., 1974. *Offshore sediment transport and equilibrium beach profiles.* Delft Hydraulics Lab., Publ. 131.

Tanaka, H. & Thu, A., 1994. Full-range equation of friction coefficient and phase difference in a wave-current boundary layer, *Coastal Eng.*, 22, 237-254.

Tennekes, H. & Lumley, J.L., 1972. *A first course in turbulence*, MIT Press.

Tran-Thu, T., 1995. *Modélisation numérique de l'interaction houle-courant-sédiment*, Ph.D thesis, Université Joseph Fourier – Grenoble, France (in French).

Tran-Thu, T. & Temperville, A., 1994. Numerical model of sediment transport in the wave-current interaction, in *Modelling of Coastal and Estuarine Processes*, Eds. F. Seabra-Santos & A. Temperville, 271-281.

Wilcox, D.C., 1993. *Turbulent modelling of CFD*. DCW Industries, La Canada, Calif.

Permissions

The contributors of this book come from diverse backgrounds, making this book a truly international effort. This book will bring forth new frontiers with its revolutionizing research information and detailed analysis of the nascent developments around the world.

We would like to thank Dr. Hyoung Woo Oh, for lending his expertise to make the book truly unique. He has played a crucial role in the development of this book. Without his invaluable contribution this book wouldn't have been possible. He has made vital efforts to compile up to date information on the varied aspects of this subject to make this book a valuable addition to the collection of many professionals and students.

This book was conceptualized with the vision of imparting up-to-date information and advanced data in this field. To ensure the same, a matchless editorial board was set up. Every individual on the board went through rigorous rounds of assessment to prove their worth. After which they invested a large part of their time researching and compiling the most relevant data for our readers. Conferences and sessions were held from time to time between the editorial board and the contributing authors to present the data in the most comprehensible form. The editorial team has worked tirelessly to provide valuable and valid information to help people across the globe.

Every chapter published in this book has been scrutinized by our experts. Their significance has been extensively debated. The topics covered herein carry significant findings which will fuel the growth of the discipline. They may even be implemented as practical applications or may be referred to as a beginning point for another development. Chapters in this book were first published by InTech; hereby published with permission under the Creative Commons Attribution License or equivalent.

The editorial board has been involved in producing this book since its inception. They have spent rigorous hours researching and exploring the diverse topics which have resulted in the successful publishing of this book. They have passed on their knowledge of decades through this book. To expedite this challenging task, the publisher supported the team at every step. A small team of assistant editors was also appointed to further simplify the editing procedure and attain best results for the readers.

Our editorial team has been hand-picked from every corner of the world. Their multi-ethnicity adds dynamic inputs to the discussions which result in innovative outcomes. These outcomes are then further discussed with the researchers and contributors who give their valuable feedback and opinion regarding the same. The feedback is then collaborated with the researches and they are edited in a comprehensive manner to aid the understanding of the subject.

Apart from the editorial board, the designing team has also invested a significant amount of their time in understanding the subject and creating the most relevant covers. They scrutinized every image to scout for the most suitable representation of the subject and create an appropriate cover for the book.

The publishing team has been involved in this book since its early stages. They were actively engaged in every process, be it collecting the data, connecting with the contributors or procuring relevant information. The team has been an ardent support to the editorial, designing and production team. Their endless efforts to recruit the best for this project, has resulted in the accomplishment of this book. They are a veteran in the field of academics and their pool of knowledge is as vast as their experience in printing. Their expertise and guidance has proved useful at every step. Their uncompromising quality standards have made this book an exceptional effort. Their encouragement from time to time has been an inspiration for everyone.

The publisher and the editorial board hope that this book will prove to be a valuable piece of knowledge for researchers, students, practitioners and scholars across the globe.

List of Contributors

Germán González Silva, Natalia Prieto Jiménez and Oscar Fabio Salazar
State University of Campinas, Brazil

Jesús I. Minchaca M, A. Humberto Castillejos E and F. Andrés Acosta G
Centre for Research and Advanced Studies – CINVESTAV, Unidad Saltillo, Mexico

S.A. Karabasov
University of Cambridge, Department of Engineering, UK

V.M. Goloviznin
Moscow Institute of Nuclear Safety, Russian Academy of Science, Russia

E. Lubarsky, D. Shcherbik, O. Bibik, Y. Gopala and B. T. Zinn
School of Aerospace Engineering, Georgia Institute of Technology, Atlanta Georgia, USA

Majdalani and Tony Saad
University of Tennessee Space Institute, USA

H. Herrero, M. C. Navarro and F. Pla
Universidad de Castilla-La Mancha, Spain

V. K. Veera, M. Masood, S. Ruan and N. Swaminathan
Department of Engineering, Cambridge University, Cambridge, UK

H. Kolla
Sandia National Laboratory, Livermore, CA, USA

X. San Liang
Harvard University, School of Engineering and Applied Sciences, Cambridge, MA, USA
Central University of Finance and Economics, Beijing, China
Stanford University, Center for Turbulence Research, Stanford, CA, USA
Nanjing Institute of Meteorology, Nanjing, China

Balázs Pritz and Martin Gabi
Karlsruhe Institute of Technology, Department of Fluid Machinery, Germany

George Vahala and Bo Zhang
College of William & Mary, USA

Jeffrey Yepez
Air Force Research Lab, USA

Linda Vahala
Old Dominion University, USA

Min Soe
Rogers State University, USA

José Simão Antunes do Carmo
University of Coimbra, Portugal

Printed in the USA
CPSIA information can be obtained
at www.ICGtesting.com
JSHW011452221024
72173JS00005B/1047